TM
109

MATHEMATICS FOR TEACHERS

Problem Solving and Other Basic Skills

MATHEMATICS FOR TEACHERS

Problem Solving and Other Basic Skills

Charles A. Reeves **Thom B. Clark**

Scott, Foresman and Company   Glenview, Illinois
Dallas, Tex.   Oakland, N.J.   Palo Alto, Cal.   Tucker, Ga.   London, England

Library of Congress Cataloging in Publication Data

Reeves, Charles A.
    Mathematics for teachers.

    1. Mathematics—1961–   2. Problem solving.
I. Clark, Thom B.   II. Title.
QA39.2.R444   510   81-23234
ISBN 0-673-16051-3   AACR2

Copyright © 1982 Scott, Foresman and Company.
All Rights Reserved.
Printed in the United States of America.

1 2 3 4 5 6-RRC-88 87 86 85 84 83 82

Production Management
and Art Direction by: *Michael Bass & Associates*

Cover Design by: *Lorena Laforest Bass*

*Dedicated to Cathy, Dillon, and Rosemarie,
and to my students*—CAR

*Special appreciation to my family,
Franny, and Zack*—TBC

Contents

1. PROBLEM-SOLVING PROCESSES — 1

    Problems — 2

    A Model for Problem-solving Processes (rational, intuitive, or autistic thought) — 3
    Set 1 — 7

    A Closer Look at Rational Processes (memory and recall, algorithmic, or analytic and synthetic thought) — 9
    Set 2 — 12

    Useful Analytic and Synthetic Processes (nine standard rational approaches to solving problems) — 14
    Set 3 — 21

2. THE THIRD R — 26

    Numeration (several ancient whole number numeration systems and the Hindu Arabic system) — 26
    Set 1 — 31

    Foundations (concrete and semiconcrete interpretations of whole number addition, subtraction, multiplication, and division) — 33
    Set 2 — 39

    Building Brick Walls (algorithms for whole number addition, subtraction, multiplication, and division) — 41
    Set 3 — 48

    Fractions—An Endangered Species? (basic concepts for fractions: symbolism and interpretations, renaming, comparison, and common denominators) — 50
    Set 4 — 56

    Number Crunching, with Fractions? (addition, subtraction, multiplication and division algorithms; accompanying properties) — 60
    Set 5 — 65

Guess That It Was Bound to Happen
(the relationship of the decimal system to fractions)   67
   Set 6   73

The Two Faces of Percent (relating percent to fractions and decimals; computing)   75
   Set 7   79

Negative Numbers (a basic model for computing with positive and negative numbers)   83
   Set 8   88

Holes in the Road (irrational numbers)   90
   Set 9   94

Another Look at Whole Numbers (primes and composites; odd and even   96
   Set 10   100

3. GEOMETRY   103

Shapes (lines, segments, rays, circles, spheres, and angles; straightedge and compass)   104
   Set 1   107

Shape Transformations (slides, flips, and turns)   108
   Set 2   114

Look-alikes (symmetry and congruence)   117
   Set 3   122

Curves (simple closed curves and polygons)   124
   Set 4   128

When Two Lines Meet (opposite angles, perpendicularity, parallelism, and traditional constructions)   130
   Set 5   136

Tessellating the Plane (regular and irregular polygons)   138
   Set 6   142

Tessellations in Art, You, and Escher (exploring tessellations that are not made from polygons as a basic shape)   144
   Set 7   148

Polyhedra (vertices, edges, and faces)   151
   Set 8   153

4. ANSWERS   156

Forms for Answers (numbers, figures, logical arguments, conclusions, and processes)   157
   Set 1   161

Reasonable Numerical Answers (estimating and rounding off)	162
Set 2	166
Communicating Answers (precision, the size of numbers, and appropriate units)	168
Set 3	171
Practicing (practicing giving reasonable answers that communicate to others using mass and temperature in the metric system)	172
Set 4	174

5. CHANCE   177

Chance (a beginning definition of the probability of an event, and its complement)	177
Set 1	181
Counting the Possibilities (the fundamental counting principal)	184
Set 2	187
The Chain Law Is in Effect (applying the chain law to unrelated events, and the modified chain law to related events)	189
Set 3	193
Through the Back Door (finding the probability of an event by using the complementary event)	195
Set 4	198
The Way Things Are (determining probabilities through empirical evidence)	199
Set 5	203

6. THERE MUST BE A BETTER WAY!   206

Calculator versus Computer (a working distinction between the two tools)	207
Set 1 (problems for a calculator)	208
Flowcharts (charting solutions to real-life dilemmas, some of which require "decisions" and "looping")	209
Set 2	214
Flowcharts that Count (counting the number of times through a loop in a flowchart and checking the logic of a flowchart with a loop by reducing the problem to one that can be traced by hand)	217
Set 3	221
Introduction to Programming (simple programs that involve no looping; coding into BASIC)	223
Set 4	227
Using a Loop (writing programs that involve looping)	230
Set 5	234

The Right Tool (choosing to use a calculator, computer, or paper
    and pencil to solve a problem)   237
  Set 6   238

7. **MEASUREMENT**   241

  Measuring Length (using meter as the base unit)   243
    Set 1   245
  Measuring Area (using tessellating figures—squares—with sides
    being subdivisions of a meter, to measure area)   247
    Set 2   251
  Measuring Volume (using cubes with metric dimensions to "fill up"
    space)   254
    Set 3   257
  Measuring Angles (degrees and radians)   258
    Set 4   261
  Measures Common to All Classrooms (some of the more common,
    nonphysical attributes frequently measured in the classroom)   262
    Set 5   267

8. **DESCRIPTIVE STATISTICS**   270

  Grouping and Classifying (preset vs. clustering technique)   271
    Set 1   276
  Presenting Data Visually (pictograms, bar graphs, line graphs,
    and circle graphs)   279
    Set 2   285
  Population Studies (random and representative samples;
    stratified random sampling)   288
    Set 3   290
  Group Measures (mean, median, and mode as descriptors of
    central tendency; range and standard deviation as measures
    of dispersion)   291
    Set 4   296
  The Normal Curve (how standard deviation determines the shape
    of the normal curve)   298
    Set 5   304

9. **PROBLEM SOLVING, E X T E N D E D**   307

  Other Solutions (finding other paths to a problem that has
    been solved)   307
    Set 1   310

Tip of the Iceberg (extending a solved problem; generating new
    problem situations)   311
    Set 2   314

Explorations (problem-solving experiences that come initially
    from informal "messing around")   315
    Set 3   320

Unique Interpretations and Perspectives (exploring different
    interpretations of given problem situations)   322
    Set 4   326

Problems for the Road (a collection of problems for the curious
    student who wants more)   328
    Set 5   328

COMMENTARY   335

GLOSSARY   367

INDEX   375

Preface

To the student:

Relax—it won't be that bad! You won't be asked to give long, detailed proofs or waste too much time practicing computational skills that you should have mastered long ago. We've replaced those aspects of mathematics with something much tougher —we're going to ask you to think! And explore! And create! And release your natural curiosity!

Why? Because that's what mathematics is mostly all about; somewhere along the way we seem to have lost sight of this fact. Elementary mathematics often gets equated with, and bogged down in, a murky sequence of skills that must be mastered. But mathematics is so much more than this, and this may be your last chance to find that out before you're entrusted with the educational future of our country.

A word of caution: the rewards of the experience we've planned for you will come in direct proportion to the amount of energy you're willing to invest. Our writing style for this text—an ongoing, informal dialogue—makes the assumption that you will carefully read, and then reread, certain portions. We are assuming that you want to become more competent with mathematical ideas and consequently that you will "risk" experimentation when certain questions are raised. The notion of your exploring mathematics with us is an important orientation for this text. Actually, we're still thinking ourselves about some of the issues we ask you to consider.

We've tried to capture and transmit to you, as best we could, our notion of the true nature and spirit of mathematics. If you enjoy the material as much as we've enjoyed preparing it and testing it on your peers, we'll count the whole effort a success. Good luck!

THEN said a teacher, Speak to us of Teaching.

And he said:

No man can reveal to you aught but that which already lies half asleep in the dawning of your knowledge.

The teacher who walks in the shadow of the temple, among his followers, gives not of his wisdom but rather of his faith and his lovingness.

If he is indeed wise he does not bid you enter the house of his wisdom, but rather leads you to the threshold of your own mind.

The astronomer may speak to you of his understanding of space, but he cannot give you his understanding.

The musician may sing to you of the rhythm which is in all space, but he cannot give you the ear which arrests the rhythm nor the voice that echoes it.

And he who is versed in the science of numbers can tell of the regions of weight and measure, but he cannot conduct you thither.

For the vision of one man lends not its wings to another man.

From *The Prophet*
KAHLIL GIBRAN (1893-1923)

To the instructor:

We endorse the full spectrum of the National Council of Supervisors of Mathematics' Basic Skills List (1977), and consequently have based this text almost exclusively on those topics. However, our coverage of the material may be different from another text covering the same topics, because of our philosophy about mathematics and how to teach it. Our perspectives have evolved from years of teaching at all levels in the curriculum.

Several tenets that structure our feelings are important enough to be mentioned:

a. Mathematics, in its true character, is an art form that must be experienced to be appreciated. Directing students through a ready-made collection of concepts, definitions, axioms, and theorems is something like guiding tourists through a mathematical museum, permitting them only to observe from a distance the polished art forms that resulted from years of earnest labor. We hope instead to engage the student in some of the thinking processes used and enjoyed by problem solvers who created those aesthetic forms and relics found in the archives of mathematics.

b. It's been our experience that this one course cannot accomplish, in a lasting fashion, what 12 years of formal schooling has concentrated on and possibly failed in—computational proficiency for the elementary education major. Therefore we merely review this topic as one of the basic skills, we hope in a somewhat different fashion from previous exposures. We trust that your teacher-education program has minimal entrance standards, or provides remedial help outside this course to students for whom a review of computation is not enough. We've attempted to avoid "teaching toward the lower end of the spectrum."

c. "Teaching" is not synonymous with "explaining." In fact, sometimes the appropriate teaching technique may be to squelch the temptation to "explain everything so it's crystal clear." Struggling with a concept many times brings us to grips with a dilemma so that a later discovery or explanation is much more meaningful. The implication of this tenet, for the text, is that we have rejected the typical writing style of "explain and practice."

d. A "teacher's manual" should provide an alternative to "lecturing on what's in the text." We hope our text can be understood outside of class most of the time, and class time can be spent enriching the material by extensions, side-tracking with personal experiences, and sharing and generating new ideas related to the experiences at hand. Instructors should view themselves as storehouses of valuable side trips that cast a new perspective on the material in the text.

e. Problem sets offer a chance not only to practice and extend and explore material that has already been presented, but also to introduce new material intuitively. Many of the dilemmas in our exercise sets will crop up later in the text discussions. We hope the class will be anxious to discuss a few of the situations. Urge students to use the Commentary on Problems section of the text, but selectively.

f. An effective writing style for this audience is an informal one. This does not imply a sloppy, noncaring approach to mathematics, but an approach more concerned with communicating ideas and thought processes than technical information. One of our primary objectives was to write a mathematics text that would be easy to read.

g. Units that are developmentally related (such as Geometry and Measurement, and Probability and Statistics) should not necessarily be placed right next to each other. We have physically separated these chapters on purpose to provide a break from the monotony of a continuous extension of the same idea.

h. This is probably the elementary education major's last chance to view mathematics as a living entity, well worth a personal investment of time and energy. This undoubtedly places a tremendous burden on the instructor because of the need to provide a model for the way mathematics can, and should, be taught. This seems to work best when the teacher openly represents the processes, questions, and difficulties that he or she experiences in solving a problem. The authors recommend excursions into your own dilemmas with the problems in the text as demonstrations of the dynamics of problem solving for your class.

The material covered in the text, and suggested in the Teacher's Hints, works for us. Perhaps it will for you also. In either case, let us know the results.

Charles A. Reeves   *Thom B. Clark*

MATHEMATICS FOR TEACHERS

Problem Solving and Other Basic Skills

1
PROBLEM-SOLVING PROCESSES

> If life were of such a constant nature that there were only a few chores to do and they were done over and over in exactly the same way, the case for knowing how to solve problems would not be so compelling. All one would have to do would be to learn the few jobs at the outset. From then on he could rely on memory and habit. Fortunately—or unfortunately depending on one's point of view—life is not so simple and unchanging. Rather, it is changing so rapidly that about all we can predict is that things will be different in the future. In such a world, the ability to adjust and solve one's problems is of paramount importance. (Henderson and Pingry 1953, p. 233)

How do you solve your problems? Are you an analytic person, or do you rely on emotion? Do you use both reason and emotion at times? Do you feel secure when you make a decision, or do you frequently change your mind later on? Do you enjoy taking risks, or do you proceed only on safe ground?

Your approach to problem solving might be described as "cautious" or "risk taking," "intuitive" or "structured," "internally" or "externally" motivated, "concrete" or "abstract," or in countless other ways. Such characteristics would probably vary from problem to problem, but taken generally, they would constitute your personal problem-solving style. In other words, you have a unique style of thinking about and solving problems. Analyzing, and perhaps adding flexibility to, your own thought processes and problem-solving strategies is an intriguing venture!

This introductory chapter explores with you some of the many ways in which we think about problems, particularly as they apply to mathematical situations. We will examine general approaches to thinking and problem solving and then analyze methods used in resolving mathematical problems, including three specific levels of mathematical thought processes used as unifying themes throughout the text.

You are encouraged to "pick up the challenge" from here on. No matter what your past experience with mathematics has been, give yourself a chance to become a better problem solver. After all, "the ability to adjust and solve your own problems is of paramount importance."

Problems

Problem solving is resolving some dilemma. Sometimes the solution to the dilemma is obvious, easily obtained, or otherwise straightforward. On the other hand, sometimes there is great difficulty generating the elusive connective tissue leading from the known information to a resolution of the problem. In fact, there are many problems, both in mathematics and in general, that have yet to be satisfactorily resolved!

Consider some of these problems whose solutions are pending:

We are having a population explosion. What will we do about it? Where can we go?

$1/2 = 0.5 = 0.4999\ldots$

$1/3 = 0.3333\ldots$

$2/9 = 0.2222\ldots$

Our decimal notation for fractions is not so good for some fractions, is it? Can we do something about it?

Is there a relationship between the ancient geometrical arrangement of rocks at Stonehenge and the position of the sun?

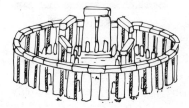

Source: From *The Book of Knowledge* (1960, courtesy Grolier Incorporated).

Can our three-dimensional universe continue to expand? Why isn't what it's expanding into considered part of today's universe?

There are thousands of important (and not so important) problems for which there are no solutions yet or only tentative ones. However, as you will see in this course of study, no matter how different from each other these problems appear to be, often there are some processes of thinking commonly used to deal with them.

A Model for Problem-solving Processes

As in Peppermint Patty's case, that "elusive connective tissue" for solving a problem may not come easily. Your frustrations in trying to solve problems may even be increased by your efforts to fit your understanding of the problem to another person's method. So it is very important to learn to rely on your own style in thinking about problems and their solutions.

Problem solving includes a variety of actions, processes, and ways of thinking ranging from highly organized and rational to haphazard and autistic. Being intuitive falls roughly between these two extremes of problem solving. The rational processes include the structured, well-known, and accepted methods of solving problems; intuitive processes, by contrast, are less formal and somewhat loosely tied to the orderly, well-structured rational procedures. Autistic processes, on the other hand, often yield novel solutions—sometimes so novel as to be called "off-the-wall" by some people. However, some problems have virtually required unusual ways of thinking about a dilemma, even to the point of fantasizing.

Source: © 1979 United Feature Syndicate, Inc.

▲ Rational   ▲ Intuitive   ▲ Autistic

Consider this problem as an example of how these three ways of thinking might deal with a dilemma:

A truck has approached an underpass on a sidestreet in lower Manhatten. Unfortunately, the top of the truck is two inches higher than the underpass allows for safe passage through.

**TAKE A MOMENT AND THINK OF SOME
                             SOLUTIONS TO THIS DILEMMA!**

Rational thought might lead to reversing the truck's direction and searching for an alternative route to the other side. *Intuitive* processes might convince us to try letting some air out of the tires, or loading some more cargo on the truck, to lower the body enough to go through. *Autistic* ideas might include:

> Airlift the truck over the overpass.
>
> Raise the overpass.
>
> Dismantle the truck or the overpass.
>
> Approach the overpass with sufficient speed to force the truck through. (After all, it's just the top two inches that won't fit!)

What were some of your solutions to this trucker's dilemma? Do they fit into one of the categories above? If you're like most of us, your first solution would be a rational one; hitting on a logical idea first probably prevented your mind from going any further and considering any other options. Further characterizing these three thought processes might now be in order.

RATIONAL PROCESSES

Rational processes have rigid structures, patterns, and routines that have "stood the test of time" in being used effectively to organize problems and find solutions. Many of these thought processes can be learned formally; our public K-12 schooling concentrates on this type of mental process. Cultivating your skills in using rational procedures can greatly increase the number of alternatives you have in organizing information about problems. Thus a fairly reliable self-control can be exerted over the rational methods of solving problems. In fact, a considerable number of rational methods frequently provide ways of resolving a wide range of dilemmas.

One rational approach to Peppermint Patty's banana problem might be to represent the problem algebraically. Such an approach might include a step-by-step analysis of the integral parts of the situation:

1. Represent "unknowns" with letters or symbols.
2. Search for a relationship between the unknowns.
3. Write an equation, expressing mathematically the relationship between the knowns and unknowns.
4. Solve the equation.

Peppermint Patty's banana problem is not a very important one, but the general analytical process outlined above is useful for a large number of similar problems, some of which *are* worth our time. Keep in mind also that this particular approach is just one of many rational ones that could have been used. If you haven't had much recent experience with algebra, you might have organized the problem another way. Later in this chapter you'll get a chance to explore several different types of rational thought methods.

Realistically, there are many problems that do not lend themselves to existing logical methods, or perhaps personally you can't seem to generate a successful method.

In such a situation, it may be necessary to generate new or more suitable strategies. Perhaps it is fortunate for us that life is not of such a constant nature that only a few problem-solving methods are all that are required of us. Life is infinitely more interesting, being variable and refusing to be totally under our control!

INTUITIVE PROCESSES

Brainstorming, formulating hunches, and daydreaming are all intuitive processes. Compared to the rational, the intuitive methods are less willfully controlled, less predictable, and not as reliable; they involve greater risk taking than the rational variety. But in order to develop new strategies for solving problems, we often need to invent or design different ways of viewing the dilemma. This might require organizing our perspective of the situation in a way that doesn't conform to previously accepted rational views of the situation. We might have to "loose our imagination" to release our intuition.

Consider again Peppermint Patty's problem. You might decide to try an intuitive "guess an answer" approach. You would test your first guess against the information in the problem, and then probably revise your guess a second and even a third time. The better your intuition, the better your succeeding answers would be. (Notice that, even with intuitive thought, you return to rational thought to check out your intuition.) If you were successful with this method on several different problems, you might even refine the whole procedure into your own "guess-test-revise-test" rational process to use on future problems. In so doing, you would have used your intuition to develop a new solution strategy, a new way of approaching future dilemmas.

• • •
• • •
• • •

PROBLEM

Connect the nine dots to the left with four connected straight lines without lifting your pencil from the paper.

TO GAIN THE FULL EFFECT OF THIS PROBLEM, TRY TO RESOLVE IT BEFORE READING THE EXPLANATION THAT FOLLOWS!

If you were unsuccessful at connecting the dots in the prescribed manner, it was probably because you unnecessarily restricted yourself to the boundary formed by the dots along the outer edges. Our minds have encountered situations like this before, and most of the time we were supposed to stay within the figure itself, even if it wasn't stated explicitly. Our rational minds filed this information away somewhere, and it cropped up here. The intuition suggesting that you extend the lines beyond the edge of the figure is helpful in breaking this self-imposed confinement.

So, how did you do? Were you successful? Did you initially restrict your exploration to the interior portion? Can you solve the problem now that you have a hint? Try it out on a friend and see what happens.

Our description of problem-solving processes thus far implies a close relationship between the rational and intuitive methods of thought. In fact, the processes support each other. The rational procedures are used to test our intuition, and using our intuition to give us a new way to view a situation may in turn result in a new rational approach to similar situations.

AUTISTIC METHODS

Autistic processes consist of such acts as dreaming and fantasizing. The product of such thinking may be "illogical" in that it doesn't fit well into rational explanations, unless some adjustment or revision is made in the current way of thinking. Indeed, most products of autistic thought will *never* fit into existing beliefs or theory, and since society generally doesn't encourage thought patterns that yield results only a small percentage of the time, we are reluctant to spend time developing our ability to think autistically. However, since so many important problems are still unresolved, even after many years of rational thought having been expended on them, we can't afford to discount all autistic thinking. We sometimes need revolutionary ideas to resolve perplexing situations.

Speaking of autistic ideas, think how tentative you would have felt if you had been the first person to:

Suggest that the sun—not the earth—was the center of the universe!

Consider changing Euclid's fifth postulate which says that "through a given point not on a line, there's only one line parallel to the given line," and create "new" geometries!

Imagine that people might be able to communicate by thoughts alone!

Discover that a common, ordinary mold could stop the growth of harmful bacteria!

Think you might be able to relieve people of pain by sticking needles in their bodies!

Autistic ideas may not enjoy a fantastic batting average, but when they do get a hit, it's usually a home run! This type of thought, when successful, is associated with a major upheaval.

SUMMARY

When dilemmas occur, old information and assumptions are reviewed, reorganized, and tried out—rational thought. If this is not successful, our minds search for new evidence resources, or explanations; we begin to use our intuition to gather new "hunches"— intuitive thought. Finally, perhaps the situation will require that we completely restructure basic assumptions, considering even ideas that seem (at the moment) illogical— autistic thought. Only after we've exhausted all these possibilities should we claim that we've given the problem our best possible effort.

SET 1

1. By moving only three coins, make the set of coins on the left look like the set on the right.

   Did you systematically (rationally) organize an approach to this problem, or was your method strictly based on intuition?

2. Revisit Peppermint Patty's banana problem. After you have tried to find the weight of the banana, analyze your thinking in terms of rational or intuitive thought.

3. Multiply 43 x 37 using paper and pencil. Would you consider what you did "rational thought" or "intuitive thought?"

4. Consider this method of multiplying 43 x 37:

   Does the answer agree with yours? Is this a rational procedure, do you suppose? If so, can you figure it out? Do you think it might even be as good as your own method?

   ```
     43
    x37
    111
    148
   1591
   ```

5. A first grade student had been taught to add single-digit "signed numbers" but had never encountered "regrouping" or "borrowing." Yet she figured out a method of solving 22 - 17. Her work looked like this:

   Would you say the student was using intuitive thought, or rational thought?

   ```
    22
   -17
    -5
    10
     5
   ```

6. Ferdinand Magellan was the first person (purportedly) to demonstrate that the Earth was round instead of flat. Discuss in a paragraph how his thinking processes might have incorporated combinations of rational, intuitive, and even autistic thought.

7. The pie to the right needs to be sliced into the maximum number of pieces possible, using only six straight-line cuts. How many pieces will that be?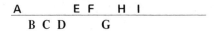

8. Place the remaining letters of the alphabet above or below the line, according to some rational process consistent with what is shown.

9. In one continuous drawing, without lifting your pencil and without retracing any line segment, recreate the figure below:

Also, try this one:

10. The solutions to some problems may be "beyond" mathematics. What's the catch to these dilemmas?

   a. Two beggars were busy bumming dimes on the street corner. The tall beggar was the short beggar's father, but the short beggar was not the tall beggar's son. How could this be?

   b. On a straight road, a man in his car heading due east drove his car for one mile without making any turns. At the end of the mile the man was exactly one mile due west of his starting point. Explain how such a thing might happen.

   c. Determine the number of grooves on a 12-inch 33-1/3 RPM record.

   d. Under what conditions might two sisters who look alike and were born of the same mother on the same day not be twins?

11. Use six real coins to form the configuration shown below. Then arrange the coins to have four coins in each line (across and down).

A Closer Look at Rational Processes

Rational thought processes can be organized into three levels of mathematical thought and action. The organization described here is hierarchical in nature; that is, a higher level is said to be intellectually more complex than a lower one, and the higher one subsumes all lower levels.

LEVEL 1: MEMORY AND RECALL

The lowest level in this hierarchy of rational thought consists of recognizing and remembering mathematical facts or concepts. Functioning at this level means merely recalling information—such things as the basic addition facts, the meaning of =, and the decimal value of π. In the cartoon to the right, Sally has failed not only to memorize the basic multiplication facts, but also to learn the conventional names for numbers. She's not operating at the first level of rational thought for multiplication and hence will have difficulty in moving to the next higher level of rational thought.

Operating at this level does not demonstrate any understanding of underlying concepts, nor is it engaging in true problem solving. At this recognition/recall level, we are mainly dealing with the question—What is it? The answer to such a question involves "static knowledge," since just recalling a fact, definition, or rule certainly doesn't imply we can do anything functional with it. Knowing that 3 x 3 = 9, 3 x 4 = 12, 7 x 3 = 21, and 7 x 4 = 28 is required for the problem 43 x 37, but knowing those facts by themselves certainly would not ensure success. Working at Level 1 simply means that we have committed certain things to memory, and can recall them at the appropriate time.

Source: © 1967 United Feature Syndicate, Inc.

The student below is demonstrating Level 1 knowledge of the Pythagorean theorem.

Source: B.C. by permission of Johnny Hart and Field Enterprises, Inc.

In any math class, no matter how advanced, there is some memory/recall work to do!

Remembering and recognizing simple facts is a necessary part of problem solving, but learning objectives must focus on more complex processes. Hence we arrive at the second level of rational thought.

LEVEL 2: ROUTINE STEP-BY-STEP PROCEDURES

Performing step-by-step processes is the next highest level of rational thought. Success at this level requires mastery of the previous level (memory of basic facts and definitions), yet this stage goes beyond simple recall.

The student to the left is recalling more than the basic facts of multiplication. She is also demonstrating mastery of a step-by-step procedure for solving any whole number multiplication problem!

To function at this level is to apply a previously learned set of steps in a fairly routine manner.

Elementary students learn routine processes—*algorithms*—for going beyond the basic facts of arithmetic, so they can handle such exercises as:

```
   24        43        29      8 ⟌ 1939
  +18       -27       x38
```

Take a moment and work the problems above. Can you tell when you are merely recalling a fact, distinguishing it from the step-by-step procedure?

In your K-12 experiences, you probably learned a procedure for solving simple equations. Here is an example of one such procedure:

Equation	Procedure	
$3x + 4 = 7$	Add -4 to both sides.	$(3x+4) + {}^-4 = 7 + {}^-4$
	Use associative property to regroup.	$3x + (4 + {}^-4) = 3$
	$4 + {}^-4 = 0$	$3x + 0 = 3$
	Rewrite simplified form.	$3x = 3$
	Divide both sides by 3.	$3x \div 3 = 3 \div 3$
	Rewrite simplified version.	$x = 1$

Due to its routine nature (after much practice), solving simple equations like the one above is considered a Level 2 process.

Can you work at the second level of thought with the Pythagorean theorem? If so, use a step-by-step procedure to find the hypotenuse of the triangle to the right.

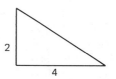

Operating at this level of rational thought is beginning to deal with static knowledge in terms of the question—What of it? These processes are predicated on their use in solving some meaningful problem at some time in the future. However, learning a step-by-step procedure is primarily involved with developing routine skills and proficiencies. We need these skills, but mainly so we can progress to the third level of rational problem-solving processes.

LEVEL 3: ANALYSIS AND SYNTHESIS

The third level of rational thought consists of mental processes known commonly as "analyzing" and "synthesizing." *Analysis* of a dilemma involves breaking it down into its component parts, examining each in turn.

When a vacuum cleaner won't suck up a pile of dirt, we might try taking the whole thing apart and looking at the individual pieces and how they fit together. Perhaps we'll find the problem—a full dust bag, a hole in the hose, a clogged pipe, etc.—by examining each part separately.

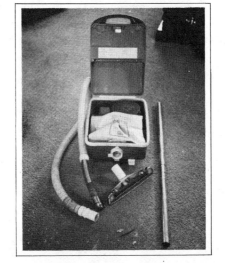

Sometimes, analyzing a problem will yield a solution, while at other times the analysis is only an attempt to get to know the problem intimately.

Another example of analysis comes from problem 7 in Set 1, which asks you to determine the maximum number of pieces of pie you could obtain from six straight cuts. Consider this breakdown of the problem:

Draw several circles to represent the pie.

Then begin drawing lines, one by one, across the circles to represent the slices, trying to get the largest number of slices with each new cut.

step 1   step 2   step 3

Note carefully in the preceding example that two different situations are possible for the third slice: we can either slice through the intersection of the two previous cuts, or slice somewhere else. We'd choose the last option, since our only concern is to produce the maximum number of pieces. The "best" third slice would produce seven pieces, then. (Count them.)

So our analysis would continue, breaking the problem into the integral parts of adding a new slice to what's already there, working toward the final goal of six slices.

Synthesizing—one of the highest forms of rational problem solving—involves putting separate pieces together to resolve a dilemma. After a vacuum cleaner has been taken apart and examined piece by piece, it must be put back together carefully so that the individual parts will perform the overall function correctly. This act of seeing how unique parts can mesh together to form a total picture is what synthesis is all about.

Perhaps somewhere in our analysis of the pie-slicing problem, we would discover a pattern that could be generalized to similar situations. This process would be a synthesis of the problem in that we would be observing that the given dilemma fits into a larger picture. Notice how this highest form of rational thought begins to encroach on intuition!

SUMMARY

In this section, we have discussed three levels of rational problem-solving processes: (1) memory/recall; (2) step-by-step, routine procedures; and (3) analysis and synthesis. "Problem solving" is a term usually reserved only for Level 3.

In the following set of exercises, you will have the opportunity to work at all three of these levels. You might find it beneficial to review your work on each problem in light of Level 1, Level 2, and Level 3 processes.

SET 2

1. Thirty-two teams are to compete in a single elimination basketball tournament to determine the NCAA champion. How many games must be scheduled?

2.  Two gamblers are playing a game in which the loser must pay an amount equal to what the other gambler has at the time.

   Player A won the first and third games, while B won the second game; they both ended with $12. With how much money did each begin play?

3. A farmer was asked by a passing stranger how many rabbits and chickens he had in his yard. Not being a very direct person, the farmer replied, "Between the two there are 76 eyes and 126 feet." How many of each kind of animal did the farmer have?

4. In a certain variety of sunflower, the seeds seem to emanate in clusters from the center of the flower according to the sequence:

   3   3   6   9   15   24   ... etc.

   Assuming that the number pattern above continues, determine the size of the next three clusters.

5. How many fence posts are required for 130 feet of straight fencing if the centers of the posts are to be ten feet apart? What if the fence is going to be a closed circle?

6. In the circle to the right, what is the length of the dotted line segment?

   (The center of the circle is labeled 'C,' and the radius is 10.)

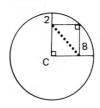

7.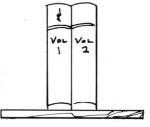

   In a certain pond, a patch of lily pads doubles its size each day, once it gets started. If the pond was completely covered on the 20th day, on what day was it half covered by lily pads?

   What fractional part of the pond was covered on the 15th day?

8. A follow-up to the pie-slicing problem: Suppose that you always slice a pie to get the most pieces, disregarding whether the pieces are of the same size! Can you determine whether a relationship exists between the maximum number of pieces and the number of slices? Without actually drawing "pies" and "slicing" them, try to determine the number of pieces you would get from 20 slices.

9. A two-volume set of history books is on a shelf in the usual manner, the books upright and side by side. The total thickness of the pages of each volume is 2 inches and the thickness of each cover is 1/8 of an inch.

   What is the distance between the first page of Volume I and the last page of Volume II in this arrangement on the shelf?

10. Construct a square whose area is approximately 7 square centimeters. By "approximately," we will mean that the area of the square you construct should not be more than 0.1 cm^2 from being 7 cm^2. (To do this, you might want to think first about the correct length of the side of such a square.)

11. If asked "What is the largest number in the world?" a first grader would probably respond with something like "one billion." Can you come up with an argument that might convince him that there is no largest whole number?

12. You are placed in charge of designing a stained glass window, and decide to base your design on 20 straight lines that cross the window, with no three lines crossing at the same point. This will give you the greatest number of colored sections possible for 20 lines (if you cut across the previous 19 cuts). How many different sections of glass will you have to cut for the window in doing this?

13. Give a rational argument that there is no whole number (i.e., no number from the set 0, 1, 2, 3, 4, 5, . . . , etc.) that will solve this equation:

   $N \times N = 90$

   Things to consider in doing this: Try to estimate a value for N that would make the equation approximately true. Find values that are too small and values that are too large and compare them.

14. Can you name at least one number between 1/3 and 1/2? Can you verify that your number is between these two numbers? List the mathematical ideas that occur to you in analyzing this problem.

15. Patterns are useful in most areas of study, and often interesting and sometimes important relationships are discovered. Analyze the following arithmetic statements for any pattern or conjectures you think might be suggested by them:

   $(0 \times 2) + 1 = 1$
   $(1 \times 3) + 1 = 4$
   $(2 \times 4) + 1 = 9$
   $(3 \times 5) + 1 = 16$
   $(4 \times 6) + 1 = 25$

   Would you venture to guess the next five arithmetic statements in this list? Why did you choose the ones you did?

Useful Analytic and Synthetic Processes

An analysis of a problem can take many forms, and a particular approach to investigating a problem is a matter of the personal style and intuition of the problem solver. The infinite varieties of all of the useful analytical processes could not possibly be described, but we can take a look at some of the rational procedures that have been in general use for a long time—most of them over a thousand years! In this section, we will consider nine of these analytical methods.

Often, more than one rational process can be applied to the same problem. The first four analyses will be applied to the first problem from Set 2: 32 teams are to compete in a single-elimination basketball tournament to determine the NCAA champion. How many games must be scheduled?

Before going ahead, review briefly your own solution to this problem.

A concrete approach to this dilemma is to represent the teams and their games with a picture of the "pair-brackets" shown to the right. Then you could merely *count* the number of games as the teams, and subsequent winners, are matched. A "concrete" analysis involves either physically or pictorially showing the situation. This analytic method has an advantage over others in that, many times, the solution can be obtained with nothing more than counting, as in this problem. It also is extremely easy to show others the solution.

A CONCRETE ANALYSIS

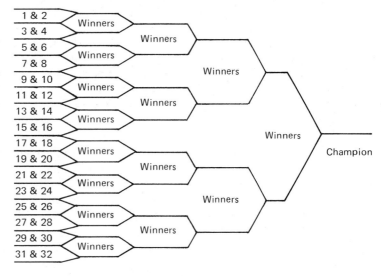

The usefulness of this method should not be slighted. Drawing diagrams of problems is an activity that's encouraged all the way from kindergarten through careers in mathematical research.

A slightly different approach to this problem is to represent the events of the problem numerically, or with words, instead of pictures:

A NUMERICAL ANALYSIS

    32 teams are matched (first round)   — 16 games

    16 teams are matched (second round) — 8 games

    8 teams are matched (third round)   — 4 games

    4 teams are matched (fourth round) — 2 games

    2 teams are matched (fifth round)   — 1 game

Now we merely add the number of games at each round and obtain the total. This sum is 31 games. This method seems almost as "concrete" as the previous one, but keep in mind that that's probably because you've already had the benefit of seeing the "blueprint" of the solution above. This analysis is definitely harder to arrive at, on its own, than the former one.

There are some limitations to reaching a solution using the first and second types of processes. Certainly they solve the problem at hand with ease, but they don't lend themselves directly to generalizations. That is, we are clearly able to solve similar dilemmas by using similar representations, but if our analysis stops here, we will not have learned any general principle that can be used directly to solve other such problems. We haven't yet used synthesis.

INDUCTIVE ANALYSIS

A different way to view this problem is to consider it as an instance of a larger situation. By solving less complicated instances of the same problem, we may be able to get a hint about how to solve this particular case. In this case, we could get some simpler problems with the same essential features of this one by varying the number of teams to start the tournament. We would solve these simpler instances by brute force (perhaps using a concrete approach), and set up our work so that any number pattern that's hidden will come into focus.

# of teams	# of games
1	0
2	1
3	2
4	3
5	4
.	.
.	.
.	.

Fortunately, in this situation there appears to be a fairly obvious pattern between the number of teams and the number of games necessary to determine a champion. Provided that this relationship continues to hold, we have a valid method of finding the number of games that must be played, given *any number* of teams that begin.

This third analytical method is often called "empirical induction." The procedure is to observe sufficient numbers of the same phenomenon and gain confidence enough to draw some conclusion for the general case. The validity of the method depends heavily on the extent to which each of the simple cases represents the general case, and on the relationship observed continuing. Therefore there are dangers in relying completely on empirical induction for drawing general conclusions. However, the method is an excellent one for exploring and predicting reasonable relationships between two or more variables. In most cases, the activity is followed by some form of "proof" that the pattern observed does, in fact, truly explain the situation.

A TRANSFORMATIONAL ANALYSIS

Sometimes a simple change or transformation, derived from the basic concepts or events of the problem, aids in the analysis. Consider the following line of reasoning for the tournament problem:

1. Each game determines one losing team.

2. In all, 31 teams must lose.

3. Since each loss requires 1 game, and 31 of the teams must lose, 31 games are necessary to determine the champion.

This process transforms the basic event—a team loses—into a corresponding event—a game is played. There is a one-to-one correspondence between each game played and each losing team. Then, to solve the problem, all that must be recalled is the number of teams that have to be eliminated, namely 31. This transformational process also provides us with the power to generalize about similar problems of this type. In fact, the transformation (a one-to-one correspondence) allows us full confidence in establishing a general rule for solving others of the single-elimination kind.

Another example of a helpful transformation comes from Problem 6 of Set 2. The problem calls for finding the length of the dotted line segment, which is a diagonal of the rectangle shown. The following transformation yields an easy solution:

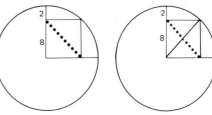

1. The dotted segment is a diagonal of the rectangle, and a rectangle has two diagonals, both the same length.
2. The other diagonal is a radius of the circle, and hence has length 10.
3. So the dotted line has a length of 10 also.

The transformation here involves substituting *radius* for *diagonal.*

Solutions derived from these and other types of transformations frequently require some experimentation. A useful new way to view the problem may come when we realize that some key concept may be substituted for another. In the first example, one *loss* was substituted for one *game*; in the last example, one *diagonal* was substituted for a *radius* of the circle.

WORKING BACKWARD

Some problems may best be approached "through the back door." Problem 2 of Set 2 is of this nature: Two gamblers are playing a game in which the loser must pay an amount equal to what the other gambler has at the time. Player A won the first and third games, while B won the second; they both then had $12. With how much money did each begin play?

This problem is highly suggestive of working backward for the simple reason that we are aware of the final result and we are asked to derive the beginning state of affairs.

From the end of game three, we can trace step by step the amount of money each gambler had at the end of the previous game:

After Game #	A had	B had	Explanation
3	$12	$12	Given this information.
2	6	18	A won game 3; doubling 6 gives 12.
1	15	9	B won game 2; doubling 9 gives 18.

So A must have started with half of $15, since A won the first game. Therefore A started with $7.50. But then B must have started with $16.50 since B had $9 after giving A $7.50.

If you solved this problem initially, was your method similar to the one described above?

Another of the problems from Set 2 is easily solved by "going through the back door." Can you find it now? The key is that you know the end result of the dilemma!

A "GUESS, TEST, AND REVISE" TECHNIQUE

Problem 10 of Set 2 required that you find the dimensions of a square whose area is 7 cm². One useful bit of information to recall (Level 1) is that the area of a square is the length of one of its sides, multiplied by itself. Or in this case, $s \times s = 7$ cm², where s is the required length.

One practical way of finding an approximation of s is to begin *guessing*. Then we would need to *test* our guess to find out whether it's "good enough." Recall that "good enough" for this problem means that whatever value we get for s, it must hold true that $s \times s$ can't differ from 7 cm² by more than 0.1 cm². If our guess is not good enough, we would *revise it* and test again.

So let's try it, and begin by making a first guess. We might start by thinking along these lines:

1. 2 is too small, since $2 \times 2 = 4$. Likewise 3 is too large since $3 \times 3 = 9$. So the value for s must be between 2 and 3. Let's guess 2½ or 2.5 first.

2. Testing: $2.5 \times 2.5 = 6.25$; $7 - 6.25 = 0.75$. So our guess is not yet close enough. To get closer to 7 with $s \times s$, we'll have to raise the estimate. Try $s = 2.7$.

3. Testing: $2.7 \times 2.7 = 7.29$, which is too large! So we'd have to revise downward from $s = 2.7$.

In continuing, we would "narrow down" the estimate for s by using this guess, test, and revise approach.

The usual method of doing a long-division problem also uses this procedure. The first step is to use a "trial quotient," and then multiply this estimate by the divisor. At this stage, the product is tested to see if the "trial quotient" was too large. If so, a revision is necessary;

Example

$$\begin{array}{r}4 \leftarrow \text{TRIAL} \\ 42\overline{)16452}\end{array} \qquad \begin{array}{r}4 \\ 42\overline{)16452} \\ 168\end{array} \text{TEST (too large)} \qquad \begin{array}{r}3 \leftarrow \text{REVISION} \\ 42\overline{)16452}\end{array}$$

if not, we subtract and then test again, but this time to see if the "trial quotient" was too small originally.

Example

$$\begin{array}{r}8 \leftarrow \text{TRIAL} \\ 58\overline{)5636}\end{array} \qquad \begin{array}{r}8 \\ 58\overline{)5636} \\ 464 \\ \hline 99\end{array} \text{TEST (too large)} \qquad \begin{array}{r}9 \leftarrow \text{REVISION} \\ 58\overline{)5636}\end{array}$$

This analytic method of solving problems relies on your intuition for its efficiency, and the problem must be one in which the guess can be tested in very certain terms as to "how good" it is. Otherwise, you wouldn't know how to revise the guess, or even when you'd obtained the correct solution. However, the method is extremely useful when a more direct method is not available or has been forgotten, and the "test condition" is given in the problem. The "rabbit/chicken" problem is a nice one to try this method on, particularly if you've forgotten algebraic methods (see Problem 3, Set 2).

A RECURSIVE RELATIONSHIP

The pie-slicing problem of Set 1 lends itself to recursive analysis. If each step of a problem can be considered an extension of the previous step or steps, we are searching for a relationship that recurs in all the steps. The table below summarizes the first several steps in the pie-slicing dilemma.

Notice that to find the total number of pieces at any step, say after the 20th slice, you would have to know the number of pieces after the 19th slice—the number after the 20th slice is the number after the 19th slice, plus 20.

A recursive analysis can be considered a "partial generalization" since you do arrive at a way to find the answer to a whole set of similar, individual problems. Having to know the results of all previous steps before reaching the stage you're interested in certainly imposes a limitation on recursive methods. However, it should be noted that the process of searching for such a relationship sometimes releases a previously hidden, direct solution to the problem. Can you find a direct solution here?

# of Slices	# of Pieces	Increase in # of Pieces
1	2	1
2	4	2
3	7	3
4	11	4
5	16	5
6	22	6
.	.	.
.	.	.
.	.	.

The sunflower seed problem in Set 2 lends itself to a recursive formula for finding the number of seeds in any given cluster. If you have not already done so, derive such a pattern and apply it to finding the number of seeds in the 12th cluster.

USING GENERALIZATIONS OR METHODS FROM PREVIOUS PROBLEMS

In the analysis of the 32-team basketball tournament problem, both the inductive and transformational analyses yielded a generalization applicable to certain related problems. The rule discovered was: "In single elimination tournaments, the number of games required is one less than the number of teams participating."

Thus, analyzing a problem and labeling it as one whose essential features fit the requirements above allows us to apply this previously learned rule. Or, if we'd forgotten the rule, perhaps the method used to generate the rule would be recalled, and could be used again. In either case, recalling a previous problem would give us a head start in solving a new problem like it.

AN INDIRECT ANALYSIS

There are many times when people do not approach problems directly. Usually, it is the nature of the problem that suggests an indirect method of attack. For example, budgetary matters are sometimes considered as in this manner:

> *You have $600 for the month, to cover all your essential expenses, and, perhaps, a four-day vacation. You know beforehand that the vacation will run upwards of $300. Should you go ahead and take the vacation early in the month, or wait and see?*

Before the decision above can be made, you would probably calculate your other essential monthly expenses, and base your holiday decision somewhat indirectly on taking care of these other commitments first. In other words, if your essential expenses were greater than $300, you would decide against the trip.

The last problem from Set 2 asked you to come up with an argument that might convince a young child that there was no largest number. We have to be careful here, because many young children can't follow a string of arguments of any nature, particularly if there's a "twist" at the end. But you might try something like this:

To the child: Pick the biggest number you know of.

Child says: One billion.

To the child: How about one billion and one?

Child says: Oh, yeah, that's bigger.

To the child: Well, try again.

Child says: Twenty zillion.

To the child: How about twenty zillion and one?

Child says: Oh, yeah.

The conversation above could be continued as long as the child's interest held up. Notice that you're planting the seed of an indirect proof that there's no largest whole number. The indirect nature of the argument is that you're essentially saying "give me the largest whole number, and I'll prove there's a larger one by adding something to what you gave me. Thus, there *is* no largest whole number."

While the example of indirect analysis above may appear somewhat trivial, don't let this method escape your attention. Actually, indirect methods of proof are extremely powerful, and have been used to crack some of the toughest problems in advanced mathematics—such as the fact that there are an infinite number of prime numbers and that $\sqrt{2}$ can't be a rational number.

SUMMARY

We have briefly examined nine rational thought processes that have a long and successful history in mathematical problem solving. While some problems seem suited for a particular approach, many problems can be solved by several of the methods or by a combination of methods. As you gain experience in using the techniques, your intuition grows and you increase the likelihood that you will solve other problems in the future. This list of nine processes certainly doesn't exhaust the rational procedures, but it does give you a good start.

The three categories—rational, intuitive, autistic—described in our discussion of how we solve problems are not discrete and disassociated methods. Several processes are often (and, indeed, must be) used together. This is particularly true of "rational" and "intuitive" processes.

In closing this introductory chapter, mention of the "creative process" may be in order. Synthesis, the highest form of rational thought, is certainly part of original, creative endeavors. And yet sometimes autistic thought is also required. Since the greatest problem solvers in history have been those fortunate few who could use rational, intuitive, and autistic processes on the same problem, why not spend time now covering methods of thinking in alternative ways, some of which go beyond rational and intuitive thought processes?

Divergent, autistic processes have not been scrutinized by history like the rational and intuitive methods have been. Therefore less is known about how to induce, or at least liberate, such methods of thought. For now, we'll just have to be content with recognizing them when they occur, realizing that they have made legitimate contributions to scientific theory throughout history.

In the problems below, you'll have the opportunity to practice some of the rational techniques covered in this unit. Keep in mind that there's no "set" way to solve any of these problems. Select a method that's in accordance with your understanding of the dilemma. Take a chance!

SET 3

1. For a national soccer tournament, 50 teams are to compete, and a team is out after its first loss. How many matches must be scheduled to find the champion?

2. Give an argument to show that there is no largest multiple of three (a multiple of three is, by definition, 3 x n for any counting number "n"; the multiples of three are 3 x 1, 3 x 2, 3 x 3, 3 x 4, 3 x 5, 3 x 6, etc.).

3. A room had both girls and boys in it. After 15 girls left, there were two boys for each girl. Shortly afterward, 45 boys left the room, leaving five girls for each boy. How many girls were originally in the room?

4. Three gamblers were playing a card game in which one person loses and the other two win each time. The loser must pay each winner the amount the winner has on hand at the time. At the end of the third game, each gambler had lost once, and in the order A, B, C. Each also had $6 at the end of the third game. Did they have the same amount of money to start with? If not, determine how much each had.

5.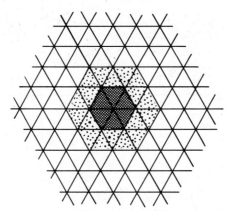

The figure to the left consists of triangles. The inner six triangles have been shaded—that is the first layer. The next layer of triangles is dotted, and has 18 members. The next layer beyond that would have 30 triangles!

The pattern has been cut off in the middle of the fifth layer. If it were continued, how many members would the tenth layer possess?

6. A hungry fox ate 100 grapes in five days, each day eating six more than the day before. How many grapes did he eat on each of the five days?

7. Find the sum of the first 1000 counting numbers.

$$(1 + 2 + 3 + 4 + 5 + \ldots + 997 + 998 + 999 + 1000 = ?)$$

8. Without using a calculator, find an approximation for $\sqrt{15}$ that's close enough so that, when squared, the difference from 15 is less than 0.01.

9. Eight baseball teams are to play in a double-elimination tournament in which you are out after two losses. How many games *must* be played to determine a champion? Is it possible that an additional game may be required?

10. A pen costs $1 more than an eraser; the two together cost $1.10. How much does the eraser cost?

11. A contractor buys a square city block to build a new development, and finds that each of the four sides can be divided into six lots (including the lots on the corners of the block). How many houses can be built?

If he decides to plant "privacy hedges" in any spot where the side of one house is right next to the side of the neighbor's house, how many hedges will he plant?

12. Name a ten-digit number such that:

    a. The first digit tells how many 0s in the number.

    b. The second digit tells how many 1s in the number.

    c. The third digit tells how many 2s in the number, etc., until the tenth digit tells how many 9s in the number.

13. Toothpicks are arranged as in the figure on the right to form five squares of the same size. By moving (but not removing) exactly two toothpicks form a figure of exactly four squares, each of the same size as the original squares. Do this so that each of the toothpicks is the side of some square in the configuration. (Note: Placing a toothpick side-by-side another toothpick is illegal.)

14. Suppose you must be at your doctor's office at 3:00 p.m., and you have several errands to run on the way. Discuss the reasoning—in terms of the kinds of thinking processes described in this section—you use to decide when you will actually leave your house to get to the office on time.

15. Study the configuration of numbers shown below (it is often referred to as Pascal's triangle) and determine the rules that seem to govern it. Use your rules to write the next three lines in the pattern.

$$
\begin{array}{ccccccccccc}
 & & & & & 1 & & 1 & & & \\
 & & & & 1 & & 2 & & 1 & & \\
 & & & 1 & & 3 & & 3 & & 1 & \\
 & & 1 & & 4 & & 6 \searrow \swarrow 4 & & 1 & & \\
 & 1 & & 5 & & 10 & & 10 & & 5 & & 1
\end{array}
$$

Can you determine the 20th row of numbers in this pattern? Try it.

16. Suppose in a study of Slavic* documents you came across the figuring recorded below:

$$84 \times 16 = \underline{\quad ? \quad}$$

H.	D.	
~~84 × 16~~		64
~~42 × 32~~		256
21 × 64		+ 1024
~~10 × 128~~		1344
5 × 256		
~~2 × 512~~		
1 × 1024		

Analyze what is going on here in terms of the "rules" which govern the sequence of steps. Determine what the rules are, and use them to carry out the same process on 92 × 12.

17. Consider the dilemma represented in the cartoon:

Source: © 1979 United Feature Syndicate, Inc.

Peppermint Patty's problem above is a formidable one to solve by brute force. A concrete approach would be to find nine books and a shelf, and physically arrange them in all possible ways, counting the ways as you go. Probably you'd soon begin to develop an organized procedure for arranging them so you could keep up with what you'd done before (it turns out there are over 300,000 ways, so you'd need a recording system).

---

*Note: This process has often been called the "Russian peasant method" for multiplying (or "halving and doubling").

After a while you'd probably repeat Peppermint Patty's remark and quit, or look for an easier method. One thing you might try is an *inductive analysis* as described in this section, that is, you would look for simpler problems with the same essence as the given problem. You would solve these simpler problems by brute force, and set up the data so that if a pattern is present, it will become obvious. If successful at finding a pattern, you would *assume* the same thing held for the given problem, and so answer it using inductive reasoning as your basis.

Applying this process to Peppermint Patty's problem we would find the number of ways one book could be arranged on a shelf—there's only one, of course. The next simplest cases would be two books, then three books, then four books, and so on. These can be solved by brute force, as shown below. Study this carefully for a moment.

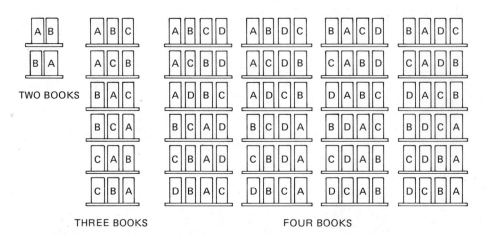

Doing the same for five books becomes pure torture, as there are 120 such ways to arrange five books on a shelf. (The authors did it and it took up a whole page. Be glad for small favors like not being asked to do it yourself!)

The table on the right shows a way to organize the data collected so far.

With a little messing around, you'd probably soon notice a relationship between the two sets of numbers. Play around with the information for a couple of minutes. If you do notice a pattern, test it on "six books."

Finally, can you help Peppermint Patty answer her question?

# of Books	# of Ways
1	1
2	2
3	6
4	24
5	120

REFERENCES

Henderson, K. B., and Pingry, R. E. *Problem Solving in Mathematics.* Reston, Va.: National Council of Teachers of Mathematics, 1953.

Polya, G. *How to Solve It.* 2d ed. Princeton, N.J.: Princeton University Press, 1945.

———. *On Understanding, Learning, and Teaching Problem Solving.* Mathematical Discovery, vol. 1. New York: John Wiley & Sons, 1962.

Wickelgren, W. A. *How to Solve Problems.* San Francisco: W. H. Freeman & Co., 1974.

2

THE THIRD R

Reading, 'riting, and 'rithmetic—the three R's—the focal points around which school subject matter has been centered for hundreds of years, is gaining even more notoriety lately than in years gone by. The third "R" has come to be associated with such terms as "computational proficiency" and "minimal arithmetic skills." While these terms are misleading and certainly limiting to a true interpretation of what mathematics is all about, the phrases do indicate society's concern that elementary students learn the survival skills of arithmetic. Most likely our society will always expect its citizens to be functional with regard to the basic number systems. Elementary students will continue to study addition, subtraction, multiplication, and division of whole numbers, fractions, and decimals.

Basic to any study of a number system, and computation in it, is an understanding of the associated numeration system. Just as written words and sentences represent ideas and thoughts, a numeration system with its own "grammatical rules" represents number ideas and relationships. A *numeration system,* then, is merely a way of naming numbers. As you might expect, different civilizations have produced different ways of naming even the most basic set of numbers, the whole numbers (0, 1, 2, 3, 4, . . .).

A rather surprising fact to most people is that many experts can't agree on which came first, the concept of number, or that of numeration. Certainly it seems logical that systems of naming concepts would not have preceded the concepts themselves, yet there's evidence to support this contention in the number-numeration issue. Regardless of who's correct about the origins of the two, it makes good sense that teaching both number and numeration should follow the "number concept first, then numeration concepts" sequence. Students should be intimately familiar with "twelveness" before they confront the numeral 12. Unfortunately, the reverse of this natural sequence occurs all too often.

The material to follow in this chapter is not new to you, but perhaps you'll gain new insights into the structure of the simplest number and numeration systems from this presentation. And perhaps you'll polish up your own "computational proficiency."

Numeration

Contrary to what many people assume because of their limited exposure to mathematics, the development of number and numeration systems was not an orderly, cooperative venture among different people and civilizations. What we see today as polished, almost intricate art forms survived a rather tempestuous, often haphazard historical development, fraught with lack of foresight and hindered by the prejudices of religious and governmental power structures. Almost incidentally and accidentally at times, the improvement of number and numeration systems centimetered ahead.

Originally, numbers were used for recording how many physical objects were present in a collection. The *counting numbers* (one, two, three, . . .) sufficed for this purpose, but you can imagine the problem of trying to represent these numbers symbolically. The members of an infinite set of things can't be named with a finite set of symbols unless some rules are established for using the symbols over and over again. A *numeration system* consists of a finite set of symbols, together with rules for using these symbols to name any number under consideration.

Notice that zero is not included with the counting numbers; for thousands of years, no one thought it necessary to have a number to represent "nothingness." This fact probably delayed as much as anything the development of efficient numeration systems, since not having some symbol for zero prevented the concept of a "placeholder." And inefficient numeration systems made calculations extremely cumbersome. In fact, without such a concept, simple arithmetic like 998 - 225 would still be done only by trained personnel who spent years becoming proficient at such tasks, like the gentlemen to the right.

"Now, with the *new* math . . ."

Source: Courtesy of Sidney Harris and *American Scientist* Magazine.

Let this somewhat rugged start not overshadow the present state of the art of mathematics. The powerful and elegant number and numeration systems we have today are truly marvelous achievements of the human mind. Many scholars and aesthetes place mathematics and music side by side as the highest art forms because both have reached stages of development that primarily depend on the ingenuity of our human mental faculties. That is, mathematics in particular does not depend any longer on observable, physical situations in order to continue its growth and be of value to humankind.

As primitive as ancient numeration systems were, we still find uses for them today. In some situations, these outdated relics are actually preferred over our Hindu-Arabic numeration system.

TWO ANCIENT NUMERATION SYSTEMS

Undoubtedly one of the very first and simplest numeration systems to arise was the tally system (⊞ ⊞ ⊞ III). And yet it's still used today in many situations. Its simplicity is obvious — there is only one unique symbol to worry about (1), and the rule for naming any given number using this symbol is "mark the symbol down as many times as you need to count to the number." The only modification to the rule is that of a horizontal use of the same symbol so that counting the number represented can be done by 5's, instead of by 1's.

The simplicity that characterizes the system ultimately limits its usefulness; large numbers are difficult to represent and read, and computation that goes beyond addition and subtraction rapidly proves extremely difficult. Yet the fact that we still use this numeration system often reminds us that we are still performing some of the same kinds of chores with numbers that our caveman ancestors did.

Roman numerals are also used today in certain contexts. Major headings in formal outlines, dates of film productions, designation of volume numbers in periodicals, and clock faces make frequent use of these numerals. The values of the symbols are given below:

I means 1	L means 50	M means 1,000
V means 5	C means 100	$\bar{V}$ means 5,000
X means 10	D means 500	$\bar{\bar{X}}$ means 10,000,000

How many unique symbols are there above? The bar used above 'V' and 'X' indicates a multiplicative principle employed to represent very large numbers. Can you determine exactly what this multiplicative principle is by the values given above?

There is also a subtractive principle at work in the Roman system. If a smaller-valued symbol precedes a larger one, the smaller value is subtracted from the larger one. So 'CD' means 500-100, or 400, whereas 'DC' means 500 + 100, or 600. (Note: the restriction that only two symbols are allowed in the subtractive process precludes symbols such as IXC, which might be interpreted as either 89 or 91.) The Roman system is basically an additive one in that the value of a numeral is found by adding the value of the individual symbols, yet this additive system has been slightly modified by subtractive and multiplicative principles.

It would be nice to find some inherent reason (such as the simplicity of the tally system) for the survival of Roman numerals in this modern age. The truth of the matter is, however, that we will continue to use Roman numerals because they can be written using a standard typewriter!

The Babylonians are generally credited with being the first to develop a numeration system that utilized place-value concepts. Their groupings were based on both *ten* and *sixty*, and even though their symbolism died out, vestiges of their groupings remain. For example, our unit of an hour for measuring time is divided into 60 minutes and each minute is further divided into 60 seconds; the circle is usually divided into 360 degrees, which is the product of 6 and 60. The Babylonians therefore provide us an example of a numeration system that is no longer in common use but is still making its influence felt.

THE HINDU-ARABIC SYSTEM

The numeration system you've been using all your life rests on two important principles: it is *positional*, and it has a standard, fixed *base* number. The base is ten, probably because our first counting devices (hands) had ten counters. Having a standard base in the numeration system is a tremendous advantage over systems whose base is not fixed. The "old English" system of measurement—with its twelve inches to a foot, and three feet to a yard, etc.—does not have a standard base, and that is the prime reason for switching to the metric system.

Having a base of ten means that numbers are uniquely expressed after the objects counted have been systematically grouped into tens, ten tens, ten groups of ten tens, and so forth. The sketch on the right shows the relative size of the first two groups, as compared to a single object the same size.

Grouping in this manner means that eventually no more than nine objects or groups will exist in each category. Once you get ten things, these are grouped into one of the next largest category. Consequently, the only unique symbols needed to describe such a grouped collection are the ten numerals 0 through 9.

The *positional* principle implies that the value of a symbol is determined by its position in a numeral. For example, the 4's in '468' and '4268' have two different values because their positioning in the total numeral differs. We read numerals from right to left, just as we do words, with our eyes meeting the largest, most important part of the numeral first. 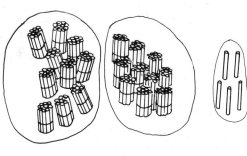 The grouping of sticks to the left indicates a situation in which the "placeholder" concept is required; a symbol is needed to show there are no groups of ten present. The correct numeral for this collection is 205, of course.

One of the most popular and effective sets of materials for showing numeration concepts at the most elementary level was developed by Zoltan P. Dienes, a psychologist who became interested in the way children learn mathematics. For base ten, his wooden blocks look like those shown to the right. Notice that, with these materials, a "Long" looks, tastes, feels, sounds, and smells like ten Units, and a "Flat" like ten "Longs." A "Block" is ten Flats, or one hundred Longs, or one thousand Units. The collection below, then, would be used to show 3457:

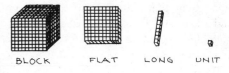

This same numeral—3457—is shown below on another common manipulative device used to teach numeration, the abacus. When compared to bundled sticks or the Dienes Blocks, an abacus is a step up the abstraction ladder. The disks used to represent different place values are physically identical, except perhaps for the color, and therefore it's not as easy for a child to remember exactly what the disks represent in relation to each other.

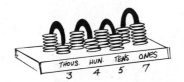

At the same time the beginning learner is using manipulative devices to internalize numeration concepts, he's trying to remember the word names that go with the symbols. And there are good reasons for the difficulty usually encountered.

WORD NAMES The word names we assign to the first hundred whole numbers point out quite clearly that not everything in mathematics follows a logical pattern. In some places, the word names are consistent—take the "sixties," for example:

60	sixty	(six tens)
61	sixty-one	(six tens and one)
62	sixty-two	(six tens and two)
63	sixty-three	(six tens and three)
⋮	⋮	⋮
69	sixty-nine	(six tens and nine)

Nice and neat, and it all makes sense. If this same system were followed for the other numbers between nine and one hundred, we'd have the word names on page 31.

10	onety	(one ten)
11	onety-one	(one ten and one)
12	onety-two	(one ten and two)
13	onety-three	(one ten and three)
:	:	:
19	onety-nine	(one ten and nine)
20	twoty	(two tens)

Throwing away your prejudices for a moment, which set of word names for 10 through 20 makes more sense, the one above, or the one you use all the time (ten, eleven, twelve, etc.)?

A "physiological accident" accounts for the fact that our counting system is based on the number ten. If our ancestors had possessed three or four fingers on each hand, instead of five, we might be using a base six or base eight system. There are several excellent reasons for choosing other numbers besides ten as a base for our numeration system (two, seven, eleven, and twelve are the most commonly suggested), but the arguments are generally beyond the scope of this book. Suffice it to say that both tradition and consensus lead us to retain ten as the "grouping number" for our numeration system.

SET 1

1. At the current price of a cup of coffee, how much money is owed by the group of people whose names appear in the list in the photograph on page 28?

2. Practice writing Roman numerals by writing symbols for these numbers:
   a. forty-four
   b. three thousand eighty
   c. fifty million
   d. nine hundred ninety-nine
   e. ||||| ||||| ||||| ||||| ||||| |||
   f. 1984

3. Suppose human hands looked like these, and that mankind learned to count in the same fashion but on six counters instead of ten. Show by drawing bundles of sticks the size of the groups that would result.

4. In the previous problem, you bundled sticks using a base six grouping; that is, it takes six individuals, or groups, to form the next largest size. To represent uniquely any collection of sticks in that system, you'd never have more than five of any particular type or group—once you had six of something, you could form one of the next size. Consequently, in base six, the only symbols necessary to write numerals are 0, 1, 2, 3, 4, and 5.

Write numerals for the first thirty-seven whole numbers, but in base six instead of base ten. When through, develop a logical word name system for those numerals, similar to the system used in our "sixties."

5. The value of the base-six numeral 231_{six} is said to be ninety-one, in our base-ten language. Or, the total number of objects in 2 groups of 36, and 3 groups of 6, and 1 single object, is ninety-one objects. Find the value of the base-six numerals below:

   a. 334_{six} b. 105_{six} c. 1100_{six} d. 555_{six}

6. Look back at the Dienes Blocks shown before this problem set. To represent the next place values, we would start over again on the shapes (longs, flats, blocks) but this time build the shapes from Blocks instead of from Units. For example, the next place value after Block would be represented by:

   This shape is a "long" again, but since it's made from Blocks, we could call it a "Long Block."

   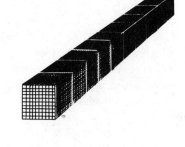

   a. Draw on your paper a "Flat Block," the next size up, and give its value.
   b. What name would follow for the next shape?
   c. Give the value of a "Long Block Block."

7. The photograph to the left shows a common set of manipulative material used in the elementary grades to teach a base other than ten. What's the base of the system, and how can you tell?

8. The two expressions below can be turned into true statements by merely moving one match in each expression. Can you do it?

a.　　　　　　　　　　b.

9. The Mayan culture did develop a numeration system that used place value and had a symbol for zero (around the 4th century, B.C.). They used a modified base 20 system, so they needed symbols for the numbers zero through nineteen. The symbols below show what they used to do this:

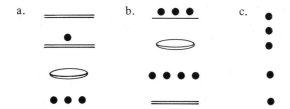

The Mayans wrote their numbers larger than nineteen in a vertical fashion, to indicate different place values. So you might see larger numbers written as shown in the example to the right. The first grouping had a value of twenty, so the next-to-the-bottom symbol (⊖) had to be multiplied by 20 to find its value.

You would expect the third place to have a value of twenty-twenties, or 400, but here the system was modified—because of their calendar with 360 days—so that the third place had a value of 18·20. In the example above, ••• is in this place and so has the value of 3·18·20, or 1080. The place values of higher positions went up by multiples of twenty, so that the value of ••• above is 13·18·20·20.

For this problem, first speculate on a possible reason — a physical one — that the Mayans chose twenty as the base of their system. Then find the values of the three numerals below:

Foundations

It's impossible to view a modern skyscraper and not marvel at the beauty and ingenuity that is the end result of years of labor. If you're not familiar with the construction industry, it may even be hard to imagine how all of those bricks could have ever been laid, or how so many different tradesmen's efforts could have been coordinated so closely to produce the final result. And yet the true strength of the structure—the foundation on which the whole edifice rests—is completely hidden from view. The most important part, and from that standpoint the most beautiful part, is hidden from sight. The foundation of a building is critical, and for that reason it's the first concern of the contractor and building inspector.

Computational proficiency in arithmetic has striking similarities to the situation described above. The underpinnings of addition, subtraction, multiplication, and division are critical to a student's success, and yet we rarely get a chance to observe these underlying factors beyond the first few grades in school. The foundation on which computational proficiency should be established consists of a crisp understanding of concrete situations that call for a particular operation. The descriptions to follow are meant to review these basic interpretations of addition, subtraction, multiplication, and division for you. If they seem to make immediate sense to you, it's probably because you have a firm foundation yourself for computation in the whole numbers.

ADDITION Forming the union of two disjoint sets is perhaps the most basic "real world" situation corresponding to addition of whole numbers. In the example to the right, a set with four dots has been combined with another set with three dots, and the resulting set has seven dots in it. The symbolism $4 + 3 = 7$ is the conventional, corresponding number statement for this action.

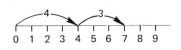

Shown to the left is another basic interpretation of addition. A jump of four units, followed by a jump of three units in the same direction, leaves us at seven on the number line. Therefore again $4 + 3 = 7$ is a reasonable statement to make.

In the concept-forming stages, these two situations are presented to students through real-world examples, and reinforced by concretely manipulating physical objects. The ideal is to have the operation of addition *follow from* a consideration of many such situations.

At the same time, the *properties* of addition become apparent if the time is taken to notice them. Addition is *commutative* $[x + y = y + x]$,* *associative* $[x + (y + z) = (x + y) + z]$, and has an *identity* element zero $[x + 0 = x]$. These properties are taught informally, of course, using only specific instances of the general rule, as demonstrated below:

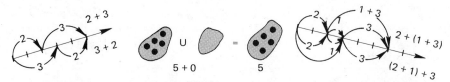

An important point to realize here is that these properties follow quite naturally from the physical interpretations of addition, and can become second nature to the student if introduced at this concrete stage. The advantages of knowing these and the other properties to follow will become clearer later on.

---

*x, y, and z will stand for arbitrary whole numbers in this unit.

Again, there are real-world situations of both the "sets" and "number line" interpretations that should precede, and call for, the operation of subtraction. Instead of "putting two sets together," as in addition, we *take away* some elements from the set we begin with for subtraction. This is more difficult to picture meaningfully than forming the union of sets, but it is attempted to the left below, for the problem 8 - 3. For the "set version," we start with a set of eight, and five are left after three are removed. So 8 - 3 = 5 seems reasonable.

The number line interpretation of this same problem is given to the right. For this concept, a jump of eight to the right is followed by a jump of three in the opposite direction. Since we end up on five, once again 8 - 3 = 5 makes sense.

SUBTRACTION

    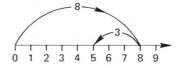

A second situation that requires the operation of subtraction and is distinctly different from the "take away" concept, involves a *comparison* of sets or numbers on a line. The real-world situation of this sort is usually identified by a question along the lines of "how many more in this group than in the other group," or "how much longer

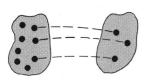

is this than that?" If the comparison involves sets, the answer is obtained by matching each member of the smaller set with one of the larger set, and counting how many are left over. For the problem 8 - 3, the sketch to the left shows how this matching might occur, and gives credibility to five as the answer.

The number line model for comparative subtraction situations is shown to the right, again for 8 - 3. In such a situation as this, the answer is gained by determining how large a jump would be required to go from the smaller number to the larger number: in this case, it would take a jump of five units, so again 8 - 3 = 5.

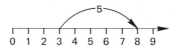

What about the properties for subtraction? Is it a *commutative* operation in that x - y = y - x? Is it *associative* so that x - (y - z) = (x - y) - z? Or is there an *identity element* for subtraction: is x - 0 = x? Of these three questions, only the last one could be answered affirmatively, since it's very easy to find whole numbers for x, y, and z that produce untrue statements for x - y = y - x and x - (y - z) = (x - y) - z. Again, the appropriate time for the learner to confront the concept that subtraction is *not* commutative or associative but has an identity element, is at the intuitive, concrete initial stages.

Several conceptual situations involve the operation of multiplication, the most prominent one being that of "repeated addition." If students encounter problem situations which require that they repeatedly add the same number to itself, before multiplication

MULTIPLICATION

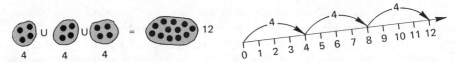

is introduced, then the formal operation itself should have been partially internalized before the symbolism is introduced. The examples above show the bare essence of this "repeated addition" concept, for the two interpretations of 3 x 4 as 4 + 4 + 4.

Another useful way of viewing multiplication uses the concept of "area of a rectangle." Even first grade students are finding the area of geometric figures by counting unit squares, and a natural extension of this concept can lead to another useful interpretation of multiplication. "The area of a rectangle that's 3 by 4" is interpreted as the answer to "3 x 4." Notice that this interpretation of multiplication complements that of the "repeated addition" concept, and the more interpretations available to the learner, the greater are the chances of recognizing a situation that calls for multiplication.

The repeated addition model is useful for demonstrating the *associative property* for multiplication:

3 x (4 x 5) = 3 x (5 + 5 + 5 + 5)

= (5 + 5 + 5 + 5) + (5 + 5 + 5 + 5) + (5 + 5 + 5 + 5)

= 12 x 5 = (4 + 4 + 4) x 5

= (3 x 4) x 5

In a similar vein, we could show that $x \cdot (y \cdot z) = (x \cdot y) \cdot z$, and therefore that multiplication is associative.

The "area" model of multiplication lends itself to showing several other properties enjoyed by this operation. Take commutativity, for example. To demonstrate that 4 x 5 = 5 x 4, all that's required is to sketch a rectangle similar to the one on the left to show 4 x 5, and then physically turn it to show 5 x 4, as in the rectangle to the right. Notice that you don't even have to know what number 4 x 5 is, to see that 4 x 5 = 5 x 4, since the area of a figure certainly wouldn't be altered by turning the figure. Likewise, we could say that $x \cdot y = y \cdot x$, so multiplication is *commutative*.

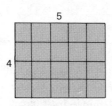
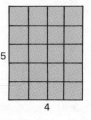

A property that links addition and multiplication can be demonstrated easily using the area model. The rectangle to the left below shows 4 x 6 or 4 x (4 + 2); that same rectangle on the right has a heavy dark line inserted to show two distinct rectangles, one which is 4 x 4 and the other 4 x 2. To find the area of the two rectangles on the right, we could add their individual areas. So we could conclude that 4 x (4 + 2) = (4 x 4) + (4 x 2). The general case here is called the *distributive property of multiplication over addition,* and states that $x \cdot (y + z) = (x \cdot y) + (x \cdot z)$.

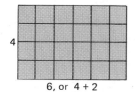
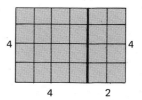

How would you show that multiplication has an *identity* element? That is, how would you show that x · 1 = x? Which model would you use? Could you conveniently use both?

The case bears repeating—the properties of multiplication arise from, and can be demonstrated with, physical situation, which should be used to build the concept of the operation. If so, then the properties will be internalized as an essential part of the operation, rather than appearing to be "imposed from on high" to the learner. It will then seem more natural to *use* the properties when they are needed.

DIVISION

Just as multiplication is sometimes interpreted as "repeated addition," division situations frequently involve the concept of "repeated subtraction." Finding out the number of times that 2 can be subtracted from 10 can eventually be interpreted as 10 ÷ 2. Again this concept is introduced and reinforced through manipulative situations as shown below. To the left, a set of ten objects has been divided into subsets, with 2 in each of the subsets. There are 5 such subsets, so 10 ÷ 2 = 5. This same problem is given

a number line interpretation in the diagram to the right. Beginning at 10, we move backwards in jumps of 2, and determine the number of jumps it takes to get back to zero. Five jumps suffice, so again 10 ÷ 2 = 5 seems logical. Both of the division situations above, which directly correspond to "repeated subtraction," are called *measurement* interpretations.

A second basic concept that requires the operation of division is called *partitioning*. Starting with a set of ten objects, suppose we partition the set into 2 equal subsets, and determine the number of elements in each subset. Again we arrive at 5 as the answer to 10 ÷ 2, through this distinctly different set concept of division. The number line model for the partition interpretation of 10 ÷ 2 = 5 is shown to the right below. In this case, we begin at 10 on the line, and make it back to zero in exactly 2 jumps of equal size. The question to be answered, then, is "What is the size of a jump of this nature?" Since the jumps are both 5 in length, again 10 ÷ 2 = 5 seems to be appropriate.

Which of the properties enjoyed by the other operations are possessed by division? Is it a *commutative* operation, i.e., is $x \div y = y \div x$? Is it *associative*, that is, does it hold true that $x \div (y \div z) = (x \div y) \div z$? How about an *identity* element (is $x \div 1 = x$)? Could you find a way to demonstrate the truth of any of the properties, using one of the models shown above?

RELATING THE FOUR OPERATIONS

The relationship between addition and multiplication has already been mentioned: a basic interpretation of multiplication is that of repeated addition. The same thing has been mentioned for subtraction and division: the measurement interpretation of division relies heavily on the concept of repeated subtraction. Are there other relationships among the four operations, relationships that are advantageous to notice at the beginning stages?

An extremely basic relationship links subtraction to addition, and division to multiplication. Subtraction is related to addition in that $x - y = z$ only when $y + z = x$. Similarly, division and multiplication are tied together in that $x \div y = z$ only when $y \cdot z = x$ (assuming $y \neq 0$). These definitions are stated formally, but it's important for a beginning learner to realize intuitively that a subtraction problem can be solved by addition, and a division problem can be solved by multiplication.

EXTENDING THE PROPERTIES

Summarizing the properties from what has been discussed in this section, we would have that:

Addition and multiplication are both commutative operations.	$x + y = y + x$ $x \cdot y = y \cdot x$
Addition and multiplication are both associative operations.	$x + (y + z) = (x + y) + z$ $x \cdot (y \cdot z) = (x \cdot y) \cdot z$
Multiplication distributes over addition.	$x \cdot (y + z) = x \cdot y + x \cdot z$
Each of the four operations has an identity element.	$x + 0 = x \quad x - 0 = x$ $x \cdot 1 = x \quad x \div 1 = x$

Notice that the authors have already made some important marginal notes to help you state these properties formally.

Eventually, the learner will have generalized these properties and relationships, and personalized the language into such things as:

I can add a bunch of numbers in any order I want to.

> *I can multiply a bunch of numbers in any order I want to.*

> *Zero doesn't add anything to another number, but it sure wipes out anything it's multiplied by!*

> *Multiplying or dividing by 1 doesn't change the number at all.*

> *Instead of multiplying two numbers outright, I can break them down into convenient sums, multiply all those together, and add the results.*

While the previous language may be somewhat rough, it does indicate that the learner has internalized the property to the point where he or she can actually *use* it appropriately. And the usefulness of the properties is what makes them worth studying, even at an early stage.

SET 2

1. The problems below are based on actual facts — a good source for such trivia is the *Guiness Book of World Records*. Don't find the answers to these problems; rather choose an appropriate interpretation discussed in this section that would be useful if you were to try to find the answer to the problem.

   a. The Peachtree Plaza Hotel in Atlanta is 733 feet tall, has 70 stories and 1100 rooms. Each story would be about how tall?

   b. A library book checked out from the University of Cincinnati Medical Library in 1823 was returned by the borrower's great-grandson 145 years later. In what year was the book returned?

   c. The fine (which was waived) for the overdue book above was $18.25 per year. How much would have been paid?

   d. The longest frog leap on record was by "Ex Lax" at the 1975 Calaveras County Jumping Frog Jubilee, and measured 17 feet, 7 inches. How much shorter than the human broad jump record of 29 feet, 2½ inches is this?

   e. The Waldorf Astoria Hotel in New York City is 625 feet tall, and has 1900 rooms in its 47 stories. About how many rooms are on each floor?

   f. Based on word count, the longest personal letter took eight months to write and contained 1,113,747 words. At the typical count of 350 words per page for textbooks, this letter from Jacqueline Jones (Lindale, Texas) to her sister would be equivalent to a text of how many pages?

g. A 42-year-old insane woman reported slight pains in her stomach. Her physicians subsequently operated and removed 2533 foreign objects, of which 947 were bent pins. How many were not bent pins?

h. Chang and Eng Bunker were Siamese twins who lived from 1811 till 1874, never being separated. At age 32, they married sisters and fathered 10 and 12 children, respectively, over the next 30 years. How many children did they have together?

2. Choose one of the interpretations of division, and explain why "dividing by zero" makes little sense.

3. If you've truly mastered the whole number properties, and built up your own set of allowable ways to manipulate computational situations, you should be able to do each problem below *in your head*. Look for easy ways to do each problem, and write only the answers down. Don't do any of the calculations if you can't do them mentally.

   a. (5 + 68) + 95
   b. (25 + 93) + (7 + 50)
   c. (5 x 13) x 2
   d. 4 x (25 x 7)
   e. 7 x (100 + 8)
   f. (6 x 97) + (6 x 3)
   g. 7 x (100 − 2)
   h. (1032 x 5) − (32 x 5)
   i. 113 + (7 + 50)
   j. 8 x (35 x 125)
   k. 25 x (4 + 3)
   l. (150 ÷ 4) + (50 ÷ 4)
   m. (48 + 317) + (45 + [2 + 5])
   n. (20 x 40) x (5 x 3)
   o. (7 x 999) + (7 x 1)
   p. 13 x 99
   q. (998 + 84) + [3 + (16 + 2)]
   r. (4 x 5) x (7 x 5)
   s. (480 x 3) + (20 x 3)
   t. 47 x 110
   u. (5 x 5 x 5 x 5) x (2 x 2 x 2 x 2)
   v. (1300 + 79) + (200 + 11) + 21
   w. 1 + 2 + 3 + 4 + 5 + . . . + 95 + 96 + 97 + 98 + 99
   x. (11 x 37) x (11 + 37) x (15 x 0)

4. The odd whole numbers are 1, 3, 5, 7, 9, etc., while the even whole numbers are 0, 2, 4, 6, 8, etc. Explore each of the following types of computations with addition and multiplication, and see if you can draw a reasonable conclusion about the statements below:

   a. An even added to an even is an _____ number.
   b. An odd added to an odd number is an _____ number.
   c. An even plus an odd number is an _____ number.
   d. An even times an even number gives an _____ number.
   e. An odd times an odd number is an _____ number.
   f. An even times an odd number is an _____ number.

   Look at the very last section of this chapter if you need more information about "odd" and "even" numbers.

5. Returning for a moment to the theme of this text, how would you characterize the mental processes you used in attacking problem 4? (Look back at Chapter 1 if you need to.)

6. For each diagram below, write the specific problem that accompanies the sketch. There may be more than one correct answer for the sketches. For the first one you might have: "From 12 subtract 3 four times" or you might interpret it as "12 − 4 (3)" or as "a 12-inch ruler could be broken into how many 3-inch pieces?"

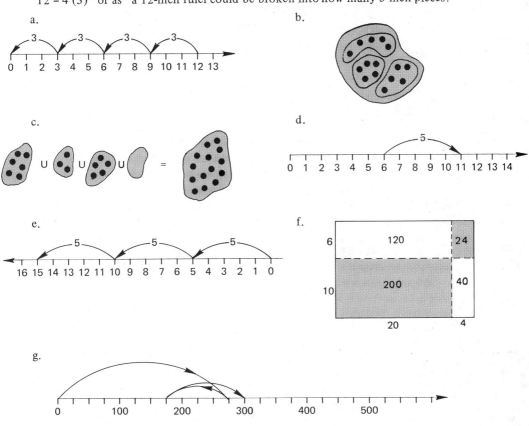

7. A checker at a grocery store doesn't pay much attention to the order in which she rings up the items being purchased. Which properties allow this to be done?

Building Brick Walls

Constructing a sturdy, attractive brick wall requires four things — a firm foundation, a pile of bricks, some mortar to hold the bricks in place, and a mason with the knowledge of how to put the bricks together to produce a pleasing structure. All four things are essential — if any one of them is missing, the wall would not be adequate.

Not unlike the construction of brick walls is the building of computational proficiency in arithmetic. To become reasonably successful in adding, subtracting, multiplying and dividing whole numbers, a student should have internalized the *basic concepts* of each operation (the foundation) and the *properties* (the mortar). Furthermore, the student must have learned the *fundamental facts* for each operation (the

bricks), and then a step-by-step process for putting these basic facts together as an *algorithm* for more complex calculating (the mason). Basic concepts and the properties for the operations were both discussed in the last section — it behooves us now to spend some time on the other essentials for computational literacy.

Consider a problem such as 3287 + 649. The fundamental facts for this problem are 7 + 9, 8 + 4, and 2 + 6. The algorithm would involve knowing that you start off by lining up the two numerals from the right-hand side, then perform the basic facts with digits that are aligned. Another essential part of the algorithm is knowing how to "regroup" or "carry" when the sum of any two single digits is ten or more.

$$\begin{array}{r} {}^{1}\;{}^{1} \\ 3287 \\ +\;649 \\ \hline 3936 \end{array}$$

Each of the four whole number operations (addition, subtraction, multiplication, and division) can be thought of in this manner. Each operation has its own fundamental facts, and each has its own algorithms for attacking problems that go beyond the basic facts. In order to build an attractive brick wall quickly and easily, the bricks must be at your fingertips. Furthermore, ways to put the bricks together must be second nature. To be proficient in computation, you must have instant recall of the fundamental facts (Level 1 knowledge) and intimate knowledge of an algorithm (Level 2 knowledge).

THE FUNDAMENTAL FACTS

The sum of any two single digits is a fundamental addition fact. For example, 2 + 4 is a basic fact, but 10 + 5 is not since 10 is not a single-digit numeral. Note that 2 + 4 and 4 + 2 are two different basic facts, yet the answer is the same for both facts. The table to the right shows places for all one hundred of the facts for addition.

+	0	1	2	3	4	5	6	7	8	9
0										
1										
2										
3										
4									9	
5										
6			8							
7										
8										
9										18

The fundamental subtraction facts are derived from the basic addition facts. Since 3 + 4 = 7 and 4 + 3 = 7 are two fundamental addition facts, 7 − 4 = 3 and 7 − 3 = 4 are two basic subtraction facts. Notice that 17 − 8 = 9 is a fundamental subtraction fact since 9 + 8 = 17 is a fundamental addition fact, yet 17 − 7 = 10 is not a basic subtraction fact since 10 + 7 = 17 isn't a basic addition fact.

We're only halfway through with the operations, and we've labeled 200 things as "essential" for a beginning learner to commit to memory! Who wouldn't rebel at such a task! Thank goodness we've got the properties and relationships between addition and subtraction to fall back on—the commutative property of addition, by itself, cuts the 100 basic addition facts in half. If learners know the answer to 4 + 5, and have internalized the commutative property, they also know the answer to 5 + 4. And help with the fundamental subtraction facts can come in the form of previously learned addition facts, if the relationship between subtraction and addition has been internalized. So even in the very early stages, the value of having intimate knowledge of the properties and relationships between operations becomes obvious.

Multiplication and division have a similar relationship regarding the fundamental facts. The product of any two single digits is a basic fact. For example, 0 x 0, 3 x 4, 4 x 3, and 8 x 5 are four different fundamental multiplication facts. But 7 x 10, 9 x 11, and 12 x 12 are *not* basic facts and hence do not have to be committed to memory.

x	0	1	2	3	4	5	6	7	8	9
0										
1										
2										
3							18			
4										
5		10								
6										
7										
8									72	
9	0									

The basic division facts (the "gozintos") come from the basic multiplication facts. Since 3 x 4 = 12 and 4 x 3 = 12 are two basic multiplication facts, 12 ÷ 4 = 3 and 12 ÷ 3 = 4 are two basic facts for division. It looks again like we have added 100 more facts to be memorized, just as with multiplication. Yet recall from a problem in Set 2 that division by zero is not defined, so we can eliminate 0 ÷ 0, 1 ÷ 0, 2 ÷ 0, ..., 9 ÷ 0 from the basic division facts. What a deal—we're left with only 190 more things to commit to memory!

Again, the task of memorizing all of the fundamental facts is made somewhat more palatable by applying the properties and relationships between multiplication and division. The commutative property cuts the memory work for the multiplication facts in half, and such things as "multiplying by zero," "multiplying by one," "dividing into zero," and "dividing by one" lead to generalizations that cut down on the work load.

Keep in mind that mastering the 390 basic arithmetic facts is a Level 1 skill—necessary, but far from the goal of developing overall computational proficiency. Instant recall of these facts is like getting the pile of bricks up close to the mason so he can work without interruption.

THE ALGORITHMS

The basic facts are standard, unchanging definitions. Anyone in a country that uses a base-ten numeration system like the Hindu-Arabic system has to wrestle with these at one time or another. There's no room for variation or creativity. However, the processes for going beyond the fundamental facts to solve a general computational problem do vary quite a bit from place to place, and even from one person to the next. The person who sits next to you in this math class, for example, might well have a different procedure from yours for solving a complex addition, subtraction, multiplication, or division exercise. A few of the more standard algorithms will be considered in this section, but these certainly do not exhaust the possibilities. Inventive people often find their own modifications of existing algorithms and use them as shortcuts whenever they can be applied.

Addition. Most of us use the same addition algorithm. We start by lining up the digits from the right-hand side so we'll be adding digits with the same place value. Impasses (basic facts with sums of ten or more) are resolved by regrouping, or "carrying." Consider the four problems below:

```
    468      329      573      385
   +347      +86      +28      +46
  71015     1189      501      433
```

Can you determine which one of the four mistakes is due to not knowing the basic addition facts? For the other three mistakes, can you tell which part of the standard addition algorithm has not been mastered?

"Adding to tens" is another useful algorithmic process when there are a good many single digits to be added, as in:

$$8 + 4 + 6 + 5 + 2 + 3 + 7$$

To use this method, you search for numbers that add up to ten (usually marking them out as you go) and count the tens in your head as you get them. Anything left over is tacked onto the total. For the previous problem, the work might look something like:

$$\underbrace{8 + \overbrace{4 + 6}^{20} + 5 + \overbrace{3 + 2 + 7}^{30}}_{10} = 35$$

Being able to add numbers in any convenient order, of course, is an extension of the commutative and associative properties of addition.

Subtraction. The most popular subtraction algorithm (in the United States) begins in the same fashion as in addition—by lining up the digits according to place value. Each digit is subtracted from the one above it, moving from right to left. If an impasse is reached (i.e., a place where the top numeral is smaller than the one underneath it), the impasse is resolved by regrouping within the top numeral. In the problem to the right, the impasse is that 8 can't be subtracted from 3; the impasse is overcome by renaming

    7 hundreds + 3 tens + 5 ones as

    6 hundreds + 13 tens + 5 ones.

Since 13 - 8 is a basic subtraction fact, everything is okay from this point on — the impasse has been resolved satisfactorily by "borrowing" a hundred and changing it into ten tens.

By contrast, the impasse above could have been worked out in another fashion, one that typifies another subtraction algorithm. Since the difference of two numbers remains unchanged if the same amount is added to both, we could add ten tens to the top number if we add the same amount to the lower one. We *do* add the same amount to the lower number, but instead of adding ten tens, we'll add one hundred as indicated to the left. Again the impasse has been resolved, as we now have 13 - 8, which is a basic fact, in the tens place.

While this algorithm is possibly more difficult to understand and explain, it certainly is just as efficient as "borrowing," and is actually the preferred method in some countries. In fact, many people (usually at least four or five in a typical college class) use the two processes—at different times —in the same problem! They're truly manipulating the system, making mathematics work the way they want it to.

```
  17 1
  3 8̸ 4̸      }  FIRST IMPASSE : BORROWING
  ² 1̸ 9 6        SECOND IMPASSE : ADDING SAME
      8                          NUMBER TO BOTH
```

Multiplication. Multiplication problems that call for an algorithm are usually started in the same manner—lining up the digits from the right-hand side—but this is certainly not necessary. The process does not require this alignment, since each digit in one number is eventually going to be multiplied by each digit in the other number, regardless of place value. In this case, we align the numerals to be orderly and systematic, lest we overlook some digit and commit an error. The right-most digit of the lower number is multiplied by each digit of the upper number, regrouping or "carrying" when necessary. Then multiply the second digit times all those on top, and so forth. When each digit of the lower number has been multiplied by each one of the top number, the individual results are added to find the final answer.

```
              2 2                1 1
              2 2                2̸ 2̸
Step 1      4 6 8      Step 2   4 6 8
            x 2 3              x 2 3
            1404               1404
                               9360
              1 1
              2̸ 2̸
Step 3      4 6 8
            x 2 3
            1404
           +9360
           10764
```

The algorithm above is fraught with places where a mistake might be made, even if the basic multiplication facts have been memorized. Students often forget to erase the numbers they "carry" when moving to the second digit of the lower number, and so the wrong numbers are "carried" on the second line. Or they "carry" correctly, but add the number carried before they multiply. Or perhaps they forget to "move over one place" or "tack on an extra zero" each time they begin a new partial product. The list is endless!

Consider the same problem as above, but this time using a process called the *lattice method* of multiplication. Before going to the next paragraph, see if you can figure out what's going on below:

Problem: 468 x 23

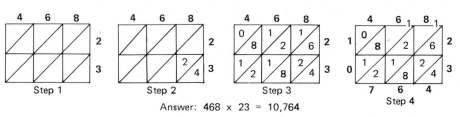

Answer: 468 x 23 = 10,764

Well, how did you do? See if you agree with, or can now follow, this explanation of the lattice method:

First, draw a rectangle the right size so that the numbers can be written on the top and side. Draw in the unit squares and the diagonals.

Next, multiply each top digit by each digit along the side, putting the product in the square corresponding to the two, with the tens digit on top of the diagonal.

Last, begin in the lower right-hand corner, and add the digits along the diagonals, carrying to the next set of numbers when necessary.

The answer can be read starting at the upper left-hand corner and moving counter-clockwise to the bottom right. Notice that in this case, we arrived at the same answer as when the usual algorithm was applied (10764).

We probably wouldn't choose this algorithm as the basic one for students to learn. However, it does remind us of the fact that other options are often available in even the most fundamental arithmetic processes.

Division. In case you've forgotten, our standard division algorithm (called "long division") is a bear for many people. Some teachers go so far as to claim that, due to the complexity of this commonly used technique, division itself should not be called "basic." However, division *is* a basic operation; it does not depend on any procedure we happen to adopt that aids us in calculating the answer.

That our standard division algorithm is more complex than the common algorithms for addition, subtraction, and multiplication is manifest in several steps of the procedure. First, we start not by aligning the numerals from right to left according to place value, but by placing the divisor to the left of $38\overline{)18962}$ the dividend, right beside it! And the digit in the largest place value of the answer is found first, a complete reversal of what goes on in the other three algorithms.

We begin the calculations involved not by using basic division facts, but by *rounding off* the divisor and dividend and *estimating* the first digit of the quotient. Then we *multiply* and check to see if our estimate was too large; if so, we revise the estimate. If the estimate was not too large initially, we *subtract* and this time check to see if the original estimate was too small; if so, another estimate is required. If everything seems okay, we "bring down another digit" from the dividend, and start the whole process again, continuing in this fashion till there are no more digits to "bring down."

This complicated process called "long division" isn't learned overnight, of course. Generally, it takes you several years to master the techniques involved, and even then you may sometimes have difficulty. Even if you learn to do every step routinely, there are times when you'll need an eraser, because rounding off a divisor and a dividend introduces a source of error in the estimating process. But the algorithm does have advantages over some simpler techniques for calculating the answer to a division problem: it is definitely shorter, and mastering the process helps "build character!"

A division algorithm called "repeated subtraction" is exactly what you'd expect it to be from the name. It's easy to explain—to obtain the answer to a division problem, you figure out how many times the divisor can be subtracted from the dividend. Students rapidly go beyond the stage of subtracting the divisor itself, and begin subtracting multiples of the divisor, keeping up with how many times they subtracted the divisor in the process. This is illustrated for the problem $8832 \div 23$.

Long Division	Repeated Subtraction	
		# of 23's subtracted
384	8832	
23 ⟌ 8832	−2300	 100
−69	6532	
193	−2300	 100
−184	4232	
92	−2300	 100
−92	1932	
	−1150	 50
	782	
	−690	 30
	92	
	−92	 4
		384

Note above that 100 23s were subtracted three times, then 50 23s, then 30 23s, and last, 4 more 23s. So altogether, 384 23s were subtracted—this is the same answer obtained using "long division" shown to the left above.

The contrast between the two methods appears obvious—long division is difficult to explain and leaves no room for variation, but it is shorter for people who can easily handle the rounding off, estimating, and checking procedures mentally. Repeated subtraction uses more paper and doesn't stretch the mental muscles as much, but it is easy to understand and stresses the relationship between division and subtraction. Which one do you prefer?

SUMMARY

Computational proficiency with whole numbers involves both "Level 1" and "Level 2" mental processes. The basic facts for each operation are introduced when the operation itself, and the properties, are being internalized. However, somewhere along the way, the fundamental facts should be memorized so that the brain can concentrate on more complex calculations that go beyond the basic facts. At this stage, the learner is trying to master a routine, step-by-step procedure that can be used on any future problems requiring the operation. The learner is trying to commit to memory a Level 2 process, an algorithm.

Two algorithms have been given a cursory exposure for each of the four operations (addition, subtraction, multiplication, and division). As you attempt the problems in the next set, concentrate on mastering the algorithms that are not already familiar to you. Besides the experience of expanding your own horizons, these procedures will help you later on as a teacher by giving you ways to enrich the basic curriculum. Alternative algorithms evoke high interest in the students, and help the learner realize that more than one process can frequently be used to solve even the most mundane of mathematical situations.

An apprentice brickmason concentrates on building standard walls that are sturdy and functional. As experience is gained, the mason develops new techniques for putting the bricks together, and eventually feels in complete control of the task at hand. With a little encouragement, an experienced mason may even build you a brick structure using a pattern that no one's ever used before!

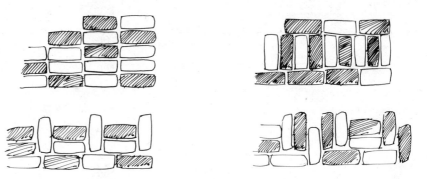

SET 3

1. Try the "adding to tens" method on the problems below:

   a. 34507
        2837
       70523
         483
        1932
      854916
        7465
   +   23

   b. 14007
        2356
       38275
         491
        7340
          16
       28494
   +   615

2. Resolve the impasse in each problem below by adding the same amount to both the top and bottom numbers — i.e., don't borrow!

   a. 4583   b. 3052   c. 653   d. 5002   e. 4570
      -2259   - 36   -74   -2091   -988

3. Another subtraction algorithm involves adding the same amount to both numbers, as above, but this time the purpose is not to avoid any particular impasse. The number you add to both is chosen so it will "round off" the lower number to one that can be subtracted "by observation!" Consider the two examples below:

Given problem:	235 - 68	Solution:	(1) Add 2 to both: 237 - 70 (2) Add 30 to both: 267 - 100 (3) "Eyeball" answer as 167.
Given problem:	3242 -579	Solution:	(1) Add 1 to both: 3243 - 580 (2) Add 20 to both: 3263 - 600 (3) Add 400 to both: 3663 - 1000 (4) "Eyeball" answer as 2663

With a little practice the method above becomes a real jewel—notice you never have to "borrow"—particularly for someone who can add in his head, but has trouble with the standard subtraction process. Try the *"rounding off"* method on the problems below, adding any number you feel is advantageous, and using as many steps as you want.

    a. 354   b. 3046   c. 5896   d. 4956   e. 530   f. 4073
        −86   −358   −3658   −988   −479   −370

4. Use the "lattice method" on these problems:

    a. 435 x 53   b. 679 x 103   c. 45 x 73

5. Use "repeated subtraction" to solve these problems:

    a. 6745 ÷ 23   b. 51643 ÷ 344

6. Many times students are taught a standard algorithm, and never encouraged to look for shortcuts that would save time and effort. The multiplication problems below are typical of this situation:

    a. 1000   b. 472   c. 67983
       x346   x333   x1001
       6000   1416   67983
       4000   1416   00000
       3000   1416   00000
    346000   157176   67983
                                                                  68050983

For each example above, what shortcut could you lead the student to observe?

7. Study each example below very carefully to find out what sort of mistake has been made. Has the student made an error in a "basic fact," or has he made an algorithmic mistake. Or perhaps neither label truly applies. Look for the clues!

    a. 369   b. 384   c. 3000   d. 37)25601
       +47   x9   −27   (quotient 6810)
       463   3483   2972   −222
                                                                    340
                                                                  −296
                                                                    441
                                                                    −370
                                                                     71

    e. 526   f. 674   g. 785   h. 34
       +846   −87   x29   x7
       1472   787   7065   358
                                                              1570
                                                              8635

8. One former fifth grader used to drive her teacher nuts by turning in multiplication problems that looked like the one to the right. The answer is correct, but the student obviously wasn't using the standard algorithm. The teacher watched her work a problem one day, and figured out her procedure. Can you?

                                        27
                                    x34
                                  238
                                  68
                                  918

9. A third-grade student was asked to subtract 27 from 43, and show her work. She wrote:

$$\begin{array}{r} 43 \\ -27 \\ \hline -4 \\ 20 \\ \hline 16 \end{array}$$

Is the answer correct? Will this process always work? Can you develop an entire subtraction algorithm based on this process of resolving impasses?

10. Use your ingenuity to figure out ways to solve the problems below:

a. 4 ⟌ 5 wks., 3 days, 5 hrs., 8 min.

c. 3 gal., 2 qts., 3 pints
                                  x 5

b.   5 yds., 2 ft., 7 in.
   −3 yds., 2 ft., 10 in.

d. 64 lb., 13 oz. + 38 lb., 12 oz. + 85 lb., 5 oz.

11. What's the largest product that can be obtained when a 3-digit number is multiplied by a 2-digit number, and none of the five digits in the two given numbers are the same?

$$\boxed{?}\,\boxed{?}\,\boxed{?}$$
$$\times\ \boxed{?}\,\boxed{?}$$

Notice how you use intuition and rational thinking to do this problem—trial and error, checking two different attempts at getting the largest possible product, and rationalizing the importance of the roles of the hundreds and tens places in the two numerals.

Fractions—An Endangered Species?

The Studebaker was a nice car during the 1950s. It was small, ran like a sewing machine, and had a streamlined appearance. Society's demands during the late fifties and early sixties ran contrary to this concept, however—large, gawdy gas guzzlers were the standard of the day —and the Studebaker died a slow, agonizing death. Today the appearance of one is quite a shock, and makes some of us reminisce about the "good ol' days." With tears in our eyes, we wonder why such a practical automobile was allowed to become extinct.

A somewhat different case is that of the dodo bird. This creature is certainly extinct, but no one has shed a tear recently over that fact. The bird was a misfit: it was small and relatively defenseless; it couldn't fly; and it certainly wasn't cherished as a thing of beauty. If the dodo bird were still here today, it would doubtlessly exist only as a curiosity that had been saved from extinction because of our innate desire to see any species continue!

The dawning of the metric system in this country has created an interesting dilemma related distantly to those above. The old "English system" of measuring (pounds, feet, gallons) does not have a standard base of ten and hence encourages our dependence on fractions—we need to express things as "halves," "thirds," "fourths," etc. But the metric system *is* based on ten, and hence encourages the use of decimals as opposed to fractions. Since the United States is gradually accepting the metric system, some experts say we will soon have no more need for fractions, so we should begin to edge them out of the classroom in the near future. Others disagree, saying that concepts like 1/2 and 2/3 are much easier to grasp than their decimal forms (.5 and .$\overline{6}$), and that fractions should therefore remain in the curriculum.

Perhaps one day we'll have an historical perspective from which we can judge the inherent worth of fractions. Perhaps they *are* more like Studebakers than dodo birds, and if we allow their demise we'll regret it one day. Some new, practical uses for fractions may even emerge in the future if we "keep them alive" for awhile. On the other hand, we may look back from our future vantage point and realize that fractions have become useless artifacts, interesting only from the historical point of view.

Most educators agree that we would be hurt more by allowing something to go out of existence and then realizing that we need it, than by keeping something alive only to find out it's unnecessary. So for our immediate interest, we can assume that fractions will still be in the curriculum for some time to come. We can't place them on the "endangered species" list yet!

SYMBOLISM AND BASIC INTERPRETATIONS

The notation used to name fractions has become fairly standard—either $\frac{a}{b}$ or a/b is acceptable, where a and b both stand for whole numbers, with the added restriction that b can't represent zero. For reasons to be discussed shortly, a is called the "numerator" and b the "denominator" of the fraction.

Ratio. One use of a fraction is to describe the comparative relationship between two sets, or the *ratio* of one set to another. When a fraction like 7/5 is used as a ratio, what is generally meant is something on the order of "For every seven of these, there are five of the other." For example, a recent British study found that the relationship between "cigarettes smoked" and "minutes taken from life span" is 2/11—that is, for every two cigarettes smoked, 11 minutes are removed from a person's life span.

Symbolism such as a:b is interpreted in exactly the same way as a/b in speaking of the ratio of two sets, and is actually preferable, since it extends without confusion to comparing three or more sets. In the beginnings of our social consciousness, for example, it was sufficient to set ratios of a/b to represent quotas for ideal black/white policies regarding employment, school attendance, etc. The recent emergence of other minority groups has forced us to consider the numerical relationship of several groups at the same time, and for such comparisons, a:b:c:d:e seems much less confusing than a/b/c/d/e. Notwithstanding our preference, however, we'll yield to tradition and occasionally refer to a:b as a/b.

Division. An eighth-grade student, if asked to interpret a fraction such as 8/3, would likely respond with 8 ÷ 3. This interpretation of a/b as a ÷ b is the one that becomes the most useful, mathematically speaking, from the upper elementary grades on. In this situation, the symbolism is used to indicate *division of the upper number by the lower number.* Besides the usefulness of this interpretation for relating fractions to decimals, this definition of a/b as a ÷ b is necessary when a student enters the realm of algebra.

$$\begin{array}{r} 2.666 \\ 3\overline{\smash{)}8.0} \\ \underline{6} \\ 20 \\ \underline{18} \\ 2 \vdots \end{array}$$

Part of a Whole. The most primitive interpretation of a fraction — the one which gives us "numerator" and "denominator," and lends itself to the concrete examples and mental imagery necessary for beginning experiences—is that of *part of a whole.* Consider the loaf of bread to the right. The whole loaf would be the unit, and it's been partitioned into 22 somewhat equal slices. The denominator of a fraction represents the number of parts into which the unit has been partitioned, hence the natural denominator to associate with the loaf of bread would be 22 (assuming that the slices were all exactly the same size). Each

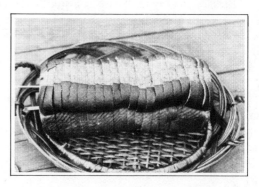

individual piece of bread would be one of the 22 pieces, and hence would be called one twenty-secondth of the loaf, or 1/22.

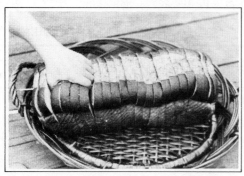

Four of the slices are being removed in the picture to the left. The numerator of a fraction expresses the number of pieces of the whole being considered, so four twenty-secondths, or 4/22, represents the part of the loaf being removed. Notice that "numerator" is closely related to "number," just as "denominator" is related to "denomination." These words associated with fractions are therefore appropriate, and do describe the relationship of each part of the symbol under this "part of a whole" interpretation. Unfortunately, many teachers don't realize that these terms aren't in the vocabulary of elementary school children, and are frustrated when they encourage students to verbally explain what's going on with a/b.

As we continue through this unit, we will lean most heavily on this "part of a whole" interpretation for a fraction. It should be pointed out, however, that any conclusions drawn would have to be consistent through each interpretation of a fraction.

OTHER NAMES FOR A FRACTION

Consider the "top view" of the loaves of bread below. In the drawing to the left, each two slices have been joined together, sandwich style, so that the loaf is now partitioned

into 11 sections instead of 22. In the second picture, party sandwiches have been made by two lengthwise cuttings down the loaf. In the photograph on the previous page, enough bread for two sandwiches is being removed—4/22 of the loaf. The sketches above use shading to indicate two different names for that same amount of bread— 2/11 from the left picture, and 6/33 from the one on the right. This "part of a whole" interpretation would therefore lead us to conclude that 4/22 = 2/11 = 6/33. But is this conclusion consistent with the other interpretations?

The work in the box to the right shows what happens when we consider the problems (4 ÷ 22), (2 ÷ 11), and (6 ÷ 33). Notice that the answer is the same for all three division problems (.1818 . . .), so the "division interpretation" of fractions would lead us to conclude that 4/22 = 2/11 = 6/33.

```
    .1818           .1818            .1818
22)4.00         11)2.00          33)6.00
   22              1 1               3 3
   ─── ─           ───               ───
   180              90               270
   176              88               264
   ───              ──                ───
    40              20                 60
     :               :                  :
```

And how about the ratio concept—would it be logical to conclude that 4:22 expresses the same comparative relationship with sets as that described by 2:11 and 6:33? The British study referred to earlier mentioned a ratio of 2/11 for the number of cigarettes smoked and the number of minutes lost from a lifetime. Could this same ratio be described by 4/22 or 6/33? Intuitively the answer would be "yes," since 2 cigarettes removing 11 minutes would be the same as 4 cigarettes removing 22 minutes, or 6 cigarettes removing 33 minutes.

The particular fraction considered above (4/22) is certainly not unique in having several different names. It was meant to give an intuitive structure for the general case of producing other names for any fraction of the form a/b. Through considering many concrete situations similar to that above, we could arrive at a method for finding other names for a given fraction. This procedure is important enough to warrant a formal statement.

> *An equivalent name for a fraction a/b is produced when both numerator and denominator are multiplied by the same number (excluding zero).*

$$\left[\frac{a}{b} = \frac{a \cdot n}{b \cdot n} \text{ if } n \neq 0\right]$$

Notice that we now have a way of producing an infinite number of names for the same fraction, since n can be any whole number other than 0.

The usefulness of finding other names for a given fraction may not be apparent at the beginning stages, but it is essential for truly understanding such things as comparing fractions, finding common denominators for addition and subtraction, reducing a fraction to lowest terms, etc. Therefore it is crucial that students internalize the concept at the very beginning stages of their exposure to fractions.

COMPARING FRACTIONS

Relying again on the "part of a whole" interpretation of fractions, it seems intuitively obvious that 3/22 should be less than 5/22, since 3 pieces of bread is less than 5 pieces of bread. This could also be interpreted as a ratio, and again as division, and the same conclusion would be drawn. Furthermore, the situation could be explored in depth and the generalization would be something on the order of:

> *If two fractions have the same denominator, the one with the smallest numerator is the smaller fraction.*

Certainly we've surprised no one at this point. Anyone who's got a good grasp of a fraction as "part of a whole" can readily see how to construct concrete examples to demonstrate the truth of the statement.

But what about the general case, that of comparing fractions with different denominators? How can we proceed to compare two fractions built from different-sized pieces, i.e., two fractions with unlike denominators? Say 2/3 and 3/4. Do we need to develop a more general rule for comparing fractions than the one above, a rule that allows us to compare fractions with different denominators?

No, we already have enough power at our disposal to compare any two fractions as to which is the larger. All we need do is rename the two fractions so that they have the same denominator, and then compare the resulting numerators! For 2/3 and 3/4, we might do the following:

$$\frac{2}{3} = \frac{2 \cdot 4}{3 \cdot 4} = \frac{8}{12} \text{ and } \frac{3}{4} = \frac{3 \cdot 3}{4 \cdot 3} = \frac{9}{12}$$

Since 8 is less than 9, 8/12 is smaller than 9/12, and so 2/3 is less than 3/4. To compare two fractions, then, all we need do is ensure that they have the same denominator and then apply the rule above. This procedure assumes, of course, that a common denominator can always be found for two given fractions. The case just used as an example—2/3 and 3/4—may have reminded you of a commonly used scheme for finding common denominators, one that has the advantage of always working. Given two fractions, *we can find a common denominator by multiplying the numerator and denominator of each fraction by the denominator of the other fraction.* So to answer our own question — yes, we can always find a common denominator for two given fractions.

The danger in discovering a procedure that will always work — like the one mentioned above for finding common denominators—is that we latch onto it and use it to the exclusion of other, perhaps more efficient, methods. Such is the case this time. In practice, there are better ways of finding a common denominator.

Suppose we have three fractions for which we want to find a common denominator, such as 5/6, 3/4, and 1/2. We could extend the method mentioned above, using $6 \cdot 4 \cdot 2$ as this denominator, and rename each fraction as shown to the left below.

SCHEMES FOR FINDING COMMON DENOMINATORS

$$\frac{5}{6} = \frac{5 \cdot 4 \cdot 2}{6 \cdot 4 \cdot 2} = \frac{40}{48}$$

$$\frac{3}{4} = \frac{3 \cdot 6 \cdot 2}{6 \cdot 4 \cdot 2} = \frac{36}{48}$$

$$\frac{1}{2} = \frac{1 \cdot 6 \cdot 4}{6 \cdot 4 \cdot 2} = \frac{24}{48}$$

$$\frac{5}{6} = \frac{5 \cdot 2}{6 \cdot 2} = \frac{10}{12}$$

$$\frac{3}{4} = \frac{3 \cdot 4}{4 \cdot 4} = \frac{12}{16}$$

$$\frac{1}{2} = \frac{1 \cdot 6}{2 \cdot 6} = \frac{6}{12}$$

Note that, indeed, this procedure works! However, a clever person might notice that 12 will also work as a common denominator for these three fractions, as shown to the right. The only advantage to using 12 over 48 as the denominator is that it produces fractions that are easier to comprehend. In similar situations, however, finding the smallest common denominator initially saves a lot of useless computation.

In reality, the only fractions we'll ever meet in real life will have relatively small denominators, and we usually revert to "inspection" to find a common denominator in such situations. This procedure can be summarized as:

> *Consider only multiples of the largest denominator, in order, till you find one that's a multiple of the other denominators. Then rename all using this number.*

To apply this to the previous example of 5/6, 3/4, and 1/2, we would consider multiples of 6 till we found a number that's a multiple of all of the other numbers. The first multiple of six doesn't work (6 · 1 = 6) since it's not a multiple of 4, but the second multiple of six does work (6 · 2 = 12) so we can stop our search here. Then we rename each fraction with 12 as the denominator. This method of "inspection" is particularly effective when all the work can be done mentally.

Again, fight the temptation to file the "inspection method" away as the ultimate procedure to select in finding common denominators. As you attempt the problems to follow, you will perhaps find different ways of renaming the fractions involved. Remain flexible in your thinking and you'll sharpen your problem-solving skills with fractions.

SET 4

1. Which of the three interpretations of a fraction is being used in each situation below?
    a. The shortest record for amputating a leg in the preanesthetic era was 14/60 of a minute, by Dominique Larrey, Napoleon's chief surgeon.
    b. A nightclub act is usually designed to last 3/4 of an hour.
    c. Burt Reynolds accepted a football scholarship to Florida State only after the head coach pointed out the obvious advantages of attending a school where women outnumber men about 3/1.
    d. Sheri got a score of 14/15, or 93%, on her last chemistry exam.
    e. Scientists conservatively estimate the odds that another intelligent life form has at some time existed in the universe as 2,000,000,000,000/1.

2. Many times elementary teachers fail to capitalize on familiar objects that seem to be somewhat naturally partitioned into fractions, as they introduce the important concepts to beginning students. From your own experiences, list some things which we quite naturally think of as being divided into:

a. Halves	f. Sevenths	k. Twenty-fourths
b. Thirds	g. Eighths	l. Thirtieths
c. Fourths	h. Ninths	m. Sixtieths
d. Fifths	i. Tenths	n. Three hundred sixty-fifths
e. Sixths	j. Twelfths	o. One of your own

3. Pick one of the interpretations of a/b, and use it to explain why the restriction $b \neq 0$ must be made for the fraction to make good sense.

4. Rewrite at least one of the two given fractions so they will have the same denominator. Try to use the smallest denominator possible.

   a. 5/12 and 1/6
   b. 5/8 and 3/20
   c. 7/6 and 3/4
   d. 5/10 and 6/4
   e. 4/15 and 3/10
   f. 3/4 and 5/9
   g. 4/6 and 8/12
   h. 6/20 and 15/24
   i. 4/33 and 2/55

5. Return to the problems above, and decide which of the original fractions is the largest.

6. Two college classes took the same final exam. In the first class, 19/26 made at least a C, while in the other class, 21/28 made at least a C. Which one of these two classes had the largest fraction to make a C or higher, or was the fraction the same for both classes since each class had 7 students with unsatisfactory grades?

7. Liquor is frequently sold in bottles labeled "fifths" and "quarts." What do you think was the original unit for these terms—i.e., a fifth of what? A quarter of what? And which should cost more, a "fifth" or a "quart?"

8.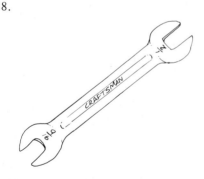
A wrench has a number printed on it — this number refers to the size of the nut it fits. Box-end wrenches like the one pictured to the left have two ends, each with a number. And if the wrenches are of the "good ol' boy" type, the numbers are fractions. A standard set of such wrenches has these numbers on them:

$$\frac{13}{16} \quad \frac{3}{4} \quad \frac{5}{16} \quad \frac{1}{2} \quad \frac{7}{16} \quad \frac{25}{32} \quad \frac{5}{8} \quad \frac{1}{4} \quad \frac{9}{16} \quad \frac{3}{8}$$

Arrange the sizes above from smallest to largest.

9. The human body holds about 5 quarts of blood. About what fraction of your blood do you donate when you answer the call from the Red Cross or other emergency health organizations?

10. As a fraction, about how thick is an ordinary sheet of paper? If you think about it for a minute, you should probably come up with a fairly accurate, yet easy, method of finding out.

11. You happen upon an accident victim and must apply C.P.R. to save the person's life. Two of their normal body functions are now up to you—to keep the heart

pumping, you compress the chest, and to supply necessary oxygen, you breathe into the lungs.

The chest should be compressed 60 times per minute, while the ratio of breaths to chest compressions is 2:15. The rescue team takes 13 minutes to relieve you. How many times should you have breathed for the victim?

12. The general rule for renaming fractions is restated below:

$$\frac{a}{b} = \frac{a \cdot n}{b \cdot n} \quad \text{for all non-zero whole numbers n.}$$

Notice that, when considered "in reverse," this justifies the process called *cancelling*, the old friend which allows us to *reduce fractions to lowest terms*. Stated somewhat loosely, "cancelling" allows us to eliminate any *factors* which appear in both the numerator and denominator of a fraction. Several legitimate examples are offered below:

$$\frac{10}{24} = \frac{5 \cdot \not{2}}{12 \cdot \not{2}} = \frac{5}{12} \qquad \frac{24}{30} = \frac{\not{6} \cdot 4}{5 \cdot \not{6}} = \frac{4}{5} \qquad \frac{36}{48} = \frac{\not{2} \cdot \not{3} \cdot \not{2} \cdot 3}{\not{2} \cdot 2 \cdot 2 \cdot \not{2} \cdot \not{3}} = \frac{3}{4}$$

The thing to remember is that only factors—numbers that divide the numerators and denominators, *with zero remainders*—can be eliminated. At least this is the general case. But under unusual circumstances, wierd things *seem* to suggest different, but erroneous rules. Look at the work below, and decide in which cases the student is correctly cancelling factors. While you're at it, check and see if the student gets the correct answer for any problems being done incorrectly.

a. $\frac{1\not{6}}{\not{6}4} = \frac{1}{4}$ e. $\frac{2\not{7}5}{\not{7}70} = \frac{25}{70}$

b. $\frac{2\not{6}}{\not{6}5} = \frac{2}{5}$ f. $\frac{4\not{9}}{\not{9}8} = \frac{4}{8}$

c. $\frac{1\not{9}}{\not{9}5} = \frac{1}{5}$

d. $\frac{xy + xz}{x^2y + x^2z} = \frac{x\not{y} + x\not{z}}{x^2\not{y} + x^2\not{z}} = \frac{x + x}{x^2 + x^2} = \frac{2x}{2x^2} = \frac{1}{x}$

13. Somewhere in the past, you've probably been exposed to graphing points of the form (x,y) on a cartesian coordinate system. On the axis system to the left, for example, the points (2,4), (4,2), (7,10), and (0,11) have been graphed. Notice that the first number of each ordered pair tells how far "out" to go, and the second number of (x,y) tells how far "up" to go to locate the dot. Notice also that (2,4) is *not* the same point as (4,2). Before going further, can you give the ordered pair for the dot above that has a square around it?

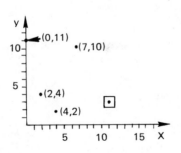

Suppose we wanted to make dots for fractions, changing a/b to (a,b), and graphing the ordered pairs on a coordinate system. To the right, dots have been placed where 3/4, 5/8, 7/4, and 0/11 would appear. Notice that now, the numerator tells us how far "out" to go, and the denominator how far "up" to travel, to locate the point.

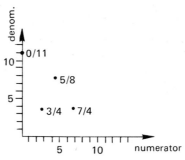

Draw an axis system similar to the one on the right and experiment with this rather strange interpretation of a fraction, considering as you do these questions:

a. What would equivalent fractions look like, say the set of fractions 1/2, 2/4, 3/6, ...?

b. Could you determine from the graph if a/b is smaller than c/d? If so, how?

c. Are there any points allowed in a normal cartesian coordinate system that would *not* be allowed here? If so, where would they lie?

14. One interesting fact about fractions is that *between any two fractions, there is another fraction*! Or, if a/b < c/d, then there is another fraction x/y such that a/b < x/y < c/d.

For example, 1/2 < 3/4. How do we proceed to find a fraction between them? First, rename them so they have the *same denominator,* and you possibly can then find a fraction between them:

$$\frac{1}{2} = \frac{1 \cdot 2}{2 \cdot 2} = \frac{2}{4} \qquad \frac{3}{4}$$

Oops! It's still not apparent, so on to the next step — rename both fractions with denominators *twice as large:*

$$\frac{2}{4} = \frac{2 \cdot 2}{4 \cdot 2} = \frac{4}{8} \qquad \frac{3}{4} = \frac{3 \cdot 2}{4 \cdot 2} = \frac{6}{8}$$

And notice that 4/8 < 5/8 < 6/8, so 5/8 is a fraction between 1/2 and 3/4. If we were still not successful at this stage, we would "double the denominators" again, and continue till we succeeded.

Find fractions between the ones below:

a. 3/4 and 5/6
b. 9/10 and 9/11
c. 4/5 and 9/10
d. 3/8 and 3/4
e. 3/4 and 4/3
f. 11/20 and 13/25

Number Crunching, with Fractions?

"Number crunching" is an engineering term for performing number computations with amazing speed and accuracy. While most number crunching these days is handled by high-speed calculators and computers, it's still possible to find a few humans practicing the art.* And while we may marvel at such performances, in reality we don't value them very highly. Rather, we realize a greater need to understand basic numerical principles and procedures for working with relatively few numbers.

Such is the case with fractions—many problems relevant to every day experiences involve simple calculations with fractions, but generally speaking the fractions involved are fairly simple and few in number. "Number crunching" with fractions is certainly possible, but it is a stage most of us never reach. The algorithms are different from those for whole number arithmetic, and seem replete with such magic phrases as "find the common denominator," "invert and multiply," and "cancel common factors." Many students can recall precisely the point in the past when mathematics suddenly became "less than a friend," and often it has something to do with the arithmetic of fractions.

Yet in many ways, computation with fractions is more intriguing than with whole numbers. There may be more rules to the game, but complex games are generally more challenging than trivial games, once the rules have been mastered. To avoid the frustration associated with learning these new algorithms, computation with fractions must be experienced slowly, concretely, and in a manner that continually relates the symbols being used to reasonable, real-world situations. But if it is handled in this fashion, computational facility with fractions is certainly within the grasp of most students. A "number cruncher" or two may even emerge from the depths!

THE ALGORITHMS Our aim in reviewing computational algorithms for fractions is to present them in their simplest forms and stress the reasonableness of the calculations we ultimately perform. If the most basic definitions and rules are internalized, computational facility will be enhanced as students add their own modifications to the algorithms.

Addition and Subtraction. Addition and subtraction of fractions seem quite logical if the denominators are the same, and the meaning of "numerator" and "denominator" are kept clearly in mind. Consider the calendar to the right. The month is partitioned into 31 equal pieces, so each day would be 1/31 of the month. Saturdays and Sundays are 5/31 and 4/31 of this month, respectively. To find the fraction of the month devoted to weekends, we would combine

```
October           1981
 S  M  T  W  T  F  S
                1  2  3
 4  5  6  7  8  9 10
11 12 13 14 15 16 17
18 19 20 21 22 23 24
25 26 27 28 29 30 31
```

---

*The Guinness Book of World Records has these listings—you are invited to try and top them:
 a. Under test conditions in 1976, R. H. Frost added 100 randomly selected digits in 32.57 seconds —about 1/3 of a second per calculation!
 b. Under test conditions in 1976, William Klein mentally extracted the 73rd root of a 500-digit number in only 2 minutes, 43 seconds.
 c. A Korean youngster – Kim Ung-Yong – performed integral calculus on television at the tender age of 4 years, 8 months.

the separate amounts of time, a concept that relates to addition. So it seems quite natural to say that 5/31 + 4/31 should result in 9/31, since there are nine Saturdays and Sundays together in this month.

If the weekends are removed from a month, the remaining days are "work days" for most of us. Finding the fraction of this month devoted to work days could be handled by subtracting 9/31 from the total month (31/31). By counting, we would find 22 such days, or 22/31 of the month. So it seems logical to say that 31/31 - 9/31 would be 22/31.

Examples such as these inevitably lead to the same conclusion about how to add or subtract fractions with the same denominator:

$$\frac{a}{b} \pm \frac{c}{b} = \frac{a \pm c}{b}$$

The most common mistake made in adding or subtracting fractions — a sure indication of rote learning without internalizing the concept — is when the two denominators are also added or subtracted, as shown to the right. A beginning learner has to stay with concrete interpretations of "denominator" till these common mistakes no longer seem reasonable. Returning briefly to the calendar situation above would make answers like 9/62 or 22/0 seem unreasonable.

$$\frac{5}{31} + \frac{4}{31} = \frac{9}{62}$$

$$\frac{31}{31} - \frac{9}{31} = \frac{22}{0}$$

?

An impasse arises when the fractions to be added or subtracted do not have the same denominator, as in the case 2/3 + 1/4. Again, we can employ a process previously found to be useful, that of renaming the numbers involved—recall that this is the most common way of resolving impasses in the algorithms for addition and subtraction of whole numbers. From the last section, we know that we can always rename fractions so they *do* have similar denominators, and in so doing we eliminate the problem. For the example above, our work might look like what appears in the box to the right. Notice that we have refrained from developing a general rule for adding and subtracting fractions, one that encompasses the possibility that the denominators are not the same initially. Defining

$$\frac{2}{3} \pm \frac{1}{4} = \frac{8}{12} \pm \frac{3}{12} = \frac{8 \pm 3}{12}$$

the algorithm with fractions that have a common denominator is all that is necessary, if the prerequisite knowledge is at hand.

Multiplication. Multiplication of fractions is generally related to, and introduced through, real-world situations that require determining a "fractional part of a given fraction." For example, consider the three pictures below showing a case of soda bottles. Each drink would be 1/24 of the case—as a sidelight, notice how conveniently several other denominators could be represented (halves, thirds, fourths, sixths, eighths, and twelfths):

In the middle picture above, one row has been removed, leaving 5/6 of the case; in the third picture, another row has been removed, leaving 3/4 of the 5/6 of the case of cokes left. The latter situation exemplifies 3/4 x 5/6 since it represents finding 3/4 of 5/6 of the case. Since 15 cokes are left in the last picture, it seems reasonable that 3/4 x 5/6 should be 15/24.

Considering many such examples would eventually lead us to conclude that multiplication of fractions should be defined as:

$$\frac{a}{b} \times \frac{c}{d} = \frac{a \times c}{b \times d}$$

In other terms, to find the product of two fractions, place the product of the numerators over the product of the denominators.

This algorithm is certainly the most popular one involving fractions—there are no impasses to be resolved! Many educators recommend that grade-school children should encounter multiplication of fractions before the other three operations, for this one reason. And the idea has merit since, unlike its whole-number counterpart, multiplication of fractions is both conceptually and computationally independent of addition and subtraction of fractions.

Division. When you stumble across a real-world situation that calls for division of fractions, it's almost a cause for celebration. There just aren't many such examples around. Yet division of fractions needs to be taught, since the algebraic manipulations required of students in later years assumes familiarity with this algorithm.

A piece of ribbon is shown to the left below, lined up beside a ruler. The width of the ribbon can be measured to be 5/8 of an inch. It needs to be cut into strips that

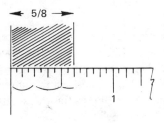

are 1/4 inch wide, so an immediate question—one that connotes the concept of division—is "how many 1/4 inch strips can be cut from a ribbon that's 5/8 inch wide?" In gaining an answer to this question, we'd be obtaining the answer to 5/8 ÷ 1/4. Examining the picture above, we could conclude that 2½ such strips could be obtained, so 5/8 ÷ 1/4 should lead us to the answer 2½.

Another name for 2½ is, of course, 5/2. So above, the answer to $5/8 \div 1/4$ should be 5/2. Or, as shown to the right below, there seems to be an obvious way to define division of fractions, a way that's consistent with this concrete example. To divide fractions, we simply divide the numerators and the denominators in the indicated fashion. And if we continued to examine real-world situations of this sort, as long as we could perform the indicated divisions and end up with whole numbers (that is, as long as $a \div c$ and $b \div d$ *are* whole numbers), this definition of how to divide fractions would suffice!

$$\frac{5}{8} \div \frac{1}{4} = \frac{5}{2} \left(\text{or}, \frac{5 \div 1}{8 \div 4}\right)$$

$$\frac{a}{b} \div \frac{c}{d} = \frac{a \div c}{b \div d}$$

The impasse frequently met is that $a \div c$ and $b \div d$ aren't whole numbers. But again, the impasse can be resolved by renaming one of the fractions. In such cases, we rename a/b as x/y so that $x \div c$ and $y \div d$ *are* whole numbers, and we can then apply our simple definition of division above. How do we produce such a fraction x/y? It's really fairly simple:

If $a \div c$ isn't a whole number, rename $\frac{a}{b}$ as $\frac{a \cdot c}{b \cdot c}$.

If $b \div d$ isn't a whole number, rename $\frac{a}{b}$ as $\frac{a \cdot d}{b \cdot d}$.

If neither $a \div c$ nor $b \div d$ is a whole number, rename

$$\frac{a}{b} \text{ as } \frac{a \cdot c \cdot d}{b \cdot c \cdot d}$$

In the last case above,

$$\frac{a \cdot c \cdot d}{b \cdot c \cdot d} \div \frac{c}{d} = \frac{(a \cdot c \cdot d) \div c}{(b \cdot c \cdot d) \div d} = \frac{a \cdot d}{b \cdot c} = \frac{a}{b} \times \frac{d}{c}$$

which you'll recognize as the infamous "invert and multiply." Notice that an acceptable definition of division of fractions can be made without referring to this sometimes inefficient "trick" for resolving an impasse.

Due to this simple algorithm being unfamiliar to most people, a few practice exercises might be in order before going ahead. Take the three problems below, and work them using the method you've always used for division of fractions, and then compare your answers and your method to that shown:

1. $\dfrac{12}{25} \div \dfrac{3}{5} = \dfrac{12 \div 3}{25 \div 5} = \dfrac{4}{5}$

2. $\dfrac{6}{35} \div \dfrac{5}{7} = \dfrac{6 \cdot 5}{35 \cdot 5} \div \dfrac{5}{7} = \dfrac{(6 \cdot 5) \div 5}{(35 \cdot 5) \div 7} = \dfrac{6}{25}$

3. $\dfrac{3}{10} \div \dfrac{2}{9} = \dfrac{3 \cdot 2 \cdot 9}{10 \cdot 2 \cdot 9} \div \dfrac{2}{9} = \dfrac{(3 \cdot 2 \cdot 9) \div 2}{(10 \cdot 2 \cdot 9) \div 9} = \dfrac{27}{20}$

Did you get the same answers as those above? Which method do you prefer? Isn't it nice to have two at your disposal so you can apply them selectively?

THE PROPERTIES

The properties that accompany computation are not imposed on top of the structure that's already there. The properties arise as reasonable characteristics of the real-world situations that give the operations their meaning. The properties should accompany early work with fractions at the informal, intuitive stages. They can later be formalized as shown below:

Addition is commutative and associative.

$$a/b + c/d = c/d + a/b \qquad a/b + (c/d + e/f) = (a/b + c/d) + e/f$$

Multiplication is commutative and associative.

$$a/b \times c/d = c/d \times a/b \qquad (a/b \times c/d) \times e/f = a/b \times (c/d \times e/f)$$

Multiplication distributes over addition.

$$a/b \times (c/d + e/f) = (a/b \times c/d) + (a/b \times e/f)$$

$\dfrac{0}{x}$ is the identity for addition and subtraction, where x is any whole number but zero.

$$a/b \pm 0/x = a/b$$

$\dfrac{x}{x}$ is the identity for multiplication and division, where x is any whole number but zero.

$$a/b \times x/x = a/b \quad \text{and} \quad a/b \div x/x = a/b$$

$\dfrac{a}{b}$ is a reciprocal for $\dfrac{b}{a}$, if both a and b are not zero, and the product of a fraction and its reciprocal is the multiplicative identity.

$$a/b \times b/a = x/x = 1/1$$

Notice how closely the properties with fractions resemble those of whole numbers. Actually, we've only added one more, the powerful idea of a *reciprocal* element for each fraction.

That the properties accompanying fractions resemble those of whole numbers is not an accident. As a matter of fact, the properties for fractions hold true because of the whole number properties. For example, we could give a verification of the properties above in the manner shown below for addition being commutative:

$\dfrac{a}{b} + \dfrac{c}{b} = \dfrac{a + c}{b}$ by definition of addition of fractions.

$\phantom{\dfrac{a}{b} + \dfrac{c}{b}} = \dfrac{c + a}{b}$ since whole number addition is commutative.

$\phantom{\dfrac{a}{b} + \dfrac{c}{b}} = \dfrac{c}{b} + \dfrac{a}{b}$ by definition of addition of fractions.

For anyone so inclined, the other properties for fractions could be proved in a similar fashion, by relying on those for whole numbers.

The properties are fairly dry things to try and memorize. Their real value comes in applying them to cut short some of the calculations met with fractions.

SET 5

1. Perform the exercises below:

    a. $\dfrac{3}{8} + \dfrac{3}{4}$
    b. $\dfrac{4}{5} - \dfrac{2}{15}$
    c. $\dfrac{7}{8} + \dfrac{1}{6}$
    d. $\dfrac{3}{10} - \dfrac{1}{4}$
    e. $\dfrac{5}{7} - \dfrac{4}{7}$
    f. $\dfrac{3}{1} + \dfrac{1}{2}$
    g. $\dfrac{2}{3} + \dfrac{4}{5}$
    h. $\dfrac{5}{6} - \dfrac{3}{4}$
    i. $\dfrac{3}{8} + \dfrac{1}{11}$

2. Practice applying the multiplication and division algorithms:

    a. $\dfrac{2}{3} \times \dfrac{4}{5}$
    b. $\dfrac{3}{4} \div \dfrac{1}{2}$
    c. $\dfrac{1}{2} \div \dfrac{3}{4}$
    d. $\dfrac{7}{2} \times \dfrac{4}{7}$
    e. $\dfrac{3}{5} \div \dfrac{4}{5}$
    f. $\dfrac{5}{1} \div \dfrac{1}{2}$
    g. $\dfrac{2}{3} \times \dfrac{0}{4}$
    h. $\dfrac{4}{7} \times \left(\dfrac{3}{8} \div \dfrac{1}{4}\right)$
    i. $\dfrac{12}{35} \div \left(\dfrac{2}{7} \times \dfrac{2}{1}\right)$

3. Try some "number crunching" with fractions:

    a. $\dfrac{5}{12} \div \left(\dfrac{2}{3} + \dfrac{1}{6}\right)$

    b. $\left[\left(\dfrac{5}{8} + \dfrac{1}{3}\right) - \dfrac{1}{12}\right] \times \dfrac{1}{2}$

    c. $\left[\left(\dfrac{3}{7} + \dfrac{1}{6}\right) - \dfrac{1}{21}\right] \div \left(\dfrac{2}{3} \times \dfrac{5}{7}\right)$

    d. $\left[\left(\dfrac{3}{32} + \dfrac{9}{16} + \dfrac{3}{8} - \dfrac{5}{4}\right) + \left(\dfrac{4}{7} + \dfrac{5}{21}\right)\right] \times \left(\dfrac{1}{4} - \dfrac{1}{4}\right)$

4. You study for ½ hour for a big test, and then watch television for ½ hour as a reward. You then study again for ¼ hour, and watch television for ½ hour as a reward. A quick review right before bed takes only 6 minutes. What fraction of an hour did you spend studying for the big test?

5. You made some popcorn and ate half of it, saving the rest for a friend. But your friend was late arriving, and by that time you'd nibbled away a third of what you'd left for her. Your friend offered to share her part with you, and you took her up on it, eating half of what was left. What fraction of all the popcorn did you eat?

6. The Bible says to tithe 1/10 of what you earn. The government takes about 1/4 right off the top too. If you tithe and give Uncle Sam his share, what fraction of your earnings remain for you to live on?

7. Three roommates sharing costs equally couldn't decide whether or not to add a fourth roommate. Their rent and electricity for 9 months totaled $1530. If a fourth person could be added without increasing these basic costs, how much less would each pay?

8. Jack, Sam, and Betty were the kids for whom the "house rule" to the right was established. Prior to this rule, the first one home would take all the coke. But no more!

    The first day Jack got home first, and took his share of the full 2-liter bottle. Sam arrived a few minutes later and requisitioned his allotment. Then Betty got home and had her part. How much did Betty drink?

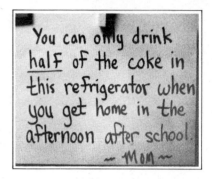

9. The kids above thought they would really like this new rule because, on the first day, everyone got plenty. What they didn't realize was that their mom had foxed them. That first 2-liter bottle was the only one she'd ever have to buy! Why?

10. Television programs are designed to allow about 1/10 of the "air time" for commercials on a 1/2-hour television program. What fraction of an hour is actually devoted to the program on such a half-hour show?

11. The book store on most campuses will give you 1/2 of the original price for a good used book, and then sell it for 3/4 of the original price. What fraction of the original price are they making as profit when they sell a used book?

12. Only about 1/50 of the high school students interested in a military academy are selected, and of those selected, only about 2/3 eventually graduate. What fraction of those originally interested finally graduate?

13. "Four-out-of-five dentists surveyed recommend sugarless gum," or so the commercial goes. But what wasn't reported was that half of those who recommended sugarless gum also stipulated that people really shouldn't chew gum of *any* sort. What fraction of the dentists surveyed were in this last group?

14. It has been estimated that, on the average, from 1/3 to 1/2 of a teacher's classroom time is spent in discipline. This means that an average teacher gets to spend between _____ minutes and _____ minutes of a 50-minute period actually teaching.

15. The class at the Riverdale school is about to watch a 30-minute, 16 mm film. Such a film has 54,000 frames. How many frames will they see per second?
    As a fraction, then, how much time is spent on each frame?

16. A recipe for a 3-layer cake calls for 1/2 cup of sugar. If you only had 1/3 cup of sugar available, would you have enough for a 2-layer cake, or a 1-layer cake, or

would you have to give up on the cake idea altogether till you could get some more sugar?

17. The local school board was having trouble finding teachers willing to work part of the summer, to help remedial students. So they offered them "time and a half" for the three weeks they would be employed. One teacher who made $1200 per month volunteered; how much could this person expect to make for that three-week period?

18. As you do the computations below, search for some shortcuts made possible by the properties of fractions.

   a. $\left(\dfrac{5}{9} \times \dfrac{3}{4}\right) \times \dfrac{4}{3}$ d. $\dfrac{2}{3} \times \left(\dfrac{5}{8} \times \dfrac{3}{2}\right)$

   b. $\dfrac{7}{8} + \left(\dfrac{3}{5} + \dfrac{1}{8}\right)$ e. $\left(\dfrac{2}{3} \times \dfrac{4}{5}\right) + \left(\dfrac{2}{3} \times \dfrac{1}{5}\right)$

   c. $\left(\dfrac{3}{10} + \dfrac{1}{5}\right) \times \dfrac{5}{1}$ f. $\left(\dfrac{7}{9} \times \dfrac{5}{3}\right) - \left(\dfrac{7}{9} \times \dfrac{2}{3}\right)$

Guess That It Was Bound to Happen

It was just a matter of time. Once our numeration system for whole numbers and the concept of fractions were well-established, it was inevitable that someone would combine the two ideas. The result was the decimal system.

A decimal point is used to distinguish the whole number part of a decimal numeral from the fractional part. The whole number part is to the left of the point, and the fractional part is to the right. The meaning of a symbol such as 91342.5687 is reviewed for you briefly below. Notice that we're still completely in base ten.

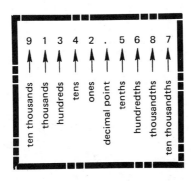

Positioning the decimal point to the right of the "ones place" gives young children problems right at the outset. They seem to sense an unwritten rule that, whenever possible, mathematical notation should be symmetrical. As evidence of this phenomenon, many times they'll respond to a question indicating that the value immediately to the right of the decimal point is also the "ones place."

Our numeration system would have been symmetric if the originators had made the "tens" and "tenths" equidistant from the symbol for the "ones place." Then the "hundreds" and the "hundredths," the "thousands" and the "thousandths," etc., would all be the same distance from the critical place, the "ones." An easy way to accomplish this would be to center the distinguishing mark for the units place. Several suggested methods are presented below:

$$913 \overset{\bullet}{4} 25687 \qquad 913 \underset{\wedge}{4} 25687 \qquad 9134 ⓐ 5687$$

Note that this would also clear up the problem of confusing a decimal point with a period, as in "The number of gallons was 17.95." (At first glance, the number appears to have two decimal points.) We will bow to tradition, of course, but not without stating our case: our numeration system for decimals could be improved!

COMPUTING Computation with decimals is consistent with computation involving both whole numbers and fractions. Addition and subtraction are concepts requiring that the numbers added or subtracted be based on a common unit. For whole numbers, the algorithm begins with "lining up the numbers from the right-hand end"; this accomplishes the task of adding digits based on similar values. With fractions, we must ensure that they have the same denominator before trying to add or subtract. The usual algorithm for adding or subtracting decimals begins with "line up the decimal points"; here again we're merely ensuring that the digits to add or subtract have the same value. For example:

$$4.278 + 56.9 + 13 \text{ becomes} \quad \begin{array}{r} 4.278 \\ 56.9 \\ +13 \\ \hline \end{array}$$

$$5.76 - 3.987 \text{ becomes} \quad \begin{array}{r} 5.76 \\ -3.987 \\ \hline \end{array}$$

Once the decimal points have been aligned, the procedure for adding or subtracting decimals is exactly like the corresponding whole number algorithm (including "borrowing" and "carrying," since we're still totally using base ten).

Multiplication and division of whole numbers does not require alignment of the digits, just as multiplication and division of fractions does not require finding a common denominator. The same concept should hold true with decimals, and it does! As a matter of fact, multiplication and division of decimals can be handled using the same algorithm as that for whole numbers, with the additional step of inserting the decimal point correctly in the final answer. For example:

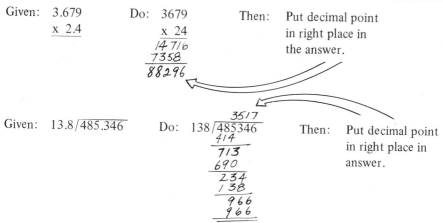

How does one decide where to place the decimal point in the answer? The algorithms you learned years ago probably include some rules like:

Count the decimals in the numbers being multiplied, and count over that many places (from right to left) in the answer.

Move the decimal in the divisor till you have a whole number, and move the decimal in the dividend equally. Bring it up.

These generalizations are fine, but if you are inclined to forget them, most of the time you can decide where the decimal should go by rounding off the numbers involved and estimating a reasonable number. In the division problem above, for example, round 13.8 off to 10, and 485.346 to 500. Since 10 goes into 500 fifty times, you could place the decimal point in the given problem so the answer would be close to 50.

We don't mean to slight the algorithms for decimals by giving them a cursory glance, but generally their development follows that already described for whole numbers and fractions. And there are some very interesting aspects of decimals that beg to be mentioned.

RENAMING FRACTIONS AS DECIMALS

The relationship between fractions and decimals is an interesting one. To rename a fraction as a decimal, use is generally made of the division interpretation of a fraction, i.e., a/b means a ÷ b. A few examples are presented for you below:

a. $\frac{3}{8}$ means 8)3.000 = .375 (24, 60, 56, 40, 40)

b. $\frac{38}{10}$ means 10)38.0 = 3.8 (30, 80, 80)

c. $\frac{13}{45}$ means 45)13.000... = .288... (90, 400, 360, 400, 360)

And a real bear:

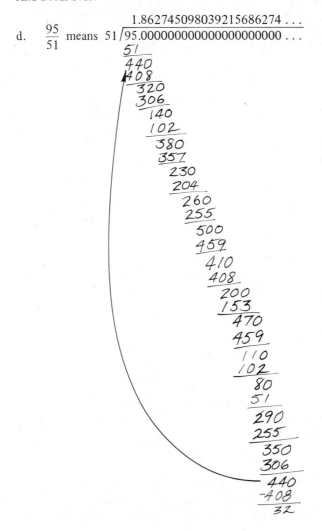

From examples (a) and (b), we would say that 3/8 = 0.375 and that 38/10 = 3.8. These are examples of *terminating decimals.* Examples (c) and (d) are referred to as *repeating decimals,* since the digits begin to have a repeating pattern after a while. The symbolism we use in such cases is typified by saying 13/45 = 0.2$\overline{8}$ and that 95/51 = 1.8$\overline{627450980392156}$. Notice from example (c) that the pattern doesn't have to begin repeating immediately to the right of the decimal, and from example (d) that sometimes we have to divide for quite a while before the repetition starts.

Which brings up an interesting question—when a fraction is renamed as a decimal, will it always be either a *terminating decimal* or a *repeating decimal*? Maybe there's some really weird fraction out there that would *never* start repeating the digits when we tried to divide the denominator into the numerator. After all, we had to go out pretty far to get 95/51 to repeat!

An intuitive investigation could begin by looking at the division process, particularly that part in which we subtract, bring down a zero after subtracting, and divide again. If we ever have a zero when we subtract, we can stop right there and we'll wind up with a *terminating* decimal. And as soon as we get a remainder we've seen before, the division process will begin to repeat itself. In example (d), 44 was the first remainder to repeat, so this is where the digits in the answer started repeating themselves. Anytime we run out of remainders and have to start dividing into numbers we've seen before, we'll have a *repeating* decimal. Do we always run out of remainders when converting a fraction into a decimal, and so always have either a terminating or repeating decimal name for the fraction?

Consider for a moment a simpler example, that of finding the decimal name for 1/7. We begin by dividing 7 into 1. Note that when 7 is the divisor, the only possible remainders are 0, 1, 2, 3, 4, 5, and 6. So how many steps would it take to be sure we had used them all up, and had to begin using them over again? If the remainder isn't zero after 7 steps, we can rest assured that the decimal has started to repeat. And in this case, as you can see above, the seventh remainder (3) *has* appeared before. So $1/7 = 0.\overline{142857}$. Going back to the previous example for a moment, we could be assured that 95/51 would have to begin repeating at least by the 51st step! (Fortunately, it only took 17 steps.) To answer the original question, then:

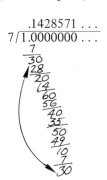

Yes, every fraction has a decimal name which either terminates or repeats!

And we can find the decimal name for a fraction by division, if we have the patience!

Sometimes it's much nicer to deal with fractions than with their decimal names, believe it or not. For example, try the addition problem to the right. Notice that even with the decimal points already lined up for you, it's awkward to apply the common algorithm. Yet it's a legitimate situation, since $0.2\overline{8}$ is the decimal name for 13/45, and $0.\overline{142857}$ is the name for 1/7. Which leads us to another interesting question: given a decimal that either terminates or repeats, can we always find a fraction as another name?

Decimals that terminate are easy to write as fractions; given a decimal such as 39.567, by merely considering the place value we know that it can be renamed as 39 + 5/10 + 6/100 + 7/1000, or as

$$39\frac{567}{1000}.$$

All such terminating decimals can be written as fractions just by going to the place-value concept.

RENAMING TERMINATING AND REPEATING DECIMALS AS FRACTIONS

The repeating decimals are a little harder to rewrite as fractions, but still it can be done. First, notice that, given a repeating decimal like 2.387777..., the decimal point can be "moved" by multiplication so that only the repeating digits are to the right of the decimal. For example, 100 x 2.387777... yields 238.7777..., and 1000 x 2.3877777... gives 2387.7777... Why would anyone want to do such a thing?

Being able to move the decimal point, as described above, is at the heart of an algorithmic process for changing a repeating decimal into its fractional name. Apparently someone long ago got the bright idea (after much trial and error, we suspect) for this conversion process, and developed the procedure illustrated below by example.

Object: Find the fraction name for 2.38777....

Procedure: Let N = 2.38777...
Then we know that 1000 · N = 2387.777...
and 100 · N = 238.777...

And taking the difference of these numbers,
1000 · N − 100 · N = 2387.$\overline{7}$ − 238.$\overline{7}$

So 900 · N = 2149

Or, N = $\frac{2149}{900}$

And all of a sudden we have the name, as a fraction, for 2.38777...

Don't be awed by the "fancy footwork" above. The main idea is to multiply the given decimal by numbers that will force the resulting decimals to begin repeating immediately to the right of the decimal point. Then subtracting one from the other will eliminate the rather annoying decimals, leaving only whole numbers.

Two other examples are provided for you below. Study them carefully for such things as "why did they multiply by that number" and "what's going on algebraically in each step?"

Given: N = 3.$\overline{7}$
Object: Write N as $\frac{a}{b}$

Process: 10 · N = 37.$\overline{7}$
N = 3.$\overline{7}$
So 10 · N − N = 34
and 9 · N = 34
Or, N = $\frac{34}{9}$

Given: N = 3.6$\overline{45}$
Object: Write N as $\frac{a}{b}$

Process: 1000 · N = 3645.$\overline{45}$
10 · N = 36.$\overline{45}$
So 1000 · N − 10 · N = 3609
and 990 · N = 3609
Or, N = $\frac{3609}{990}$

Intuitively, then, we have an answer to the earlier question posed about rewriting terminating or repeating decimals as fractions:

Yes, any decimal which terminates or repeats can be written as a fraction.

Which brings up still a third interesting question. Are there any decimals which neither terminate nor repeat, and therefore might not have a fractional name?

SET 6

1. Practice computing with decimals on these problems.

    a. (5.37 + 2.8 + 14) - (2.5 x 3.6)

    b. (11.06208 ÷ 2.76) + (5.2 - 3.881)

    c. [(49.5 x 3.8) + 27.76] ÷ 50.2

2. Perhaps you can do the problems below in your head. If so, you're using extensions of the commutative, associative, and distributive properties, which is fine, since they apply to decimals as well as whole numbers and fractions.

    a. 15 · ($1.98)   b. $4.98 + $6.75 + $7.02 - $8.25

    b. 63.9¢ x 20   d. (.5) · (20 x 76.98)

3. Change these fractions to decimals.

    a. 5/11   b. 63/25   c. 52/19   d. 9/7   e. 10/40

4. Change these decimals to fractions.

    a. 2.0$\overline{4}$ b. 5.6$\overline{21}$ c. 138.3654   d. 71.656656656...   e. 4.8

5. In problem 10 of Set 4, you expressed the thickness of a sheet of paper as a fraction. What would this thickness be as a decimal?

6. The photograph to the right shows the menu for a fast food restaurant in a small town in California. The food's not that great, but it sure is cheap!

    How many beef tacos could you get, if you only had a dollar bill to spend (forget the tax)?

7. In many public schools elementary teachers are given less than $50 a year to purchase consumable supplies for the class. Assume such a class has 27 students. Each student would then be entitled to how much of the money to be spent on him?

    A school year usually has 180 school days. How much money per day, per student, would the teacher have to spend?

8. A common automobile tire has a radius of 14 inches. How many inches would its circumference be? (You probably recall the formula C = 2πr, where C stands for circumference, r stands for radius distance and π is approximately 3.14.) A typical warranty for a good tire is 40,000 miles. How many times would such a tire turn completely around in traveling 40,000 miles? If the tread were one inch thick originally, and worn off exactly at the end of the 40,000 mile period, what part of an inch of tread was lost each time the tire turned around?

9. A basketball rim has a radius which measures 9 inches. The circumference of a basketball measures about 31 inches. When such a ball swishes through the hoop, how many square inches of clearance are there? You might recall that area of a circle is given by A = πr².

10. Sensory signals travel through your body at about 260 miles per hour. About how long would it take for your brain to realize you dropped a brick on your toe? How long would it take for you to start hopping around the room, crying out in your misery?

11. The world's fastest camera can take pictures at a rate of 600 million frames per second. How long does each frame take on such a camera?

12. Levis jeans sell for around $18.99 at an average store, but you can get them for around $10.99 at a "second store." Jack took orders from his fraternity brothers one term — 15 brothers agreed to buy one pair each at $15.99, and 3 others wanted 2 pair each at that price. He bought them at the second store. Without counting tax, how much profit could he expect?

13. Find out how much a carton of cigarettes costs at today's prices. A person who smokes a pack a day from age 20 to age 60 will spend how much (at today's prices) on this habit?

14. Several types of checking accounts are popular around a college campus. One type is where the checks cost you more (perhaps 10¢ per check), but there's no service charge. Another is where the checks cost less, but there's a service charge.
    Find two different types of checking accounts for your campus, and analyze them to see where the "break point" is, i.e., the point at which one begins to hold the advantage over the other. (This depends on the number of checks written per month, for the two types described above.)

15. During the seventies, staying in shape became the "in thing" to do. It seemed everywhere you went, you'd meet joggers, tennis players, etc. Even in the junior and senior high school athletic programs, increased attention was turned to the "life-long sports." Rather than seeing how strong or fast you were, the main questions asked were of the "how good is your heart" or "are your lungs in shape?" variety.
    Right now, without a whole lot of work, you can test your cardiovascular fitness using a well-known version of the Harvard Step Test. First, locate a sturdy bench or chair of the appropriate height, using the table to the right. You will also need a metronome, or watch with a second hand, and someone to take your pulse (unless you can take your own, when your heart is pounding).

Your Ht.	Bench Ht.
Less than 5'	12"
5'0" – 5'3"	14"
5'4" – 5'9"	16"
5'10" – 6'1"	18"
Over 6'1"	20"

Then exercise according to this procedure:

a. Step from the floor onto the bench and down again 30 times per minute, for four straight minutes. If you get too tired to go on, you can stop earlier, but it will lower your score.

b. As soon as you're through, sit quietly and take your pulse for thirty seconds, starting 1 minute after you finish (P_1); then again 2 minutes after you finish (P_2), and then once more 3 minutes after you finish exercising (P_3).

c. Compute your recovery index (RI) using this formula:

$$RI = (100 \cdot S) \div 2 \cdot (P_1 + P_2 + P_3),$$

where S stands for the total number of seconds you exercised.

RI	Rating
60 or less	Poor
61–70	Fair
71–80	Good
81–90	Very good
91 or more	Excellent

Since this method measures how quickly your heart returns to normal after exercise, the lower $P_1 + P_2 + P_3$ is, the better shape you're in. But the lower the sum above is, the higher RI would be since RI is $100 \cdot S$ divided by this sum, and dividing by a lower number raises the answer. So the better shape you're in, the higher your RI score should be. The table to the left gives you a rating system you can use on yourself.

The Two Faces of Percent

Pick up a newspaper and start perusing any page you wish. How far do you get before you stumble across the word "percent?" Probably not very far. Weathermen, salesmen, social scientists, financiers, sportswriters, doctors, educators—the list goes on and on—are immersed in the language of percent. Certainly part of its popularity is due to a psychological factor; its use seems to represent authority, or hard, unarguable evidence. For some reason, percent has become one of the most widely used topics mathematics has produced. Thus it does have a positive face.

But ask a school child about percent, and you're likely to see its other face. This topic seems to be one of the greatest hurdles that school children meet in mathematics. In a recent national study,* only 35 percent of 13-year-olds (seventh and eighth graders) knew what percent 30 was of 60, and less than 10 percent of this age group could compute 4 percent of 75. Comprehension doesn't improve much with age either; for 17-year-olds, the proportion of correct responders to the two items above crept up to 58 percent and 27 percent, respectively. Even many adults aren't intimately familiar with this topic, as evidenced by the fact that the federal government recently passed legislation to protect consumers from credit card companies and loan associations that were advertising interest rates of 1½ percent per month, and not giving the yearly rate with which the public was familiar. The general public was unable to change a monthly interest rate into a yearly one, and the wolves were having a feast!

*The National Assessment of Educational Progress, 1978, Suite 700, 1860 Lincoln St., Denver, Colorado 80295

Why all the difficulty handling a topic that seems so obvious to many people, and is undoubtedly placed in such high esteem by society in general? As much to blame as any other factor is that, prior to the introduction of percent, a child has learned arithmetic as a meaningless set of symbols, divorced from physical interpretations. If there's no mental image available for a fraction like 4/100, or the decimal equivalent 0.04, then saying that 4% = 4/100 = 0.04 just creates a larger dilemma, because it introduces more meaningless symbols. Adding to the problem are the generalizations that we force—things like "to change a decimal to a percent, move the decimal point two places to the right (or was that to the left?)." More rules to forget!

But assuming that a student does have the prerequisite meaningful interpretations of fractions and decimals, the road seems to smooth out considerably. The new symbolism makes good sense, and the usefulness of this new tool soon becomes obvious. It turns its friendly face toward us.

THE BASIC INTERPRETATION OF PERCENT

"Percent" comes from the Latin words "per centum," which mean "of a hundred" or "out of a hundred." It seems quite natural, then, that 4 percent should be interpreted as 4 out of 100. Or 50 percent as 50 out of 100.

But we already know that 4 out of 100 and 50 out of 100 can be interpreted both as fractions and decimals.

$$4\% = \frac{4}{100} = 0.04$$

$$50\% = \frac{50}{100} = 0.50$$

The examples above are typical of how "percent" should be interpreted.

Conceptual difficulties arise for many people when the percent under consideration is smaller than 1 percent or larger than 100 percent, or involves a fraction of a percent. For example, 0.3 percent, 214 percent, and 37¼ percent are typical of situations that are more difficult than most that involve percent, and yet these concepts are quite common in reporting information of interest to the general public — interest rates, discounts, and business gains or losses are just a few cases in point. Perhaps the difficulty can be traced to not returning to the basic idea that establishes percent, and following through in a step-by-step fashion, as shown in the examples to the right. Study this process and familiarize yourself with these sometimes troublesome expressions. If the basic meaning of percent as "out of a hundred" is kept in mind, these often difficult, but common, situations can be handled without too much trouble.

$$0.3\% = \frac{.3}{100} = \frac{3}{1000} = 0.003$$

$$214\% = \frac{214}{100} = 2.14$$

$$37\tfrac{1}{4}\% = \frac{37.25}{100} = \frac{3725}{10000} = 0.3725$$

THE THIRD R 77

There are several fairly common situations that involve computations with the concept of percent. The first is merely converting a decimal to its equivalent form as a percent, and the second is changing a fraction into a percent. By far the most common situation involves finding a certain percent of another number, and this is the third type of calculation to be taken up in this section. Care must be taken at this stage to keep clearly in mind the basic meaning of the word "percent" (out of a hundred) and avoid relying on any rules you may have picked up over the years.

COMPUTING WITH PERCENT

Decimals to Percent. The position of the hundredths place in our decimal system is two places to the right of the point. Therefore to convert a decimal to a percent, we only have to remember that anything to the left of this place can be interpreted as "so many hundredths" or "out of a hundred." For example,

.08 means "8 hundredths" or "8 out of a hundred," or 8%
.42 means "42 hundredths" or "42 out of a hundred," or 42%
3.58 means "358 hundredths" or "358 out of a hundred," or 358%
.847 means "84.7 hundredths" or "84.7 out of a hundred," or 84.7%
.0019 means ".19 hundredths" or ".19 out of a hundred," or .19%

Study the last three above carefully before going ahead, as they're typical of the type that give many people problems. In particular, notice that .19% is *not* the same thing as 19%! (Why not?)

Fractions to Percent. Many times a given fraction can be renamed with denominator 100, in our heads, and thus can be easily changed into a percent directly. Such is the case with fractions like 1/2, 1/4, 3/4, 1/10, 3/10, etc., and even some relatively unfamiliar ones like 3/20 and 7/50. Whenever possible, this method is preferred as it seems to lend itself more to retaining the meaning of percent. Score yourself on the fractions to the right. Can you change all ten of them into percents, in your head? If so, you've got a good grasp of both "renaming fractions" and percent. If not, don't despair; there's another way to change a fraction into a percent, and you don't have to be a whiz at mental arithmetic to apply this method.

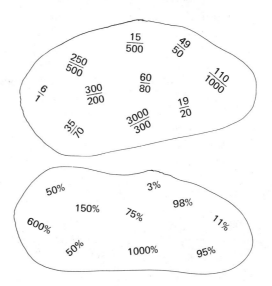

The second way to change a fraction into a percent involves first changing the fraction into its equivalent decimal name by dividing the denominator into

the numerator and then converting this decimal into a percent, as shown in the previous section. For example, to change 3/8 to a percent in this fashion, start by dividing 8 into 3, as shown to the left. The decimal .375 is obtained, and is changed into 37.5 percent as the final step. Since this method of renaming a fraction as a percent will *always* work, our minds are inclined to latch onto the procedure and use it to the exclusion of the other method. We must, of course, fight this temptation in ourselves, and certainly discourage it in elementary students. It characterizes us as preferring to compute, rather than to think!

```
    .375
 8/3.000
   24
   ‾‾
    60      and .375
    56      equals 37.5%
    ‾‾
     40
     40
     ‾‾
```

Consider these problems, which were previously encountered as fractions. Which of the two methods would you use on each of the four situations to change the language into that of percent?

    a. Four out of five dentists surveyed

    b. Of those selected, only about 2/3 graduate

    c. Television programs are designed to last 1/10

    d. Sherry got 14 out of 15 correct on her test

Most adults would be able to rename the fractions in (a) and (c) above as percents by mentally converting them to fractions with denominator 100, the preferred method. Division is usually required for problems (b) and (d), and "rounding off." Note however, that even though the fraction in (b) requires division, many people are so familiar with this particular fraction (2/3) that they can take a shortcut and rely on memory.

Calculating a Percent of a Number. The remaining frequently encountered computation that involves percent is that of finding a certain percent of a given number. If the sales tax is 4 percent and you buy an item that costs $48.99, the sales tax you pay would be figured by a problem of the "4 percent of $48.99" nature. Or if you borrow $1000 at a yearly interest rate of 10½ percent, at the end of a year the interest you owe can be calculated by finding 10½ percent of $1000. If a store has a sale with 25 percent off every item, you can find the amount you would save on a $235 sofa by computing 25 percent of $235. If we interpreted these situations as fractions instead of percents, it would be easy to conclude that they were characteristic of situations that require multiplication.

Given that we multiply in such situations, then, should we convert the percent to a decimal, or to a fraction, before we multiply? This decision is again based on personal preference, and depends both on the numbers involved and on your own computational skills. Sometimes the problem can be done mentally, if the percent involved is a familiar one, and the other number is "easy to work with." To find 25 percent of $200, for example, you would almost automatically think of "1/4 of $200 is $50." Or 10 percent of 560 people as "1/10 of 560 is 56." But to find 37 percent of $324, you might as well go ahead and pull out the pencil-and-paper! Thirty-seven percent is not that familiar to most of us, and $324 is not particularly easy to work with. So we'd choose to go ahead and calculate .37 x $324. Again, as in the last example, converting the given percent into a decimal and then multiplying, will always work, but as before we must not become too dependent on it as the only method. Which of the problems below could you do mentally, and which would require paper-and-pencil?

    a. Find 75% of 8000 people.

    b. Calculate 20% of $47.

    c. Compute 8¼% of $2500.

    d. 13% of 475 families is how many families?

Probably you could do (a) above mentally, and perhaps even (b). But unless you're a computational whiz, (c) and (d) would require some arithmetic—something on the nature of .0825 x $2500 for (c), and .13 x 475 for (d). Could you do them all now?

    The next problem set has some exercises in it similar to the ones encountered in this section on computation, plus a few more to show you ways in which percent will enter your future life. From your past experiences in mathematics, and with the sorts of problems you've seen so far in this text, can you estimate about what percent of the percent problems you'll feel successful with after you're through?

SET 7

1. Write the fractions and decimals below as percents. Rename them mentally, if possible.

   a. 0.164
   b. 5/12
   c. 16/20
   d. 0.0004
   e. 3.97
   f. 91/75
   g. 14.6
   h. 2000/200
   i. 85/50
   j. 0.4
   k. 0.001
   l. 9/16
   m. 11/4
   n. 3/30
   o. 33/50

2. Calculate where necessary below; otherwise, do the problems in your head.

   a. 10% of 94
   b. 125% of $40
   c. 99% of 1500 chickens
   d. 50% of 33 students
   e. 200% of 14 pounds
   f. 73% of $15
   g. 1% of 983
   h. ½% of 200 snakes
   i. .10% of 300 eggs
   j. 10% of 300 eggs
   k. 1% of 4 million farmers
   l. 50% of 60% of 40 jobs

3. Dade County (Miami), Florida, has a "balanced curriculum" outline for the teachers to follow, to ensure that a student is exposed to a well-rounded education. For a third grader, the following time requirements must be met:

math	45 min./day	literature/lang.	30 min./day
reading	60 min./day	music	30 min./alter. days
writing	30 min./day	physical educ.	30 min./day
art	60 min./week	science	100 min./week
health/safety	30 min./day	social studies	150 min./week

The school day runs from 8:20 a.m. till 2:40 p.m. each day except Wednesday, when school is dismissed one hour early for "teacher planning."

Calculate the percent of the school day a student spends in each activity above, and then find the percent of time left over for such things as announcements, lunch, going to the restroom, etc.

4. According to recent research findings, human beings learn through their five senses in these proportions:

Tasting	1.0% of the time
Touching	1.5% of the time
Smelling	3.5% of the time
Listening	11.0% of the time
Seeing	83.0% of the time

If our schools are to make a greater impact in helping the human species "learn," what changes might be suggested by the information above?

5. The waiter brings your group its tab, which runs $48.80 including tax. You volunteer to get the tip, and have no paper and pencil handy to calculate the extra 15%. What would you do?

6. After graduation you borrow $2500 at a simple interest rate of 9%. How much interest would you owe in a year? If you were in the enviable position of being able to pay back both the loan and the interest in one year, how much would you have to fork over?

Suppose you decided to pay the loan and interest back at the end of 10 months instead of waiting the entire year, and the bank agreed that you'd only have to pay interest on the time you had the loan instead of the full year. What do you think would be a fair amount of interest for you to have to pay back?

7. Banks loaning money at a certain discount rate is another process that's gaining in popularity. With this procedure, you don't go in and borrow a set amount—instead, you agree to pay back a certain amount at the end of the time period. The discount is then figured, and the amount of money you take home with you is the total you'll pay back, minus the discount.

Suppose you go into a bank and agree to pay them back $2500 in a year at a discount rate of 9%. How much money do you get to take home with you?

8. You and a friend are planning a holiday visit to a different part of the country during spring break. Your friend agrees to take her car—what percentage of the gasoline do you think you should pay?

9. Due to a lack of funds, teachers on a 12-month contract are cut back to 10 months with a corresponding reduction in salary of 2/12 or 16-2/3%. A year later (to keep from going to court) the teachers are offered a raise of 16-2/3% over their present salary. If they accept the offer, will the teachers have regained their loss? (Hint: be careful. This is an example of changing the base on which a percentage is figured.)

10. Your parents offered to buy you a new car when you graduated from high school — anything that cost around $7000. But you were going to college in a large city with mass transportation that made a car useless, so you opted to have them buy you one when you graduated from college instead. How much would this same car cost four years later, if the inflation rate over those years averaged a conservative 10%?

11. In an aerobic conditioning program, you are tested on a treadmill to see how much work you do to maintain an ideal target heartbeat rate. The target rate is generally 75% of your maximum heartbeat, which is approximately 220 beats per minute minus your age. Write an equation expressing the target heartbeat in terms of the age of the individual, and then use this formula to find your target heartbeat.

12. Use the page number above to calculate the percentage of this text that you have completed so far this year.

What's your initial reaction to the cartoon below? Most of us would laugh a little, and nod our heads in agreement. Particularly emphatic would be the head movements of primary-level teachers who spend a great deal of time demonstrating such subtraction impasses to their students. And yet sometimes we're struck by an even greater truth . . . .

Source: © 1957 United Feature Syndicate, Inc.

A TRUE, BUT NOT SO UNUSUAL, SHORT STORY

The first-grade class was being tested at the beginning of the year to determine each student's readiness in arithmetic. The teacher's aide had five pebbles in his hand, and was individually asking each child questions like "if I put three more in my hand, how many will I have?" Eventually it was Kim's turn, and he was the first student to answer all of these addition facts correctly.

The aide decided to test him further and began asking Kim some of the subtraction facts, such as "if I removed four pebbles from my hand, how many would there be?" Kim continued his successful ways, getting all of the subtraction facts right—even the hard one about removing all five pebbles.

"How far could he go with this," the aide began to wonder. "Suppose I had only these five in my hand, but had some magic way of removing seven pebbles—how many would you say I had in my hand then, Kim?" His immediate response—"negative two"—was not a fluke. After more testing, it was determined that Kim did know a lot about negative numbers on entering the first grade.

Kim was a bright student, no doubt, but there were several others in this class with the same potential. Where did he get this head start on the rest? Not an advanced kindergarten program, or a tutor, or even his parents could take the credit. It came from watching television quiz shows where contestants could go "in the hole."

Will there be any Kims in your classes? Most definitely! Will you be able to spot their potential, and plan a program to help them reach it? We hope so! At least give them the chance, when they try to subtract a larger number from a smaller one, to demonstrate the greater truth of the situation!

Negative Numbers

For thousands of years, mankind got along quite nicely without the concept of negative numbers. The natural way of thinking about "amount" as only a positive attribute prevented what—at the time—was such an autistic idea as "negative amount" from emerging.

Positive numbers are sufficient as long as we can point to a definite starting point (to be labeled "zero") for what we want to measure, and nothing can be imagined as preceding that starting point. Physical attributes are of this nature—height, weight, number of freckles, number of grains of sand, the area of geometric figures, etc. It's impossible for us to imagine "negative weight," for example. If these sorts of things were the only measurements we needed to make, certainly the positive numbers (plus zero) would suffice. For such situations, a number "ray," with zero at one end, can be used as a physical model.

For other measurements we need to make, picking a starting point and calling it "zero" is quite an arbitrary thing. Time, for example. What point in time could you establish as "zero," and would other points do just as well? For any point you select as the starting point to measure this concept, certainly there was time that preceded that point!

Temperature is another fairly good example—although theorists claim there really is an absolute zero, it's so far removed from anything we'll ever experience that it wouldn't be practical to use as the starting point for a measurement scale. And any other point for "zero" would certainly have lower temperatures than that point.

Financial worth is another such situation. Before the concept of owing money, there was no need for negative numbers to show someone's net worth. Unfortunately for most of us, negative numbers *are* well established in our country's financial fabric — as a matter of fact, our entire economy is based on "deficit spending" on both the national and personal levels. The common phrases "being in the black" and "being in the red" are still heard today, and come from the turn-of-the-century accounting practices of recording numbers above zero with black ink, and those below zero either in red ink, or with a circle around them.

To summarize, there are certain concepts that need to be measured on a scale that is *open on both ends*. We pick an arbitrary starting point and call it "zero," and use our usual numbers on one side of the starting point and call them the *positive numbers*. On the other side of zero, we use the same numbering system, but label the numbers as *negative numbers,* as shown below:

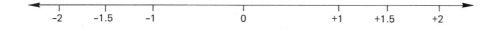

For every positive number on the right-hand side, there's a corresponding negative number on the left, the same distance away from zero. Or, for each number ^+n, there's a corresponding number ^-n that's the same distance from zero.

MATHEMATICS FOR TEACHERS

Notice that the new symbols introduced above (⁺n and ⁻n) have "plus" and "minus" signs attached to them, slightly raised to establish that they're not being used to indicate addition and subtraction. For example, ⁻ is the same thing to ⁻348 as the symbol 'B.C.' is to the year 348 B.C. The signs used with the numbers are just a part of the numeral, and in no way indicate an operation has occurred. (Note: frequently the positive numbers are written without the "plus sign" if no confusion could result.)

Now that these new numbers are available to us, how are we going to compute with them? What algorithms can we develop that would seem reasonable in light of the situations for which they came into being in the first place? Keep in mind in the discussion to follow that the algorithms for working with negative numbers spring from what should happen in real-life situations that require the use of negative numbers.

THE ALGORITHMS The average person's checkbook yields a familiar model for discussing positive and negative numbers, and thus gives us a hint about how we should define the operations of addition, subtraction, multiplication, and division. Consider this page from a checkbook register:

BE SURE TO DEDUCT ANY PER CHECK CHARGES OR MAINTENANCE CHARGES THAT MAY APPLY

DATE	CHECK NUMBER	CHECKS ISSUED TO OR DEPOSIT RECEIVED FROM	AMOUNT OF DEPOSIT	✓	AMOUNT OF CHECK	BALANCE 0 00
8/25 79		CASH DEPOSIT	⁺400 00			⁺400 00
9/1 79	1	STATE FARM INS.			-161 52	
9/3 79	2	MASTER CHARGE			-38 91	
9/10 79		CHECK FROM GRAMMA	⁺250 00			
9/11 79	3	COURTESY CARS			-500 00	
9/12 79	4	STATE OF DELAWARE			-27 43	
9/12 79		AUGUST PAYCHECK	⁺500 00			
9/18 76	5	J.T. SMITH			-14 00	

The balance can be either positive, zero, or negative, at any time. Deposits are considered positive numbers, since they increase the balance, and checks are always negative amounts, since they reduce the balance. Notice that the signs to distinguish the positive and negative numbers have been included above, even though we generally omit them since there are separate columns for deposits and checks.

Before going any further, balance each step in the checkbook register above. The final balance at the end should be $408.14, and in two places you should observe negative balances (examples of deficit spending). If successful, you are probably well grounded in the meaning of positive and negative numbers.

Addition. *Combining the previous balance with either a check or deposit* corresponds to *addition* of positive and negative numbers. Above, this interpretation would yield the following statements:

THE THIRD R   85

```
0 + ⁺400 = ⁺400                    ⁺449.57 + ⁻500 = ⁻50.43
⁺400 + ⁻161.52 = ⁺238.48           ⁻50.43 + ⁻27.43 = ⁻77.86
⁺238.48 + ⁻38.91 = ⁺199.57         ⁻77.86 + ⁺500 = ⁺422.14
⁺199.57 + ⁺250 = ⁺449.57           ⁺422.14 + ⁻14.00 = ⁺408.14
```

Note in particular above what happens when two positives are added, or two negatives are added. And equally important, what happens when a positive and a negative are combined?

Answering the questions above gives us a hint as to what our addition algorithm should be.

> *If two numbers with the same sign are added, they can be added as whole numbers and the common sign used in the answer. If two numbers with opposite signs are added, the difference in the two is used, along with the sign of the larger.*

Subtraction. *Subtraction* can be interpreted as *the mathematics involved in correcting a mistake* in the checkbook. For example, suppose that the $500 deposit recorded for September 12 was never actually made. What would you do to correct for this mistake, when you found out about it? Or suppose you found out that J. T. Smith was not going to cash your check. How would you compensate for such an incorrect entry?

Since these situations involve making corrections, they correspond to subtraction with positive and negative numbers. Correcting for the incorrect deposit would be interpreted as $- {}^+500$, or subtracting a positive number. And fixing the check that wasn't cashed is interpreted as $- {}^-14$, or subtracting a negative number.

As life would have it, we usually aren't so fortunate as to know about these false entries until we've gone much further ahead in the checkbook register itself, implying, of course, that the old eraser trick is useless. We have to adjust the present balance by doing the thing that seems most reasonable. To compensate for the wrong deposit entered, we would naturally decrease the balance that same amount; to fix the check entry, we would increase the balance that same amount. Or, in terms of positive and negative numbers,

correcting entries
- ⁺500 is the same as + ⁻500 (decreasing the balance $500)
- ⁻14 is the same as + ⁺14 (increasing the balance $14)

It seems reasonable to conclude that subtracting a positive number is the same thing as adding a negative number (and we know how to do that from the previous section). And subtracting a negative number yields the same answer as adding a positive number.

> To subtract a positive or negative number, first change the sign of the number, and then add according to the previous algorithm.

With a little practice, the procedure above becomes fairly routine, thus qualifying as a Level 2 thought process.

Multiplication. *Multiplication by a positive number* can be interpreted as *repeated entries of either checks or deposits.* For example, if your paycheck of $1000 per month was automatically deposited for you, for a whole year, the overall effect of this would be to have increased your account by $12,000. Mathematically, this would correspond to the statement $^+12 \times {}^+1000 = {}^+12,000$. Or perhaps a grandmother decides to give each of her 10 grandchildren a check for $5. The overall effect of this repeated action would be to reduce her balance by $50. In terms of multiplication by a positive number, we would have $^+10 \times {}^-5 = {}^-50$.

How would *multiplication by a negative number* be interpreted? Perhaps as *making corrections to a repeated mistake.* If you were out of the country when the monthly paychecks above were supposedly deposited to your account, and didn't find out till later that a mistake had been made and they had not been deposited, you could correct your balance by reducing it $12,000. Or if the grandmother above found other gifts for her grandchildren and got the checks back, she could adjust her balance by increasing it $50. Mathematically speaking, these would be interpreted as:

$$^-12 \times {}^+1000 = {}^-12000 \text{ (decreasing the balance \$12000)}$$
correcting entries
$$^-10 \times {}^-5 = {}^+50 \text{ (increasing the balance \$50)}$$

Notice above the only really surprising part of multiplication of positive and negative numbers—a negative times a negative yields a positive number!

> If two numbers have the same sign—either both positive or both negative—their product is positive. But if their signs are different, the product is a negative number.

While the statement above is removed from the examples that led to it, the "repeated entries" and "repeated corrections" ideas do justify the generalization, using the checkbook model.

Division. About this time you're probably beginning to wonder how in the world we're going to "stretch" this checkbook model so as to interpret *division* of positive and negative numbers. It can be done, but it seems a little farfetched. So we'll save the checkbook model of division as an exercise.

Instead, we can appeal to the relationship of multiplication and division, and use the algorithm just discussed. Recall that "a ÷ b = c" means that "a = b x c." For example, 12 ÷ 4 = 3 since 12 = 4 x 3. Using this idea gives us a means of determining the quotient in such problems as $^+45 ÷ {}^-3 = \square$. Turning it into the corresponding multiplication problem, we would have $^+45 = {}^-3 \times \square$, and using the algorithm for multiplication of positive and negative numbers, we know the answer must be $^-15$. So $^+45 ÷ {}^-3 = {}^-15$.

Use this same technique to solve the problems given in the box below before going ahead.

$$^+56 ÷ {}^-8 = ? \qquad ^+56 ÷ {}^+8 = ?$$
$$^-56 ÷ {}^+8 = ? \qquad ^-56 ÷ {}^-8 = ?$$

Your answers above should be $^-7$, $^+7$, $^+7$, and $^-7$, moving clockwise from the upper left. How did you do? If you missed the signs for these problems, you need to return and carefully work out the corresponding multiplication situations before going ahead.

We could make a generalization about division of positive and negative numbers, very similar to the one for multiplication above. This will be saved for the upcoming problem set also, along with the checkbook model for division.

THE PROPERTIES

As you might suspect from the fact that we've expanded our previous number systems by adding the negatives, all of the former properties of positive numbers apply:

Addition and multiplication are commutative.
Addition and multiplication are associative.
Multiplication distributes over addition.
Zero is the identity for addition.
One is the identity for multiplication.
Each number (except zero) has a reciprocal for multiplication.

The only new property we get from having negative numbers at our disposal is the powerful concept of an inverse element for addition. That is, given any positive or negative number, there's another number which we can add to it and get zero. The inverse for $^-73$ is $^+73$; the inverse of $^+32.897$ is $^-32.897$; the inverse of $^-¼$ is $^+¼$. If the word "inverse" bothers you, think of these numbers as "opposites" of each other.

Extensions of the properties above turn out to be quite advantageous when we begin computing with positive and negative numbers. It helps to know that these numbers can be added in any convenient order, for example. Consider the problem to the right. The clever thing would be to add $^+4$ and $^-4$ together first (to produce zero) and then $^+2$ and $^-2$ (to produce zero again). The final result is, of course, $^-8$ since it's the only thing left. Extensions of the commutative and associative laws for addition allowed us to rearrange the sequence of numbers to be added in any convenient order!

$$^+4 + {}^-8 + {}^+2 + {}^-4 + {}^-2 = ?$$

Another example of the usefulness of these properties would be in solving the problem to the left below. Notice that the distributive property of multiplication over addition would allow you to rearrange the problem as 18 x ($^-7 + {}^+6$), which is easy to compute as 18 x $^-1$. Here again, a person who has internalized the properties has a tremendous advantage over someone who sees them only as "things to be memorized."

$$(^+18 \times {}^-7) + (^+18 \times {}^+6)$$

The problems in the next set are meant to review the various interpretations discussed earlier, and extend the concept of positive and negative numbers beyond these interpretations. You will also have a chance to use the properties just mentioned to cut short some of the computation involved in problems that deal with positive and negative numbers.

SET 8

1. Write nine statements that would describe each step in the checkbook register below, in a fashion similar to that given in this unit.

BE SURE TO DEDUCT ANY PER CHECK CHARGES OR MAINTENANCE CHARGES THAT MAY APPLY

DATE	CHECK NUMBER	CHECKS ISSUED TO OR DEPOSIT RECEIVED FROM	AMOUNT OF DEPOSIT	✓	AMOUNT OF CHECK	BALANCE +316 84
1/3 80	131	UNIVERSITY (tuition)			-250 00	
1/3 80	132	BOOK STORE (books)			-56 00	
1/3 80	133	UNIVERSITY HOUSING (DORM)			-318 00	
1/3 80	134	A & P (food)			-28 46	
1/4 80		CASH FROM DAD	+400 00			
1/6 80	135	SPIEGEL CATALOG CO.			-27 85	
1/10 80		GIFT FROM GRAMMA	+50 00			
1/12 80	136	SEARS			-13 48	
1/12 80		J.C. PENNY			-6 27	

2. The last four entries above were negated. You stopped payment on the checks because the merchandise was defective, and your grandmother forgot to sign the check she sent you as a gift, and you were too embarrassed to remind her of it. Write four statements that will correct these entries.

3. Practice your computational skills on these problems:

   a. $^+7 + ^-9 - {^+3} + {^+2} = ?$
   b. $^-13 + {^+4} - {^-6} + {^-7} = ?$
   c. $(^+6 \times {^-5}) \times {^-4} = ?$
   d. $(^-3 \times {^-7}) \times {^-5} = ?$
   e. $^-155 \div {^-5} = ?$
   f. $^+60 \div {^-3} = ?$
   g. $(^+13 - {^+4} + {^-6}) \times (^-300 \div {^+5}) = ?$
   h. $(^+750 \div {^-25}) \div (^-10 - {^-4}) = ?$

4. Take stock of your total net worth by listing your assets as positive numbers, and your liabilities as negative amounts, and then adding. List only the major items ($25 or more) and estimate as accurately as possible where necessary.

5. Use the properties to make short cuts in the problems below. Perhaps you can do them all in your head!

   a. $(^-94 \times {^+16}) + (^-6 \times {^+16})$

   b. $^-1.01 \times {^-14}$

   c. $(^+38 + {^-24}) \times (^-657 - {^+428}) \times (^+43.2 - {^+43.2}) \times (^-3)$

   d. $^-3 + {^+9} + {^-18} + {^+3} + {^-5} + {^-4} + {^+6} + {^-9} + {^+4} + {^+18}$

   e. $^+1 + {^-2} + {^+3} + {^-4} + {^+5} + {^-6} + \ldots + {^+99} + {^-100}$

6. "Hearts" is an interesting card game. The object is to wind up with the fewest points (low score wins, as in golf). Each of the 13 hearts you get counts as 1 point, the queen of spades is 13 points, but the jack of diamonds is $^-10$ points. So you want to avoid getting any hearts and the queen of spades, at the same time trying to get the jack of diamonds. (The only exception is if you get all of the hearts and the queen of spades, in which case you can give each of your opponents 26 points.)

   Cathy, Jim, Thom, and Andy were playing one night, and decided to quit after one more hand. At the time, their scores were $^+48$, $^-17$, $^+13$ and $^-18$, respectively. After the last hand, had been played, the scores were $^+49$, $^-18$, $^+16$, and $^-5$, respectively. Who got the jack? Who got the queen of spades?

7. "Absolute zero" (temperature where, theoretically, molecular action ceases) is $^-459.67°F$. Suppose we did decide to drop our scale back to that starting point, so we wouldn't have to worry about negative numbers on the temperature scale. Give the resulting temperatures for the typical numbers we now use shown below:

$^-10°F$	Cold winter in Chicago
$24°F$	Cold winter day in Texas
$32°F$	Water freezes (at sea level)
$98.6°F$	Normal body temperature
$180°F$	Hot cup of coffee

8. William Setek poses the following problem in *Fundamentals of Mathematics* (Glencoe): A numismatist (coin collector) was examining a collection of coins, and found one dated 384 B.C. What could you conclude about this coin?

9. Many people are aware that B.C. stands for "Before Christ." Some people apparently think that A.D. stands for "After Death"—which is not true. Can you give a practical reason in terms of the calendar why A.D. should not stand for "After Death"? (After you've thought about that you might check with a dictionary on its real meaning, if you don't know.)

10. Return to the checkbook model, and explore trying to find an interpretation that would correspond to division of positive and negative numbers. If you think you are successful, give your interpretation for these problems:

$$^-30 \div {}^+5 \qquad {}^+30 \div {}^-5 \qquad {}^+30 \div {}^+5 \qquad {}^-30 \div {}^-5$$

Then generalize your solution by completing the following statement: If two numbers have the same sign, their quotient is a _____ number. If two numbers have opposite signs, their quotient is a _____ number.

Holes In the Road

Source: Photo courtesy of Larry Graham.

Most of us can avoid cavernous sinkholes like the one shown to the left. They're so large as to be obvious to a driver, under normal conditions. Consequently they do little damage to us, beyond some inconvenience associated with a detour.

Small potholes in a road are not so obvious, however, and therefore do much more damage to an automobile (and its passengers) in the long run.* The problem is a serious one, as evidenced by the enormous share of revenues allocated to highway departments for road maintenance. There seems to be an inverse relationship between the size of potholes and the amount of actual damage done, which is reasonable because smaller holes are harder to see, and there are more of them.

The number line itself was once reported to have "holes" in it, but they were so tiny that most scholars rejected the fact that they existed at all. Once fractions had been accepted as legitimate numbers, most scholars felt that the "road was smooth and well-paved" in the sense that any point on the number line could be named and

---

*In January of 1980, a Denver man (Bob Gilbert) reported having collected 67 hubcaps that rolled into his front yard from passing cars that hit a pothole—all of this in less than a week!

identified as one of these "new" numbers. But there were some nagging inconsistencies that refused to be squelched, and led some mathematicians to keep actively searching for some of these elusive "holes" in the line, points that could *not* be called fractions. They became convinced that there were some "potholes" out there that had not yet been filled in with names. And this question was doing as much damage to their security as established scholars, as "real" potholes were doing to their chariots.

One of the inconsistencies they noticed was brought about by geometric investigations. With whole numbers, it was easy to see a relationship between the side of a square and its area. In the three figures to the left, for example, the *length* of the sides are 1, 2, and 3, from left to right. The *area* can be found by counting unit squares, i.e., squares like the one to the far left. The areas of the three figures then, are 1, 4, and 9, respectively, leading to the generalization that:

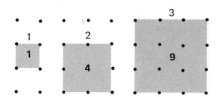

The area of a square is the length of a side multiplied by itself. Or, A = s · s, where A stands for area, and s for the length of a side.

The formula above certainly isn't new to you, and it wasn't new to the ancient scholars.

Furthermore, when fractions were introduced, they also seemed to abide by this same generalization. Look at the figures to the right. The first is a unit square, but the second has sides of length 1/2, and an area of 1/4 (and note that 1/2 x 1/2 = 1/4). The next square has sides of length 2/3, and the area is 4/9 of the unit square. The last square to the right has sides of length 3/4, and the area is what you'd expect if the "area rule" above were true for fractions also.

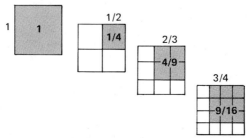

So it was natural to go ahead and assume that the formula for the area of a square was valid, no matter what numbers were used for the length of the sides. The inconsistency came about when a square like the one to the right was considered. Its area could be found by counting "half squares," and turned out to be 2. So what was the length of the sides of such a square? According to the accepted formula, this length would have to be a number such that, when multiplied by itself, the result was 2.

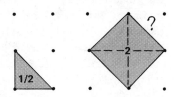

Try as they might, the ancient scholars were unable to find a fraction that, when multiplied by itself, yielded 2. To be sure, they could get "very close" to 2, and for most, this was enough. They assumed that this number they were searching for so diligently must, in fact, be a fraction that would one day be discovered. This elusive number was called $\sqrt{2}$ till its name as a fraction could be determined.

The Pythagoreans (around 300 B.C.) came up with a convincing geometric argument that $\sqrt{2}$ could not be a fraction, and this was eventually formalized into a classic "proof-by-contradiction" by Aristotle. This not only proved that $\sqrt{2}$ wasn't a fraction, but also standardized a totally new method of proving statements true. This "proof by contradiction" lives on today as an essential tool for mathematicians.

In a proof-by-contradiction, an initial assumption is followed by an argument that uses only accepted rules of logic. If the end result contradicts something you know to be true, the only error could have been in the initial assumption. Therefore the initial statement must, in fact, be false (and that's what you really intended to show all along). Note how strange this method of proof seems the first time it's encountered. You start by assuming the exact opposite of what you really think is true! In Aristotle's day, no one had ever formalized this "backdoor" approach to proofs, but as is the case with many "autistic" thoughts, hindsight makes it appear so obvious.

A polished-up version of Aristotle's proof is presented below, as an example of one of the classic proofs whose elegance is appreciated even today. Just keep in mind that it is one of those museum artifacts from which the false starts and exploratory trials have been removed.

Take it away, Aristotle!

Assume that $\sqrt{2}$ is a fraction. Then $\sqrt{2}$ has a name of the form a/b where a and b are whole numbers (b ≠ 0) and have no common factors. So a/b is the "reduced form" of the fraction.

$\sqrt{2}$ = a/b means that $(\sqrt{2})^2$ = $(a/b)^2$ = (a/b)(a/b). Or, we know that 2 = a^2/b^2, which means that $2b^2 = a^2$. Therefore a^2 is an even number by definition. This forces a itself to be an even number,* so a = 2c for some number c by definition of an even number.

But since $2 \cdot b^2 = a^2$, we have that $2b^2 = (2c)^2 = 4c^2$, and so $b^2 = 2c^2$. But in an argument like the one above, this means b^2 is an even number, hence b is itself an even number, or b = 2d for some number d.

Since a and b both have 2 as a factor, they have a common factor, which contradicts the assumption that a/b was in reduced form. So $\sqrt{2}$ can't be a fraction!

---

*Look back at problem 4 of Set 2, if this bothers you. The definition of an even number used here is covered in the next section of the text.

Aristotle demonstrated that our number line is not "well-paved" with fractions. There are some "holes" in our number line, places that cannot be labeled as fractions. These "new" numbers are called *irrational numbers*, of which $\sqrt{2}$ is one. Other familiar irrational numbers are such things as π, $\sqrt{3}$, $\sqrt{5}$. It is now known that, when the irrationals have been added to the fractions, the number line has all of its holes filled in and is therefore "complete."

It is now also known that there are *more* irrational numbers out there than fractions, a somewhat startling fact for most people. A discussion of how one infinite set can be larger than another is beyond the scope of this text, but it certainly is interesting to contemplate, even on the intuitive level.

IRRATIONAL NUMBERS AND DECIMALS

Since irrational numbers truly exist and can't be written with fractional names, you might begin to wonder if they can be written as decimals. Certainly they could not be written as either terminating or repeating decimals, because any such decimal can be renamed as a fraction. Are there any other possibilities: are there some decimals that neither terminate nor repeat?

Yes, there are plenty of decimals that neither terminate nor repeat. Some have patterns (4.565665666 . . .), and some do not (π = 3.14159265 . . .). But each of these decimals corresponds to an irrational number. For most practical purposes, the irrational numbers are approximated by terminating decimals, accurate to a desired degree regarding whatever calculations must be performed. But it should be kept in mind that we do this for convenience only. The inadequacy of our decimal system to provide precise names for irrational numbers is certainly not aesthetically pleasing. This points out again that the number concept is independent of the symbols that represent the numbers.

Approximating irrational numbers to a desired degree of accuracy is not all that hard, particularly in this day of the hand-held calculator. Even a method as elementary as "guess, check, and revise" works fairly well. Suppose we wanted to find a decimal that approximated $\sqrt{2}$ fairly closely. Recall that $\sqrt{2}$ is a number that, when multiplied by itself, yields 2. Keeping this in mind, we would want to produce a decimal which, when multiplied by itself, came close to 2. We could start such a search somewhere between 1 and 2, since $1 \cdot 1 = 1$ (which is less than 2) and $2 \cdot 2 = 4$ (which is more than 2). So we might try 1.5, which turns out to be a little too high since 1.5 x 1.5 = 2.25 (which is a little more than 2, but we're "getting warm"). Then we'd probably try 1.4, check it out, and if it wasn't close enough to satisfy us, we'd revise the estimate and continue this process till we were either satisfied or exhausted! The table on the next page shows the first few of these suggested trials. Fill in the next two entries, and see how close you are.

The method above can be used to approximate a fairly large number of irrational numbers, but for many of these nonfractions we'd need some relatively sophisticated calculating tools and principles that are not appropriate to our purpose.

APPROXIMATING √2

Guess	Guess x Guess = ?	Difference from 2	Revise
1.5	1.5 x 1.5 = 2.25	+.25	lower
1.4	1.4 x 1.4 = 1.96	−.04	raise
1.45	1.45 x 1.45 = 2.1025	+.1025	lower
_____	_____	_____	_____
_____	_____	_____	_____

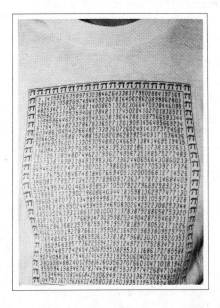

The challenge is there for some people nevertheless—for example, computers have now been used to find both $\sqrt{2}$ and π accurate to over a million decimal places! (It takes a computer almost two days even to print out such a thing.)

SET 9

1. On the right are some dots 1 cm apart. Find two dots that are exactly $\sqrt{5}$ cm apart. (Hint: Find a square whose area is 5 cm².)

2. Use the process illustrated above to determine a decimal approximation for $\sqrt{5}$, the length you found in problem 1.
   Make guesses, check the guesses by squaring the number and comparing to 5, and then revise the guess, until either the square of your guess is no more than .01 cm off, or until you're blue in the face!

3. Use a "proof-by-contradiction" argument to show that the sum of a fraction and an irrational number is another irrational number. The proof is started for you below:

   Proof: Assume that the sum of some fraction, a/b, and some irrational number, N, is another fraction, say c/d. Then:

   $$N + a/b = c/d.$$

   (The rest is yours!)

4. Use a "proof-by-contradiction" argument to show that the product of a fraction and an irrational number is again an irrational number. (The proof is very similar to that in Problem 3 above.)

5. Find an example to illustrate each of these statements:

   a. The difference of two irrationals is not always an irrational number, and

   b. The quotient of two irrational numbers is not always an irrational number.

6. Isaac Newton was one of several mathematicians who provided algorithms for approximating irrational numbers of the form $\sqrt{n}$ (for any whole number n that is not a perfect square). Much in the fashion demonstrated in the text, he used a sequence of progressively better guesses to do this. One algorithm he is often credited with is an improvement over our somewhat haphazard "guess, check, and revise" procedure. The improvement is in the "next guess" step. Newton takes all the fun out of the guesswork by literally telling us what to guess on the next step. But his method does produce estimates that rapidly approach the desired accuracy, no matter how bad the first guess is!

   Let n represent some whole number for which we want to approximate $\sqrt{n}$, let G be our "current guess," and N be the "next guess." Newton suggested that we obtain N each time by applying this formula:

   $$N = (G + \frac{n}{G}) \div 2$$

   The table below summarizes the first two steps of Newton's method for $\sqrt{2}$. Verify each step using a hand calculator, if possible, but paper and pencil otherwise. The initial guess is 1.5. Continue this process in the table for the two additional steps shown.

Guess	Guess^2	Difference in guess^2 and 2	N
1.5	2.25	.25	1.416667
1.416667	2.006945	.006945	1.414216
1.414216	___	___	___
___	___	___	___

7. Use Newton's algorithm above to approximate $\sqrt{7}$ and $\sqrt{11}$, till you have an estimate whose square differs from 7 or 11 by no more than 0.000001.

8. Stop thinking about irrational numbers for a moment, and return to the numbers we call "repeating decimals." You'll remember that any repeating decimal can be renamed as a fraction. You may have to remind yourself about how to proceed to find fractional names for $0.\overline{9}$ and $.\overline{29}$, so do this before continuing.

   Does permitting the notation $0.\overline{9}$ present any problems when we are striving for a unique decimal representation for our numbers? Should we establish, by agreement, any special rules about repeating 9s in decimal notation?

9. Look back for a moment at the T-shirt shown before this problem set. What digit occupies the 1000th place to the right of the decimal point when π is approximated as a decimal?

Another Look at Whole Numbers

The simplest numbers discussed in this chapter—the whole numbers—have been studied since ancient times, and comprise one of the oldest problem areas of interest in mathematics. Very often the whole numbers and many of the subsets of whole numbers are studied in the area of mathematics known as number theory. Central to such studies is the very special subset of the whole numbers called the *prime numbers*.

The *prime numbers* are said to be the "building blocks" for the whole numbers. You might visualize the whole numbers as chunks of ice. A *prime* block—the sort an eskimo would choose to build an igloo— is one that can't be cracked into smaller blocks. To the right, the

block labeled '13' can't be cracked apart by an ice pick, but '12' can be, as either 6 x 2, 3 x 4, or 12 x 1. Are there other ways to write 12 using multiplication in this fashion?

When we do this we usually say that we are representing a number in *factored form*; that is, in the example 12 = 6 x 2, the numbers 6 and 2 are called *factors* of 12. Similarly, 3 and 4 are factors of 12, as are 1 and 12 itself. So 12 can be represented in several ways in factored form. One of the characteristics of *composite numbers* is that there is more than one way to represent each one in factored form. The number 36 is a composite since it has 5 factored forms—can you find them all?

As suggested above, if we discount commutativity of factors, some whole numbers have only one factored form. For example, 1 x 13 is the only factored form for 13, and 1 x 5 is the only factored form for 5. Such numbers are called *prime numbers* —stated more succinctly,

> **A prime number is a whole number which has exactly two whole number factors, each different from the other.**

Keep in mind throughout this section that the term *factor* refers to a whole number which can be used in a multiplication statement to yield another whole number. In this sense, *factor* and *divisor* of a whole number mean the same thing; that is, a factor of a number is a divisor of the number.

The smallest prime number is 2—it has exactly two factors different from each other (1 and 2), and a problem in the upcoming set will demonstrate that the smaller whole numbers *zero* and *one* don't fit the definition to qualify as primes. *Four* and *six* and many other whole numbers do not fit this definition either, since they have more than two factors.

4 has the three factors 1, 2, and 4

6 has the four factors 1, 2, 3, and 6

Because many numbers have more than two factors, a special name, mentioned earlier, is used to describe them—*composite*.

> **A composite number is a whole number which has a limited number of divisors, but more than two, each different from the others.**

The chart below shows a partial list of primes and composites, with one form of the composites indicated in parentheses. Intuitively you would probably guess that there is no largest prime number and there is no largest composite number. The mathematician Euclid (3rd century B.C.) is credited with a formal proof that the set of primes is infinite. They seem to occur in an irregular fashion also, a fact which has intrigued mathematicians for hundreds of years. In fact, one unsolved problem in number theory is to find a formula which will generate the set of primes.

Primes	Composites
2	
3	
	4 (2 x 2)
5	
	6 (2 x 3)
7	
	8 (2 x 2 x 2)
	9 (3 x 3)
	10 (2 x 5)
11	
	12 (2 x 2 x 3)
13	
⋮	⋮

The special factoring indicated in the chart illustrates an important use of primes, one which helps us to understand the intimate relationship between primes and composites. And even though the study of prime and composite numbers still today provides research mathematicians with fertile soil to plow, the topic also finds itself to be quite useful in the elementary grades.

A topic invariably introduced in the intermediate grades (4-6) is that of *prime factorization.* This process can be thought of as taking a block of ice that can be broken down into smaller pieces, and doing so as far as possible. Use is frequently made of a *factor tree,* as shown below three times for 24. Notice that the process for

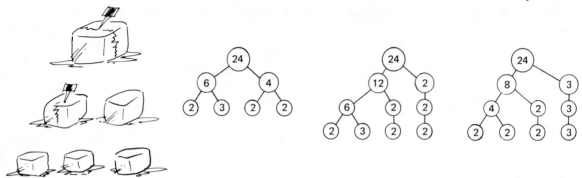

breaking a composite number down as far as possible is not a unique one, and depends on the way you begin—above, 24 was broken down into 6 x 4 to the left, 12 x 2 in the middle, and 8 x 3 to the right to show three ways to begin. But the end result is the same—notice above that, when 24 is written as the product of only prime numbers, it has to be written as three 2's and a 3, multiplied together. This example illustrates the truth of a statement so important it's been called The Fundamental Theorem of Arithmetic:

> **Each whole number greater than one is either a prime or a product of primes, and this product is unique except for the order of writing the factors.**

USE OF PRIME FACTORIZATION

Probably the most common use of prime factorization, at least in elementary arithmetic, involves finding the *greatest common factor* (GCF) and the *least common multiple* (LCM). The greatest common factor of two or more numbers is exactly what the name implies—the largest number which is a factor of them all. Remember that a factor is a divisor, so this means that the GCF of two or more numbers is the largest number which is a divisor of each of the numbers being considered. The GCF of 40, 24, and 48 is 8, since 8 is the largest divisor of all three numbers 40, 24, and 48. Try on your own to determine the GCF of 36 and 28.

The other term used above, *least common multiple,* has a related meaning. *Multiple* is probably well known to you, but just in case it isn't the following description should help:

> **A number which can be written as the product of a second number and some counting number, is said to be a multiple of both of the other numbers.**

For example, one multiple of 3 is 15, since 15 is 3 x 5. This example also shows that 15 is a multiple of 5; not shown but nevertheless true is that 15 is also a multiple of itself, and 1. So, it is probably clear that every whole number has an infinite number of multiples.

The *least common multiple* likewise means exactly what the name implies—the smallest number that's a multiple of each of the given numbers. Since 240 is the smallest number that's a multiple of 40, 24, and 48 (try to find a multiple smaller than 240 if you don't believe this), 240 is the least common multiple of these three numbers.

Prime factorization helps in locating the greatest common factor and least common multiple for several numbers. Consider the prime factorization of 24, 40, and 48 shown to the right. Notice that the only factors common to all three numbers are three 2's. So 2 x 2 x 2, or 8, is the greatest common factor—divisor—of these three numbers.

$$24 = 2 \times 2 \times 2 \times 3$$
$$40 = 2 \times 2 \times 2 \times 5$$
$$48 = 2 \times 2 \times 2 \times 2 \times 3$$

The least common multiple "pops out" from the example above also. Since the number we're searching for has to be a common multiple, it has to be a multiple of the largest number (48 above). All we need to do is multiply 48 by the factors that appear in the other two numbers that aren't already present in the prime factorization of 48 itself. Looks like '5' is the only such number in the example above, and so 48 x 5 (or 240) is the smallest number that's a multiple of each of the three given numbers.

We've shown by example how to find the greatest common factor and least common multiple of several numbers, using prime factorization, and the next exercise set will give you the chance to practice this skill. Before leaving the subject, however, a word on the correct emphasis for this topic is in order for someone who might be called on to teach it to elementary students.

The concept of least common multiple is taught to elementary students when they are learning to find common denominators in adding and subtracting fractions; similarly, greatest common factor is used to reduce fractions to lowest terms. With the deemphasis on computation of fractions in the elementary grades, and our concommitant concern with having students understand what they are doing, elementary students are meeting simpler fractions than ever before. This means that, in most cases, elementary students can find common denominators and reduce fractions to lowest terms simply by "inspection," without having to go through a rigorous procedure for these steps. Remember this when you begin to teach elementary mathematics and these topics crop up in the text you're teaching from.

However, also keep in mind that many of the students will need to have learned, at a concrete level, the process of determining the least common multiple and greatest common factor by the time they encounter algebra. These same procedures appear over and over again as students climb the mathematical ladder into more complex learning, and they're off to a much better start if they've internalized the procedure at a concrete level. So for many of your future students, using the prime factorization method for finding the least common denominator and greatest common factor is an appropriate topic to emphasize. As with many topics, you will want to consider what to emphasize with a student on an individual basis.

ODD AND EVEN NUMBERS

Another way of characterizing all of the whole numbers is to "split them in half"— numbers that have 2 as a factor are called *even* and the others are called *odd*. Note from the first few even numbers shown to the left that 2 is the only *even prime* number since all others greater than 0 are divisible by two. Also note that 3 is the smallest *odd prime* number, but that not all odd numbers are prime (9, for example, is *odd* and 9 = 3 x 3).

$$E = \{0, 2, 4, 6, 8, \ldots\}$$
$$O = \{1, 3, 5, 7, 9, \ldots\}$$

Recognizing odd from even numbers has very practical applications in elementary arithmetic, particularly when it comes to finding the prime factorization of a number. As long as the number or any of its factors is an *even* number, you can always divide it by 2. In actual practice, many people begin searching for the prime factorization of a number by dividing it repeatedly by 2 as many times as possible, then moving on to the next prime—3—and repeating the process. They "work up the ladder" of primes in this fashion, and eventually have the prime factorization of the number. For this reason, there are some "divisibility tests" that have been worked out for the smallest prime numbers. Some of these will be brought to your attention in the next problem set.

SET 10

1. Find the sum of the first 1000 odd whole numbers.

   (1 + 3 + 5 + 7 + ... + 1993 + 1995 + 1997 + 1999 = ?)

2. Find the sum of the first 1000 even whole numbers.

   (0 + 2 + 4 + 6 + ... + 1994 + 1996 + 1998 = ?)

3. Write factor trees to find the prime factorization of:
   a. 36   d. 86   g. 121
   b. 30   e. 76   h. 115
   c. 126 f. 90   i. 64

4. Use the results you obtained in problem 3 above to find the least common multiple and greatest common factor for these sets of numbers:
   a. 36, 76   c. 30, 86
   b. 126, 90, 115   d. 121, 64

5. Use the results obtained above to help you perform these computations:

   a. $\dfrac{5}{36} + \dfrac{3}{76}$ c. Reduce $\dfrac{90}{115}$ to lowest terms.

   b. $\dfrac{7}{30} - \dfrac{3}{86}$ d. Reduce $\dfrac{90}{126}$ to lowest terms.

6. One of the most famous unsolved problems in mathematics is referred to as "Goldbach's conjecture." It goes something like this:

   > *Every even number greater than 2 can be written as the sum of two prime numbers.*

   Four can be written as 2 + 2, for example, and six as 3 + 3. Verify that this statement is true at least for the even numbers up through 40 by finding two primes that add up to each of these even numbers.

7. In the section immediately preceding this problem set, mention was made of having "divisibility tests" for the smallest primes so that the prime factorization of a number could be obtained fairly quickly. Study these below:

   Two divides any number that ends in 0, 2, 4, 6 or 8.

   Three divides any number if it divides the sum of its digits. (3 divides 678 since 3 divides 6 + 7 + 8).

   Five divides any number that ends in 0 or 5.

   Seven divides any number if, when the right-most digit is doubled and subtracted from the remaining digits, the result is divisible by seven.

   Eleven divides a number if the sum of its odd-numbered digits from the left minus the sum of its even-numbered digits, is divisible by 11. [11 divides 63547 since (6 + 5 + 7) − (3 + 4) is 11, and eleven divides 11].

   Note that the processes for determining divisibility by 3, 7, and 11 above might need to be repeated several times on a given number before a final decision is reached.

   Illustrate each of the tests above by showing two large (five or more digits) numbers, one which is divisible and the other not divisible, by the prime number.

8. When students practice writing the prime factors of numbers quite a bit, they rapidly formulate a question of this nature:

   > *To find the factors of a number, how far up the prime ladder do I have to go before I've checked them all out? Do I need to go all the way to the number itself to be sure I've got them all?*

   And in fact, their suspicions that they don't need to climb the prime ladder all the way to the given number is correct. To find the factors of 61, for example, they wouldn't need to check out all the primes up to 61.

   Consider this question yourself and see if you can find an answer. Maybe you don't need to go past the halfway point, and perhaps not even that far.

REFERENCES

Vergara, W. C. *Mathematics in Everyday Things.* New York: Harper & Brothers, 1959.

Reisman, F. *A Guide to the Diagnostic Teaching of Arithmetic.* 2d ed. Columbus, Ohio: C. E. Merrill, 1978.

Eves, H. *An Introduction to the History of Mathematics.* New York: Holt, Rinehart & Winston, 1964.

Setek, W. M. *Fundamentals of Mathematics.* 2d Edition. Encino, California: Glencoe Publishing Company, 1979.

3

GEOMETRY

The humble beginnings of geometry took place far back in history. Some of the earliest records we have show crude attempts to measure, and describe, and reproduce the physical world of which early man was such an integral part. In the figure to the right, the drawings represent the world of three dimensions on a two-dimensional surface. Notice that some of the simplest geometric entities (line segments, etc.) were already in use, even though they had certainly not yet been couched in formal terms.

As human life evolved and became more complex, so did the need for geometric concepts. Establishing land boundaries, constructing dwellings, mapping territory, and producing primitive tools and weapons all placed demands on geometry, eventually (after thousands of years) producing a formal body of knowledge that could stand by itself without reference to its intuitive origins. So even though the subject was conceived "early on" in history, the gestation period was a long one. Not until Euclid's time (about 300 B.C.) did the subject become the discipline we know today.

What you experienced as "high school geometry"—with its emphasis on definitions, theorems, postulates, proofs, constructions, etc.—is actually only one stopping point in the evolution of geometry. Parallel lines, perpendicularity, obtuse angles, and π are all very familiar (but formal) terms that grew out of the intuition of early man.

Intuitive beginnings... Geometry related to the real world —— *Geometry evolves into a formal body of knowledge in itself (Euclidean)* —— *Other formal geometries are developed*

Your Level 1 and 2 knowledge of geometry (recall of the Pythagorean theorem, the value of π to two decimal places, or how to construct a line parallel to a given line from a point not on the line) may have suffered some over the past few years due to lack of use. All in all, however, you're probably not in such bad shape, since the geometry we'll be using—and what you'll be teaching to elementary students in a few years — lies somewhere between the "intuitive beginnings" and the formal Euclidean geometry on the line above. As we proceed, never lose the viewpoint that the basic concepts have their roots in intuition.

Shapes

When we look around us, a large variety of objects in our environment have interesting shapes. Some of the most fascinating occur naturally in fixed, regular forms.

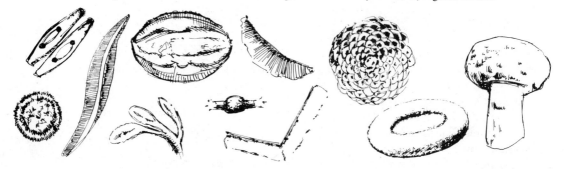

The world of nature seems to be a geometry lesson unto itself!

The functional or evolutionary aspect of the shape of naturally occurring objects is interesting to consider. A chicken egg, for example, has an obvious functional shape —any other just wouldn't do! The seed of a maple tree, for example, has a wing shape that causes the seed to carry to the soil at the right angle to dig its way into the surface for sprouting. A female grasshopper has a very long, slender "tail" that helps her dig into rotten logs in order to lay her eggs. The "flattened spherical" shape of the planets is directly related to the way in which they were formed billions of years ago. It's enough to make us wonder if everything has a shape related to its past or the job it performs, and our perspectives are just too short-lived to realize it!

Can you identify shapes by looking around you? What about human-made objects such as boxes and books (rectangular solids), phonograph records and coins (disks), tires and donuts (tori), cans and flagpoles (cylinders), and so on? Do they all have shapes that are related to function, or do you think mankind was somewhat arbitrary in selecting these shapes?

Along with color, shape is one of the fundamental attributes by which we classify things. Even four- and five-year-old children show significant development toward discriminating shape characteristics of objects. Given the pairs of objects below, for example, young children seem able to intuitively notice and crudely describe similarities and differences in shape:

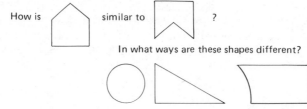

The answers you would get would be simplistic and somewhat crude, but many of the basic notions of "shape" would emerge with a little help.

How sophisticated are you with the language of "shape" at this point? Could you take a set of figures like the ones below, and discuss the similarities and differences in the shapes present?

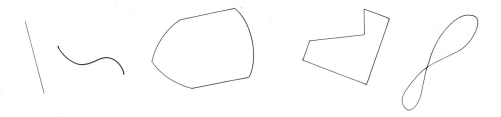

THE BASIC SHAPES

The fundamental concepts of Euclidean geometry are "point" and "betweenness." Intuitively, a point is any location in space, and is so tiny it has no dimension, and consequently no "shape."

Using two points (such as A and B below), and all the points between them, we get a *line segment*. Note that a line segment begins to have the attribute of "shape"; it can be labeled "straight." And it has "distance" associated with it, since it's a one-dimensional figure that is limited by its end points. Actually, you've probably heard that a line segment is "the shortest distance between two points."

This is technically incorrect (a line segment is a set of points), but this description does link, in an intuitive fashion, a line segment and its length. Physical models of line segments are all around us. Toothpicks, pencils, fishing poles, and pins all remind us of this geometric figure.

If a line segment is extended infinitely in one direction, the resulting figure is called a *ray*. A ray still has a shape in that it's still "straight," but continuing infinitely in one direction prevents us from attaching a meaningful length to it. So a ray has shape, but no length, as its attributes. The light coming from the sun is a good example of a ray in that this light has a definite starting point but continues forever in whatever direction it takes initially (assuming an infinite universe)!

If we continued to extend a line segment, but this time in both directions, we'd have a *line*. Again, the shape of a line is straight, but it makes no sense to attach a number to it and use this number to describe its length. The light coming from two laser beams placed back to back gives a physical model for this very important geometric figure.

Getting away from "straightness" as a shape, we arrive at a *circle* as another fundamental geometric figure. Formally, a circle is "the set of points on a flat surface that are all the same distance from a given point (the center)." We would quite naturally say that the shape of a circle is "round." And if we had not limited outselves to a flat surface, we would have arrived at another very basic geometric figure, a *sphere*. What are some physical models of circles and spheres?

A fundamental, two-dimensional figure that is neither straight nor round is the "angle." An *angle* is a figure composed of two distinct rays that share the same end point. The common end point is the *vertex*, but again it would make no sense to try to describe the length of an angle, since it extends infinitely. But angles do have shapes. We might describe an angle's shape as either "sharp" or "dull," depending on the manner in which the two rays meet at the vertex. From your past experience with geometry, you probably

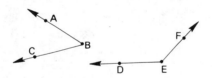

realize that some rational thinker of long ago found a way to describe how sharp or how dull an angle is, by attaching a number to it. Although a discussion of how to "measure an angle" will be saved for another chapter, it's interesting to note at this time that when we do measure an angle, we're actually measuring the *shape* of the angle, and not the figure itself.

THE BASIC TOOLS

In the classical Euclidean geometry promoted by the Greeks, only two tools were permitted for formal constructions and demonstrations—an unmarked straightedge for constructing line segments, and a compass for making circles or parts of circles. The challenge was to construct as much as possible, using as little as possible!

You should purchase a compass if you don't already have one, and find a sturdy unmarked straightedge. If your skills with these tools are a little rusty, a little practice with these two tools might be in order before tackling the next problem set. You'll also need a good sharp pencil and a pack of tracing paper.

SET 1

1. Do you know why
   a. a bird's wing is shaped like 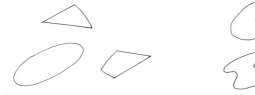 instead of like ?
   b. a football is shaped like 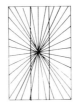 instead of like ... ?
   c. a dish is round instead of being a square or a triangle?
   d. a hunter's arrowhead is shaped like ... , but a target arrowhead is shaped like ... ?

2. Describe a basic "shape difference" in the figures to the left below and those to the right below.

3. In the configuration to the right you will probably perceive one "effect" that line segments can have on other line segments. Consider the two vertical lines and analyze real and imagined effects of the other segments on them.

4. Use only your basic tools (straightedge and compass) to do the following tasks:
   a. Make an angle that has exactly the same shape as the one below (no tracing allowed).
   b. Make two angles, one twice as sharp as the one below, and the other twice as dull (i.e., half as sharp) as the one below.

5. Use only your two basic tools to reproduce the figures below on your paper. The only points you can trace onto your own paper are those labeled P_1 and P_2.

   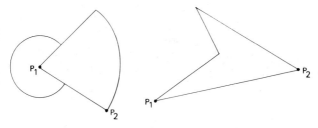

6. Experiment further with your two basic tools, and produce an abstract design that integrates various shapes. Enhance its features with shading or colors, if you wish.

7. The figure below needs to be rotated around point P. The arrow shows where we want to take point A, and <APA′ gives the shape of the rotation (i.e., the shape of the angle through which the figure is to be turned). Cut a "pie section" out of cardboard exactly the same shape as angle APA′, and then use it to rotate the rest of the figure in the same manner.

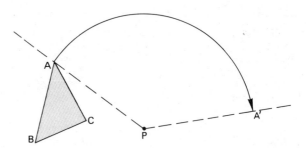

8. Since ancient times, the "golden ratio" rectangle has been frequently used in architecture because it's supposedly very pleasing to the eye. Of the rectangles below, one has its sides in this ratio. Which one do you find has the shape "most pleasing" to your eye?

Shape Transformations

Young people find their shadows fascinating. The light source projects, or transforms, the peripheral outline of the three-dimensional body onto a two-dimensional surface in a somewhat distorted fashion. An intriguing game is to cast the shadow in such a way that a recognizable figure is produced by the distortion, as in the "hand/dog" figure below.

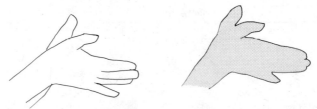

A highway map provides another example of a transformation of a three-dimensional "real" figure onto a plane. In this case, a network of roads is projected onto a sheet of paper, and even though much detail is eliminated in the transformation, sufficient information is retained to enable a traveler to drive over unfamiliar ground.

The picture on a television screen is another transformation, usually of a three-dimensional setting onto the surface of the picture tube. Notice that, in this case, the image of the transformation is actually cast onto the curved surface of the screen rather than onto a strictly two-dimensional plane.

Photography involves several transformations—a three-dimensional figure is transformed to a two-dimensional surface (the film), and then to a different two-dimensional surface (the final picture). An overhead projector provides another familiar example of transforming figures from one two-dimensional surface to another.

In the preceding examples, we would say that the transformation *distorted* the original figure, because it changed either its original size or shape. Some of the simplest, yet most interesting, transformations involve staying in the same plane and creating no distortions in producing the image of the figure. In such cases, it appears that the original figure was somehow "moved intact" without changing the relative position of the points, in producing the image. Such transformations have come to be called "rigid motions," a term that intuitively describes the actions taken. Let's take a closer look at these types of transformations.

RIGID MOTIONS

There are three types of rigid motions that move a plane figure to another position in that same plane. To get an intuitive idea of these three transformations, take a sheet of tracing paper and trace the outline of each black figure below, then manipulate the tracing until it matches up with its grey image.

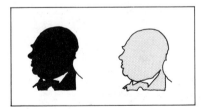

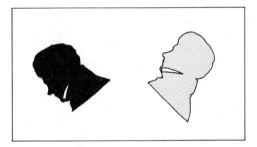

In the first example above, you probably found that you only had to *slide* your figure to match it up with the grey image. In the second example, you had to *turn* the tracing paper to make them match up. In the last example, the tracing paper had to be *flipped over* to transform the black figure to its grey image. Intuitively, these are the three "rigid motions" in the plane.

MORE ABOUT SLIDES

The results of "slide actions" are seen all around us. We slide furniture, and occasionally even our feet around a dance floor. Another name for this action is *translation*. Consider the translations of the figure to the right. The original shape is black, and it has been translated to three different images (all grey). If you trace over the original figure, and slide it to the three images, what you'll find is that the figure was translated the same distance in each case; only the *direction* of the slide was changed to produce the three different translations. This gives us a hint about one of the factors that must be specified in showing a given translation of a figure: the direction.

On the other hand, the direction by itself certainly isn't enough to identify the translation of a figure. In the drawing to the left, the black figure has been translated to two different grey images, both in the same direction. But the *length* of the slide has been changed, yielding the two different translations. So another factor—the length of the slide—must also be specified in describing a slide action.

Putting these two factors together, a slide action can be specified by showing the length and direction of the movement. Generally, an arrow (or *slide vector*) is accepted as showing these two essential features of a translation. The steps below demonstrate how a figure could be translated to its image point by point, by using the specified slide vector over and over again.

Step 1. The figure and slide vector given.   Step 2. Begin applying the slide vector.   Step 3. Apply slide vector repeatedly.

In reality, we certainly wouldn't attempt to move all of the points of a figure by applying the slide vector over and over again—we'd probably move a few of the critical ones (say, the endpoints) and "fill in" the rest according to where we know they'd be. Tracing paper helps tremendously in such a task.

Before going ahead, see if you can determine which of the four drawings below are translations, and which aren't, and why.

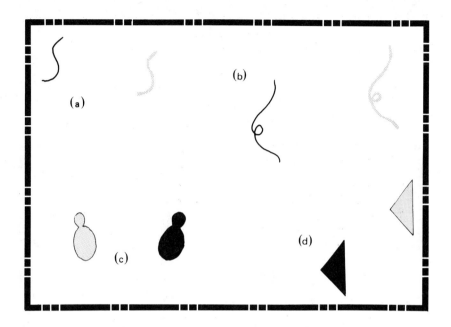

All of the four examples above are translations, except c, which is an example of the second type of rigid motion mentioned earlier, a *turn* or rotation.

Earlier, you confronted the figures to the right, and tried to make the black silhoutte match up with the grey one. At the time, we mentioned that a turning motion was necessary to accomplish the task. You might have intuitively agreed at the time, but the point about which the turn was to be made was never given. It has been supplied this time, however—trace the black figure to the right, and then stick a pin through the paper, holding it at the dot. Then turn the tracing paper till your tracing of the original matches up with the image. Note that you can go either clockwise or counterclockwise, but no matter your preference, each point in the original figure is turned through an angle the same shape as any other point on that original figure.

MORE ON TURNS

This last point is demonstrated more clearly in the example to the right. Triangle ABC has been rotated about point R, to produce the image triangle A'B'C'; note that angle ARA' has exactly the same shape as angle BRB'. Take your straightedge and lightly draw in the angle CRC'. Then use your compass to check that this angle has the same shape as the other two.

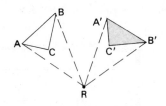

Use your tracing paper and pin to determine which of the four rigid motions below is not a rotation about the dot:

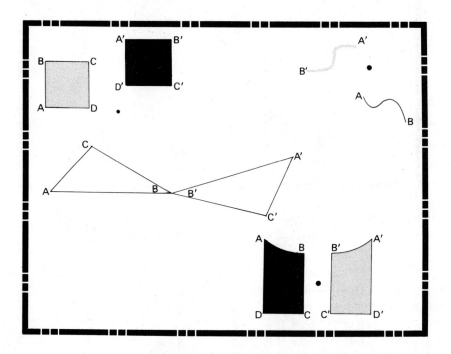

Again, all of the above rigid motions are turns about the given point, except the last one. The last example above is not a rotation about any point in the plane; it's an example of the third type of rigid transformation, a *flip*.

MORE ABOUT FLIPS The image produced by flipping a figure rigidly in the plane is quite similar to an image produced by a mirror. So *flips* are frequently referred to as *reflections* in the plane.

Earlier you considered the transformation shown to the left, in which the black figure was flipped to produce the grey image. All that has been added at this point is the "flip line," or "line of reflection." This is also the place where you would put a mirror, if you wanted to produce this same image by a true reflection.

Suppose you want to reflect the figure to the right using the line of reflection shown. One method would be to trace the figure and line of reflection and then fold the paper along the line. The figure can then be traced onto the other side of the line of reflection (or *fold line*). Do this task now, and compare your final result with the drawing of this rigid motion presented below.

If your figure and image look like the one to the left, you were successful. If not, try the task again until you are able to complete it successfully. Then for a little more practice, use this same procedure to determine which of the rigid motions below are reflections about the given fold line.

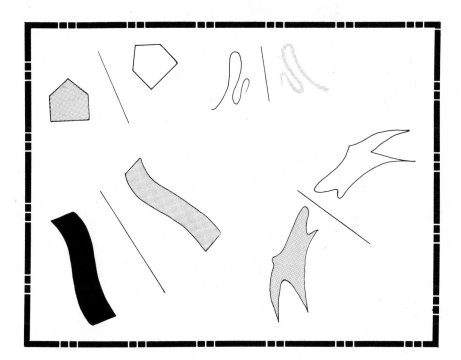

Right! Each of the transformations above is a *reflection* about the given line.

One last little bit of practice might be in order before you tackle the next problem set. In the drawing on the next page, the black triangle has been translated, rotated, and reflected to produce the three grey images. Can you pick out the slide transformation, and lightly draw in a vector producing this image? The reflection line was accidentally left in. Lightly draw in dotted lines to take the three corner points of the triangle to their respective image points. Oops! The "turn point" was not erased either. Use your tracing paper and pin to turn the original figure until it matches one of the images.

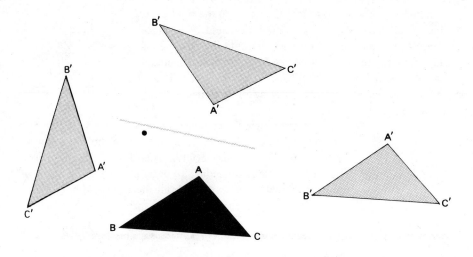

SET 2

1. Look in a mirror. The image you see is a transformation from _____ dimensions to _____ dimensions.

2. To the left you will see six transformations. Which are "rigid" transformations, and which are not rigid because of distortions?

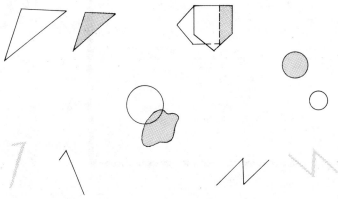

3. Each of the transformations below can be achieved by a single slide, flip, or turn. Decide which rigid transformation is operating in each case. If you're having trouble, use your tracing paper in the same manner as you did in the previous pages. For any that are rotations, approximate on your paper where the "turn point" is, as best you can. For a flip, sketch in about where the reflection line is. And for a slide, draw in the vector.

GEOMETRY 115

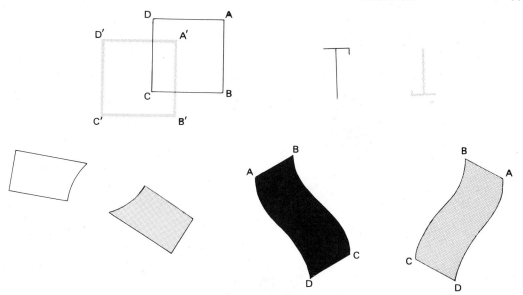

4. Trace over the figures below, and then find their images using the slide vector shown.

5. Trace over the figures shown, and then use your pin to turn each one around the given point. You choose "how far" to rotate each one.

 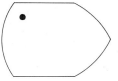

6. Trace over the figures shown, and then reflect them through the given line to find the images.

7. An exploration. Given that the transformation below is a rotation, experiment with your compass and straightedge and see if you can come up with a procedure for finding exactly where the "turn point" is.

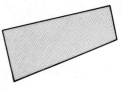

8. Another exploration. Using the same two basic tools, can you find a method for determining the reflection line, given the transformation below?

9. Start with the figure shown, tracing it on your paper. First, slide it using the vector given. Then reflect the *image* from the slide through the line shown. Finally, rotate it counterclockwise through an angle having this shape using the dot given.

Look-alikes

Cover up half of each figure below with your hand—do you like what you see as much as when the entire picture is revealed? Is half the picture as pretty as the entire thing?

Two things probably happened above. First, you most likely covered each figure in such a way that the part left was a mirror image of the part you covered, even though certainly you didn't have to do it this way. There seems to be a natural dividing line that runs through each picture. Secondly, you probably didn't find only half a picture as pleasing as the whole thing. For some strange reason, we seem to appreciate it when nature produces two identical, intricate designs on the same surface. Can you think of other naturally occurring phenomena with this same characteristic?

Symmetric is the word we apply to geometric figures in which one side is the mirror image of the other. More precisely, a plane figure is symmetric if it has a line of

reflection through the figure so that, when transformed, exactly the same figure is produced. Where is this special fold line in the butterfly? In the leaf? Is there more than one such line in either of them? Such a line of reflection, if it exists, is called a *line of symmetry* for the figure. How many lines of symmetry are there for the snowflake (careful!!)?

Did you find all six of the lines of symmetry in the snowflake? Each line of symmetry has been marked for the figures shown at the top of the next page. Study them carefully for a moment:

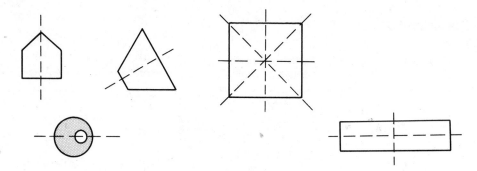

Note in the last figure above that the diagonals of the rectangle are not lines of symmetry. The picture to the right shows the image of this figure, if it were reflected about the diagonal. Notice that the image doesn't match up with the figure itself. So the diagonal shown is not a line of symmetry. The rectangle is said to be symmetric with respect to the other lines of symmetry, however.

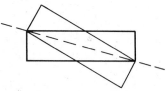

Find all of the lines of symmetry you can for the figures below. For any that give you trouble, trace the figure on another sheet of paper and fold it to see if the image and original figures match up, point by point.

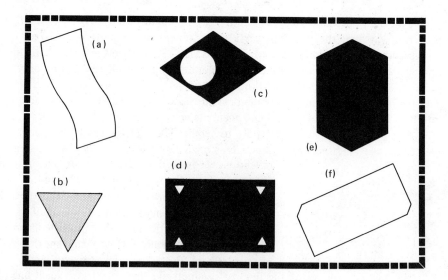

Figures (a) and (f) have no lines of symmetry; hence the figures are said to be *asymmetric*. Figure (b) has three lines of symmetry, (c) has one, and (d) and (e) each have two.

It's interesting to note that nature has a way of including humans in this set of symmetrical objects, if you accept the technically incorrect premise that we have two halves that "match up." Stand in front of a mirror and see if you can tell that your image is really not symmetrical. Probably you can't. So include yourself in nature's list of "naturally beautiful" creatures!

Now science seems to be "one up" on nature. Science claims to be able to produce a completely new person exactly like another. This process—called cloning—has been used successfully on lower order animals for years, and geneticists claim that only moral and legal issues have prevented them from repeating this process with humans. Imagine meeting someone who not only looked like you, but biologically *was* you!

In a less exciting vein, geometers have been cloning for years.

COMPOSITIONS

Given a figure in the plane, you know how to slide it, turn it, or rotate it. As you might have guessed, we can perform a chain of those rigid motions to produce other figures exactly like the original one (except, perhaps, for their orientation).

The following diagram is an example that uses first a translation and then a reflection. This chain, and others similar to it, are often called *compositions*.

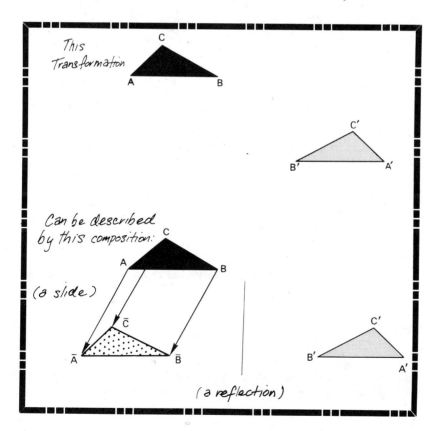

The diagram indicates a two-stage composition, but the composition equal to the original transformation is not unique. That is, many other compositions will yield the same result. Another such is:

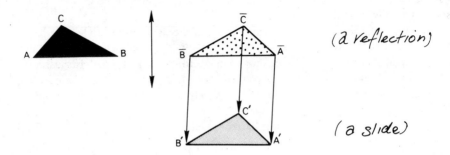

As you might suspect now, we could even conclude that there are an infinite number of compositions equivalent to a given rigid transformation in the plane.

Three and even more stages are possible for a composition. Below is an example of a three-stage chain of rigid motions. Follow these with your tracing paper and pin, just to be sure that the artist didn't make a mistake in the drawing:

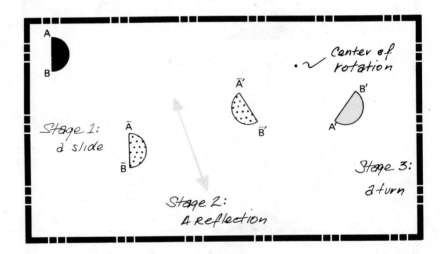

Look at all the clones above!

An interesting question to consider at this time is the reverse of putting together a long chain of rigid motions. We know that the original figure above can be transformed to its final image by using three rigid motions. Can the same result be achieved by using only *two* such simple rigid motions, or perhaps even *one*?

Below you can see a two-stage composition that accomplishes the same transformation. Maybe you can even find a single rigid motion equivalent to this two-stage one, but the authors have been unable to.

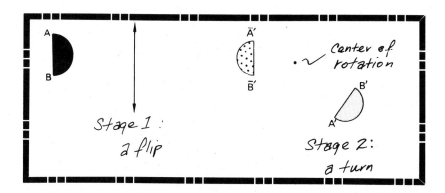

So the previous three-stage composition can, in fact, be reduced to the simpler, two-stage one above. An intriguing adventure—to be taken up in the next problem set—is to try to design a three-stage composition that *cannot* be reduced to two stages!

Occasionally we yield to traditional terminology. When identical geometric figures are produced by one of the three rigid motions, or a composition of these rigid motions, we say that the original figure and its image(s) are *congruent*. If two figures are congruent, then either can be produced by the other by using only slides, turns, or reflections in the plane.

Which of the examples below have congruent figures inside the box?

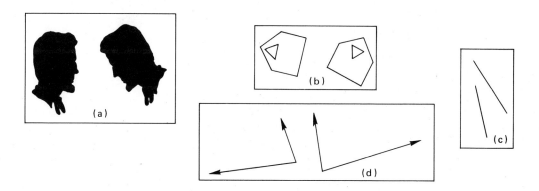

How did you do? Examples (a) and (d) above have congruent figures, but (b) and (c) do not. If you missed any above, use your tracing paper and perform a composition of rigid motions to see why.

SET 3

1. To the right is half of a person's face. Figure out a way to supply the other half of the picture, using symmetry. Can you tell who this truly autistic thinker is?

Source: Courtesy of Ontario Science Centre

2. How many lines of symmetry in each figure below?

  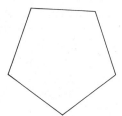 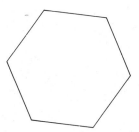

3. Count the number of lines of symmetry in each snowflake below:

4. Trace the figure and image below on another sheet of paper. Then very accurately find two different compositions that will take the figure to its image.

GEOMETRY   123

5. Design, if you can, a transformation that *requires* a two-stage composition, starting with the figure below.

6. Using the figure above, try to design a transformation that can't be done in fewer than three stages.

7. For each figure below, a point is marked. To the right of the figure, the image point is marked. Use the single transformation mentioned below for each, and produce a figure congruent to the original one.

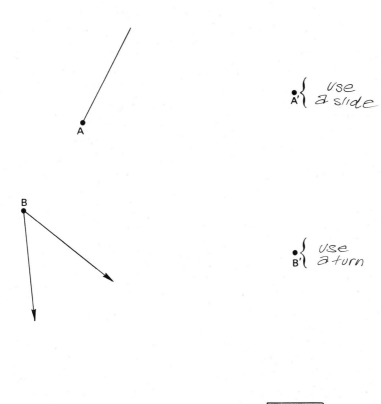

Curves

The curve below is a circle that was left out in the rain. It got wrinkled, but fortunately not twisted. So to return it to a circle, all we'd have to do would be to stretch it back out carefully, and let it dry.

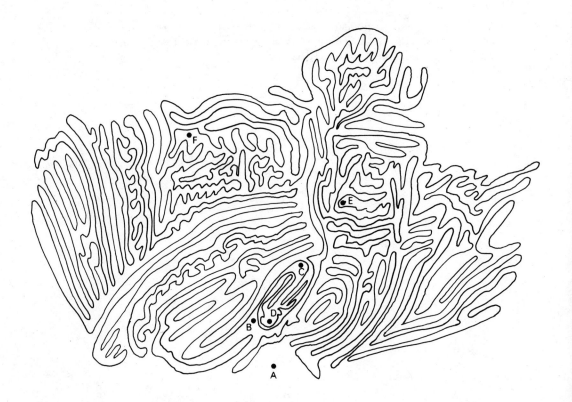

If points A, B, C, and D were kept in their positions relative to the curve as this "stretching back into a circle" was performed, they would be on the outside of the circle. If you have trouble visualizing this, take a pencil and lightly run it from A to B to C to D, and notice that nowhere will you be forced to cross the curve—nowhere will you have to enter the inside of the circle. With it all wrinkled like it is now, it's hard to tell the inside from the outside, isn't it? Look at points E and F. Can you find out whether they're on the inside or outside?

Curves like this, which are untwisted and could be stretched back into a circle, are called *simple closed curves*. Witness these examples of simple closed curves:

GEOMETRY 125

Intuitively speaking, simple closed curves are:

1. **Curves**, since they can be drawn with one continuous movement of a theoretically sharp pen. (A "theoretically sharp pen" is one which would leave an ink mark with no width to it.)
2. **Closed**, because in tracing the path with the pen, you always wind up back at the point where you started without retracing any part of the curve.
3. **Simple**, since the curve doesn't cross over itself.

For each example below, tell which of these three conditions is not met, and thus why each figure is not a *simple closed curve*:

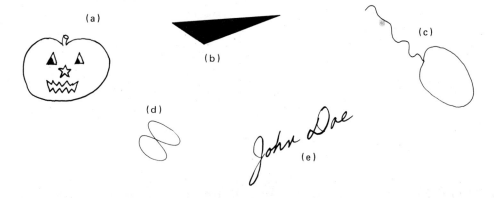

Figure (a) above can't be made with one continuous movement, while (b) is not made with a "theoretically sharp pen." Figure (c) is not closed, and (d) is closed but crosses itself. Figure (e) crosses itself, is not made with one continuous movement of the pen, and is not closed.

A mathematician named Jordan (pronounced jor-dáhn) secured his place in history by making one of those statements that appear obvious once they're made:

The Jordan Curve Theorem: *Any simple closed curve separates the plane into three disjoint sets of points – the curve itself, the interior region, and the exterior region.*

The interior (inside) and exterior (outside) portions of the plane determined by a simple closed curve are called *regions*. Unlike curves, they are true two-dimensional shapes or surfaces; you couldn't draw them with a "theoretically sharp pen" as you could a curve. The inside region is *bounded*, but the exterior region is *unbounded*, since it extends forever in the plane.

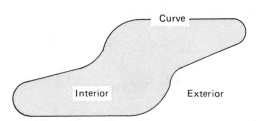

Some special types of simple closed curves prove very useful, so useful, in fact, that they warrant a closer look.

POLYGONS When a simple closed curve is made up only of line segments, the curve is called a *polygon*. The figures to the left below are polygons, while those to the right are not. Study them carefully before going ahead.

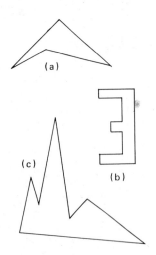
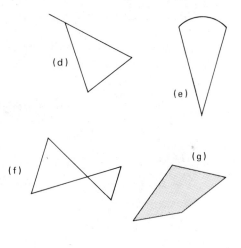

Can you tell why each of the figures on the right does not qualify to be a polygon? Figure (d) isn't closed, (e) isn't made totally out of line segments, (f) isn't simple, and (g) isn't even a curve; it's a region.

Polygons are named according to the number of sides they have. A *pentagon* has five sides, since "penta" comes from a Greek word meaning "five." Can you figure out a name for the figures to the right below, using the list of "hints" to the left?

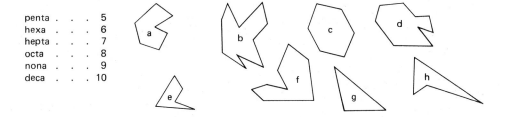

penta . . . 5
hexa . . . 6
hepta . . . 7
octa . . . 8
nona . . . 9
deca . . . 10

The names, in order, for the figures above are: heptagon, decagon, hexagon, nonagon, pentagon, octagon, triangle, and quadrilateral. Surprised about the last two? Should they be called "trigon" and "quadrigon" to continue the pattern?

Regular polygons are those in which all of the sides, and all of the angles, are congruent shapes. The simplest of the regular polygons are shown below:

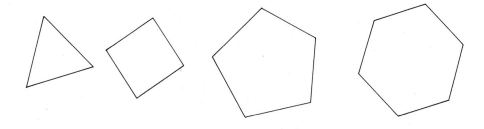

The last two of the four shown have standard names: *regular pentagon* and *regular hexagon*. The first two are seen so often they have again been awarded special names: *equilateral triangle*, and *square*. Can you think of other names for the first two, names that would follow the pattern suggested by the other two?

Intuitively speaking, a regular polygon has a center point, and this point can be used to draw a circle that touches all of the corner points, or *vertices*, of the polygon. Shown to the right is a regular nonagon, and the center of the nonagon has been marked. Put your compass on the center point, and open it up just the right amount to make the circle that would touch all nine vertices. An interesting question to be taken up in the next problem set is "how can we find the center point of a regular polygon, so the surrounding circle can be drawn?"

SET 4   1. Label each figure below with all of the appropriate words from this unit, if any apply:

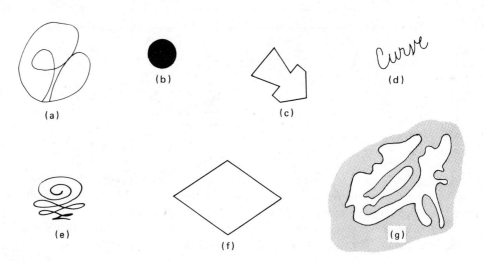

2. The teacher to the left is demonstrating "simple closed curves" by drawing an interesting one for the class. Draw one yourself that resembles a real-world object, and is more intricate than this one.

3. Only one part of a simple closed curve is shown below. You do know, however, that A is on the outside. Can you tell where point B is? How? Can you tell where C is? How?

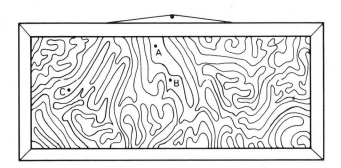

4. The *diagonals* of a polygon are the line segments that connect any two nonadjacent vertices. All of the five diagonals of the pentagon to the right have been sketched with dashed lines.

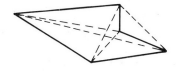

Organize your thoughts on how to proceed, and then try to answer this question: What is the numerical relationship between the number of diagonals and the number of sides (or vertices) of a polygon?

5. Look at the "palm side" of your thumb (the fingerprint). Does it appear to be a simple closed curve?

6. Devise a method for finding the "center point" of a *regular polygon*, using only your two basic tools.

7. The steps below show how you can construct an equilateral triangle, using only your straightedge and compass.

Step 1. Draw a line segment and open compass to match.

Step 2. Swing an arc, using compass as it is, from one endpoint.

Step 3. Swing an arc from other endpoint, and connect intersection with first two endpoints.

Practice this a few times, and then consider this question: Which other regular polygons could you construct using only the two basic tools?

8. In defining a *regular polygon*, mention was made that they must have both congruent sides and congruent angles, the implication being that neither by itself assures the other. Just having congruent angles doesn't mean the sides will be congruent, and having congruent sides doesn't mean the angles will be the same shape. Each of the two figures below demonstrates one of these statements. Which figure illustrates the truth of which statement?

9. Use the method described in Problem 7 to construct an equilateral triangle, and then Problem 6 to find its center point. Make a large circle that touches the 3 vertices of the triangle. After that's done, use the compass over and over again with a smaller setting each time, till you have ten concentric circles, the smallest one completely inside the triangle. Is the triangle still a polygon? Does it still look like one?

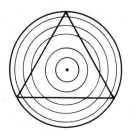

When Two Lines Meet

The four angles formed when two lines intersect have been studied for thousands of years, particularly since mankind began to construct dwellings. The first materials available (sticks, etc.) were naturally shaped like line segments, and the way in which these segments met each other had a lot to do with how sturdy the buildings were.

Consider the manner in which the pairs of lines meet each other below. Are any two of the four angles congruent? How can you tell? How could you *prove* it?

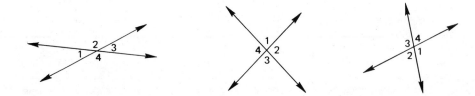

If your intuition didn't fail you in looking at the angles formed by the lines meeting above, you noticed that the congruency seems to run in pairs. Two pairs of angles are congruent in each of the three examples. These congruent pairs are called *opposite angles*, which refers to the angles that share a common vertex but not a common ray. Certainly three examples would not be a good number on which to base an inductive conclusion, but the same situation has been tested by thousands of others down through the ages, and to our knowledge (in the Euclidean sense anyway) no one has ever found an exception to the statement:

When two lines meet, "opposite" angles are congruent.

If the emphasis in this course were deductive reasoning, the statement above would be proved rigorously. Since we're concentrating on informal geometry, notice that you could use tracing paper and one of the three fundamental motions — a rotation — to show that opposite angles are congruent.

The special case of two lines meeting in such a way that all four angles produced are congruent is of particular interest. A sketch of two such lines is shown to the right. Utility has made this angle quite popular. For example, if a very heavy mailbox had to be supported from the ground with a fairly skinny pole (resembling a line segment), which of the configurations below would *you* choose?

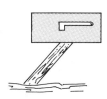 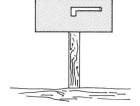 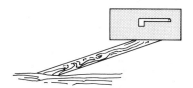

Engineering students could certainly justify—in terms of the center of gravity, torque applied at the fulcrum, etc.—what we perhaps see intuitively. The post most likely to support the mailbox meets the ground in this unique fashion. The center picture above is the "right" one to support the weight. It's easy to see, then, why people called this sort of angle a "right angle."

A right angle *is formed when two lines meet in such a way that all four of the angles are congruent.*

When two lines meet in such a way that the angles formed are right angles, the lines are said to be *perpendicular*.

Suppose someone plopped an ordinary, garden-variety angle down in front of you, and asked whether or not it was a "right angle." Or more specifically, consider those on the next page. Can you tell if any are right angles?

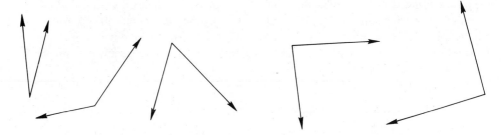

Certainly the three to the left aren't right angles; inspection tells us that. But how about those to the right. Too close to tell? Actually, they aren't either, but the important point is how to judge an angle as being "right" or not, if it's too close to "eyeball"? Can you figure out a way, using the previous definition and your two basic tools?

More often than not, rather than judging whether lines are perpendicular or not, we'll be asked to construct them so that they are. For example, consider the situation below:

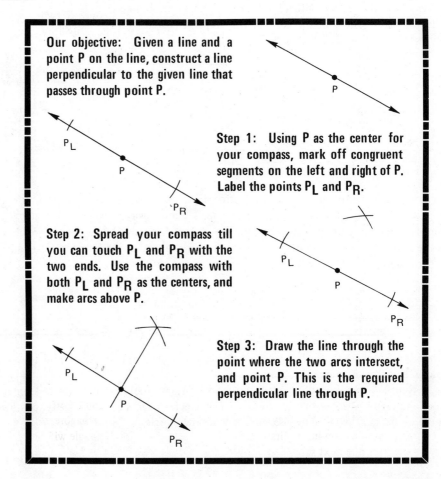

Our objective: Given a line and a point P on the line, construct a line perpendicular to the given line that passes through point P.

Step 1: Using P as the center for your compass, mark off congruent segments on the left and right of P. Label the points P_L and P_R.

Step 2: Spread your compass till you can touch P_L and P_R with the two ends. Use the compass with both P_L and P_R as the centers, and make arcs above P.

Step 3: Draw the line through the point where the two arcs intersect, and point P. This is the required perpendicular line through P.

If you're unfamiliar with this traditional construction, go through the steps on another sheet of paper, using your compass and straightedge. Be sure you can perform this procedure—which is pretty close to a Level 2, step-by-step process—before going further.

Another traditional construction problem of this nature is given below. Again, if you're unfamiliar with the procedure, follow it closely with your own compass and straightedge.

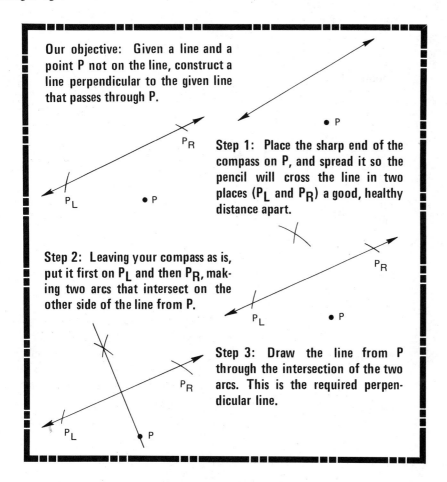

Our construction of perpendicular lines is now complete.

HISTORICAL NOTE

The Egyptians (c. 4000 B.C.) apparently used a special triangle of sides 3, 4, and 5 units to create "right angles" in their construction of pyramids and other elaborate tombs of their leaders. Yet they evidently were unaware of the Pythagorean theorem mentioned in a cartoon in the first chapter. Constructing such a triangle will be one of the problems in the next set.

Now that we've discussed the various ways that two lines can meet—either with all four angles congruent, or with only "opposite pairs" congruent—are we through? No, the other case demands to be heard also.

WHEN TWO LINES DON'T MEET

It's possible to draw two lines in the same plane that will never meet. Consider the two sketched to the right. If the drawing were extended indefinitely, the two lines would never meet. In such a case, we say that the lines are *parallel*.

Again, we might ask whether there's a way to judge when two lines in the same plane are parallel. In most cases inspection will suffice. In the following examples, however, obviously the two lines on the right aren't parallel, but the pairs on the left might be too close to call. A method is needed for judging situations like this.

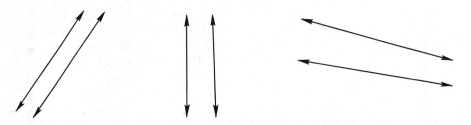

The two lines below (L_1 and L_2) are parallel, and are crossed by another line called a *transversal*. The angles labeled A and A', B and B', C and C', D and D', respectively, are said to be *corresponding angles*. Are these pairs of corresponding angles congruent? Use a slide transformation to decide. This should give you a hint about how to judge whether two lines are parallel. Test your idea on the pairs of lines below.

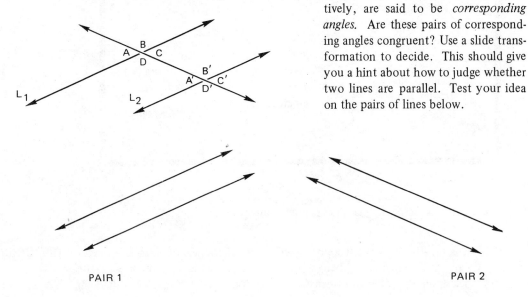

PAIR 1

PAIR 2

Historical evidence again would lend weight to the intuitive conclusion you probably reached:

Two lines in the same plane are parallel if, when cut by a transversal, corresponding angles are congruent.

Again, there's a fairly traditional construction problem associated with this concept of parallelism. Go through this with your basic tools, if it's not familiar to you.

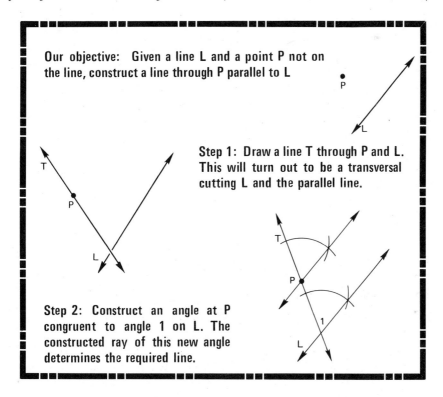

Our objective: Given a line L and a point P not on the line, construct a line through P parallel to L

Step 1: Draw a line T through P and L. This will turn out to be a transversal cutting L and the parallel line.

Step 2: Construct an angle at P congruent to angle 1 on L. The constructed ray of this new angle determines the required line.

The construction problem above completes our work with parallelism, and also our discussion of how two lines can meet, or not meet, if they're in the same plane.

ET 5 1. Use only your two basic tools, and write lightly in the book, to determine if the two angles at the top of the next page are right angles. (You can only do this approximately, of course!)

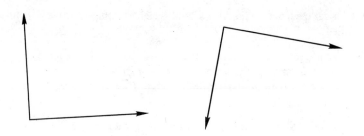

2. Use the two basic tools to determine if the two lines below are parallel. (You can only do this approximately, of course.)

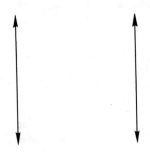

3. Mark a line segment on your paper the same length as the one to the right. Then construct a square using the line segment as one of the sides. Do this by constructing only two right angles. Is it possible to do this by constructing only one right angle?

4. Experiment with your basic tools, trying to find a way to find the exact midpoint of a given line segment. Once you've developed a process, use it on a line segment exactly the length of the one in Problem 3 to divide it in half.

5. Using a corner of this page as a right angle, pick a convenient spread for your compass, and then mark off three units along the bottom of the page (going away from the corner you chose). Then going along the outer edge, mark off four units. Lightly draw in a line segment to "finish off" this 3, 4, 5 right triangle. Measure with your compass; is the third side really five units in length?

6. Which of the line segments below (L_1 or L_2) is longer?

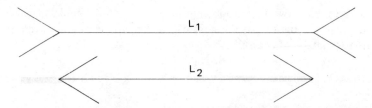

7. In the last problem set, you were asked to consider ways of constructing various regular polygons using only a straightedge and compass. Probably you didn't find a way to construct a regular pentagon. Euclid played with this particular problem himself for quite a while, and finally devised a fairly clever way using the "golden ratio" mentioned earlier.

   The German artist Albrecht Dürer (1471–1528) also found a nice method, which is presented below. Construct a regular pentagon following this procedure:

Step 1: Use your compass to make a circle with center C. Draw a line (diameter) through the circle and point C. Let P be the place where the line hits the circle. Use your method from Problem 4 to find the midpoint of $\overline{PC}$. Label it M.

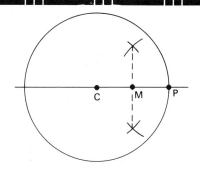

Step 2: Construct a perpendicular to $\overline{PC}$ at point C, and let X be the point where this perpendicular hits the circle.

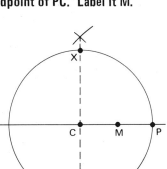

Step 3: Use your compass to mark off on $\overline{PC}$ a line segment $\overline{MZ}$ such that $\overline{MZ}$ is congruent to $\overline{MX}$.

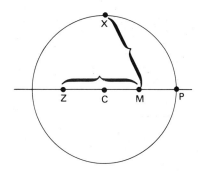

Step 4: Spreading your compass exactly as far as $\overline{ZX}$, mark off points around the circle. Then connect your marks with line segments. You should have a pentagon!

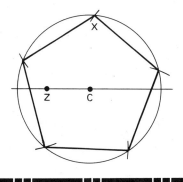

8. Can you extend Problem 7 so that you know how to construct a regular decagon (10 sides)? Going from a 5-sided figure inscribed in a circle to one of 10 sides shouldn't be that difficult.

9. In reference to his method, Dürer stated that if, when $\overline{PC}$ was bisected to find the midpoint M, the bisecting line were perpendicular to $\overline{PC}$ and hit the circle at point G, the length of $\overline{MG}$ would give an approximation useful in making a regular heptagon (seven sides). Try it on your own paper, and see how close it is!

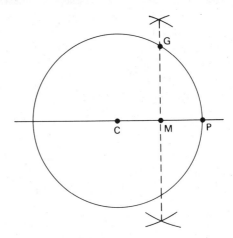

Tessellating the Plane

The Dutch artist M. C. Escher (1898–1972) is well known for his woodcuts and lithographic works of repetitive patterns and periodic drawings. By altering certain basic polygons placed side by side, he covered flat surfaces without any overlapping and without leaving any space uncovered. Escher created many interesting patterns used for *tessellating the plane*: one is shown to the right. Though Escher claimed to have little formal training in geometry, he undoubtedly discovered many principles about shapes and relationships among them as he developed his drawings and woodcuts. Of course, the art of tessellating is an ancient one, forming the basis for mosiacs found among many cultures all over the world.

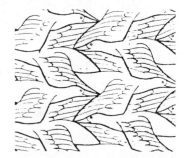

Source: M. C. Escher, "Day and Night," Escher Foundation, Haags Gemeentemuseum, The Hague.

GEOMETRY **139**

The art of tessellating a plane is merely an elaboration of the simple task of tiling a floor. Tiles are often congruent squares or rectangles, and of course these shapes will 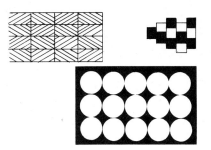 cover a floor *without any overlapping and without leaving any space uncovered.* Other basic shapes can be used to "tile a plane," but notice from the picture to the left that a circle is not one of these. There's no way to cover a flat surface with circles without leaving holes or overlapping. So a circle doesn't "tessellate the plane."

Eventually, we'd like you to get to the point where you know how to produce as interesting a figure as one of Escher's. But this process involves altering very basic shapes that tessellate the plane. Therefore before we enter the "creative" side of tessellating, let's look at some of the more common figures, and try to discover some things that might help us later on.

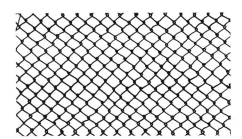 

In trying to find simple geometric shapes that tessellate the plane, our minds somewhat naturally jump first to polygons, and further to regular polygons. So we could begin our investigation with the simplest of the regular polygons, and go from there.

Trace over the figures below and make cardboard cutouts. Then manipulate the figures on a sheet of paper, tracing around the outside. Record your results—those that tessellate and those that don't—in the chart below on the next page.

REGULAR POLYGONS

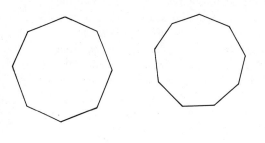

Type	Tessellate?
Equilateral triangle	
Square	
Regular pentagon	
Regular hexagon	
Regular heptagon	
Regular octagon	
Regular nonagon	

After completing the investigation above, you will realize that certain regular polygons tessellate the plane, but others do not. Just having congruent sides and congruent angles, then, doesn't mean the figure will tessellate. Now let's open the investigation a little wider, and consider the possibilities of irregular polygons tessellating the plane.

IRREGULAR POLYGONS Intuitively, it seems that other conditions than having a set number of sides would be needed to guarantee that a polygon would tessellate the plane. In general, this may well be true. Yet two types of polygons will tessellate the plane with no prior conditions other than the number of sides.

Any triangle, for example, will tessellate. Consider the one to the right. It doesn't seem unique in any way; yet it can be rotated, translated, and reflected in a routine manner to cover the plane. So in fact, this particular triangle could be used to "tile a floor."

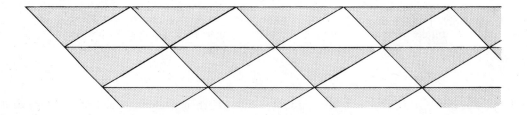

But don't take our word for the fact that any triangle tessellates the plane. Cut out some cardboard ones and experiment with them yourself. Perhaps you can find one that doesn't tessellate!

The other irregular polygon that always tessellates is the quadrilateral. Consider those below.

The figures below show how the previous quadrilaterals could be manipulated so that they cover the plane.

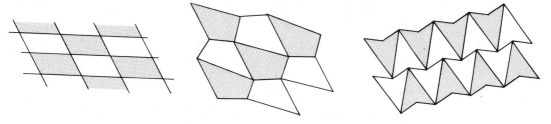

Your skepticism is still invited. These three cases shouldn't convince you that any quadrilateral tessellates the plane—make some out of cardboard and see if you can find a counterexample.

Polygons with more than three or four sides have to be somewhat special in order to tessellate. The pentagon, hexagon, and heptagon below, for example, can't be turned or twisted or slid in a routine fashion to accomplish the task.

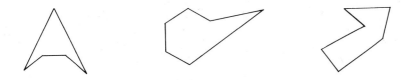

And yet there are plenty of pentagons, hexagons, heptagons, etc., that *do* tessellate. Witness those below:

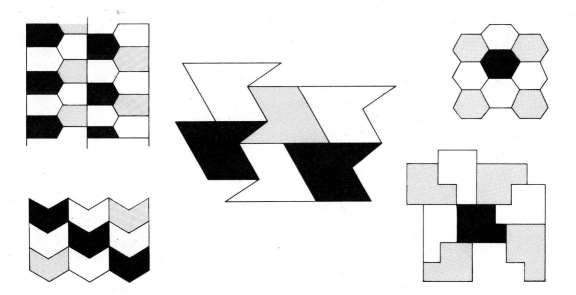

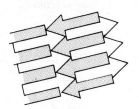

Also, take note of the fact that you could combine several shapes in each of the tessellations above (like two arrows to the left) and have tessellating polygons with enormous numbers of sides.

Before continuing, you should experiment with both modifying the shapes above to produce other figures that tessellate and developing some completely on your own. As you explore this idea of irregular polygons and the possibility of their tessellating the plane, consider some of the factors we've discussed earlier, and see if these lead you to any conjectures. For example, you might want to consider:

Will some combination of right angles in a polygon ensure that a tessellation is possible?

Does requiring the polygon to have some sides parallel to each other force a tessellation?

Will a polygon symmetric about some line always tessellate?

Is there some combination of having congruent or parallel sides that will enable a polygon to tessellate?

What if certain angles of a polygon can be made to fit completely around a point, with no overlap—will such a polygon tessellate?

Perhaps you can add some of your questions to the list above. The next problem set should help you synthesize this material, and understand the tessellation of basic shapes.

SET 6

1. Find five examples of tessellating shapes that appear in the world around you. The shapes can be from man-made objects or from nature. Sketch the shape, and tell where it was found.

2. Make a cardboard cutout of the triangle below, and use the cutout figure to make a tessellation on another sheet of paper. Try to organize your manipulations, and develop a sequence of rigid motions (slides, flips, turns) that will routinely make the triangle tessellate.

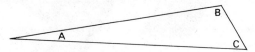

3. Make a cardboard cutout for the quadrilateral to the right. Then use the configuration on the next page to give you a technique for forcing the given quadrilateral to tessellate. Extend the drawing more than has been done here, to the

point where you have gone around the figure you start with at least twice.

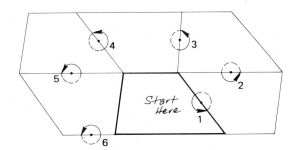

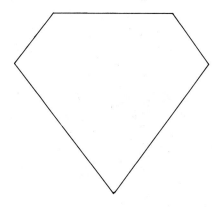

4. Make a cardboard cutout of the figure to the left, and use it to tessellate the plane.

5. Design a tessellating polygon of more than seven sides. Make a large enough pattern using your polygon to convince someone that it does, in fact, cover the plane with no overlaps or holes.

6. The figure below is quite well known as an example of something that seems to "jump back and forth" as you stare at it. The illusion was created by merely shading, in a certain way, a figure that tessellates. What type of polygon forms the basis for this figure?

Make a tessellation using a very simple polygon (as is the case above) and color or shade it, trying to produce a "jumper."

7. The figures to the left and right show two different tesselations of the same basic polygon. Use the hexagon in the center in a similar fashion to produce two different tessellations.

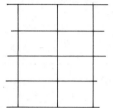 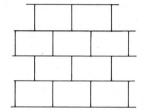

8. How many sides does the basic, tessellating polygon have in the photograph to the right?

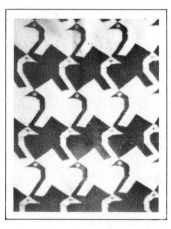

Tessellations in Art, You, and Escher

Source: M. C. Escher, "Eight Heads," Escher Foundation, Haags Gemeentemuseum, The Hague.

How did Escher create such interesting tessellations? Did he use an organized approach, or did he depend on inspiration alone to produce something like the drawing to the left? And can we hope to do the same thing? If so, how?

After virorous explorations with tessellations, it rapidly becomes obvious that several factors are at work.

There are endless ways to organize an approach to tessellating the plane, and undoubtedly Escher used some of those to begin his drawings. But creativity and inspiration play an equally important role in "seeing what's there" and turning it into a series of faces, or animals, or whatever.

In this section, we will provide you with several examples of ways to organize an exploration of tessellations, and then send you off on your own to experiment. You may not be able to produce an Escher-type drawing immediately—or ever—but then again, perhaps you can. In any event, you'll probably find that trying to produce one is an intriguing adventure, exhilarating in itself.

In the examples to follow, the first step is to select a simple polygon that tessellates the plane and then specify the composition of rigid motions (slides, flips, and turns) to be used. The basic polygon will then be modified by changing the sides into curves, in such a manner that the curves will "match up" when the rigid motions are performed. After tessellating the plane with the modified figure, shading will be used to "bring out the highlights," turning the shapes into recognizable figures.

Source: *Tessellation,* Chris Roberts, Elementary Education major, University of Wyoming, 1978.

FIRST EXAMPLE OF STRUCTURING A TESSELLATION

The basic shape to be used in this example is a rectangle, and the rigid motions to cover the plane are the translations that would produce this tessellation as a result:

Here are some steps that would produce an interesting tessellation from the basic one above:

Step 1: Put a curve on one end, and then slide that curve in a rigid fashion to the other end.

Step 2: Make a curve S_1 going from one corner, along the top side, to the midpoint of that side. Slide the curve rigidly to get curve S_2 as shown.

Step 3: Follow the same process as above, but work first on the bottom and this time move to the top, producing S_3 and S_4 as shown.

Step 4: Erase the sides of the original rectangle, leaving the modified figure. Use this figure repeatedly, in the same manner as the rectangle was used to produce the tessellation above.

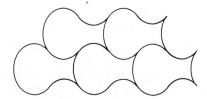

Step 5: Shade in the figures to represent recognizable shapes.

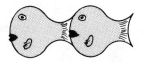 or  or ?

Notice the variables in this example: the rectangle, the rigid motions to produce that particular tessellation of the rectangle, the curves, and the shading.

SECOND EXAMPLE OF STRUCTURING A TESSELLATION

The basic shape to be used in this example is an equilateral triangle, and the rigid motion is a repeated rotation around the vertices of the triangle. This tessellation would result:

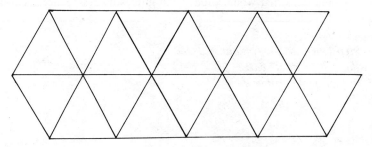

Again, here are some steps that would produce an interesting tessellation from the basic one just shown:

Step 1: Take an equilateral triangle and label the vertices A, B, and C. Draw a curve S from A to B.

Step 2: Rotate the curve S about point B until endpoint A coincides with C. Name the transformation S'. Then S' is a "new" curve with endpoints B and C.

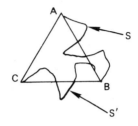

Step 3: Find the midpoint of the line segment from A to C, and label it M. Draw another curve S" with endpoints A and M.

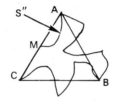

Step 4: Rotate S" about point M so that its other end coincides with point C. The result is the figure to the right.

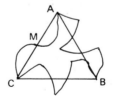

Step 5: Erase the triangle, and then tessellate the plane by continually rotating the figure around points A, B, and C.

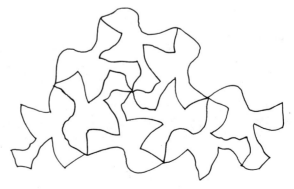

Step 6: Shade in the figures according to what you "see"; one such result is shown to the right.

Note again the variables: the polygon chosen, the rigid motions that produced the tessellation, the curves selected, and finally, the shading procedure. In the next problem set, you'll have the chance to exercise your own preferences for these variables.

PREPARING TO DO YOUR OWN TESSELLATIONS

From past experience, several things will help you with the routine steps in tessellating a plane. First, purchase (or make from a sturdy sheet of cardboard) a template for drawing the basic polygons you'll use to cover the plane. If you make your own, construct the polygons on the sheet of cardboard with your two basic tools, and cut them out carefully with a sharp razor blade.

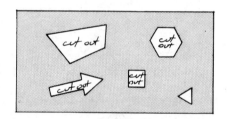

Another hint is to use paper that's thin enough to be seen through easily. An original curve can then be reproduced quite accurately by tracing over it on the thin paper; the rigid motions of sliding and turning the original curve can also be handled in this fashion. Having light come from behind the drawing to be traced is also quite beneficial. During the daylight hours, doing your drawing against a window will serve this purpose.

SET 7

1. The shape to the right was used to create the second tessellation shown in this unit. Trace over the shape, cut it out, and try to determine the basic polygon and rigid motions used to tessellate the plane before the modifications produced the shape above. (Hint: the shape and motions used were the same as in one of the two examples in this section.)

GEOMETRY **149**

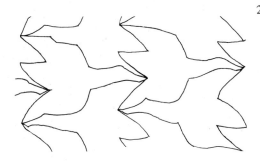

2. The basic shape seen in Escher's *Day and Night* is shown to the left. See if you can find something other than what Escher found in the shapes. Trace the basic tessellation, but use your own shading to produce a "new" tessellation.

3. The figure to the right tessellates the plane. Trace over the figure several times, and cut out the tracings and fit them together to "see" the tessellation.

    Once you know how to twist and turn the figure to create the tessellation, do so, and then use shading or coloring to bring out what you see hidden.

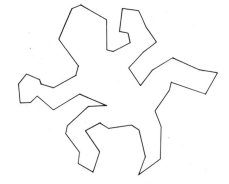

4. Create your own tessellation.

An intriguing question that has occupied the minds of great thinkers down through the ages is, "What, if anything, are the basic elements from which all existence is structured?" Long ago, scientists theorized what we know today to be true: everything in the universe is made from the same basic stuff. The variety within our cosmos is produced only by varying the relative proportions of these few basic elements. This very limited number of elements (put together in different ways) produces everything in existence.

Before the time of Christ, the Greeks theorized that there were only four basic elements—fire, earth, air, and water—used to structure the universe. The five three-dimensional shapes to the left represented earth, water, the universe, air, and fire, respectively, from top to bottom. The Greeks were incorrect about the four basic elements, of course, but there's an interesting reason they chose the figures to the left as the symbols for these four basic elements and the universe. They knew a rather startling fact about these five shapes, something that made them unique in the wide and varied world of three-dimensional figures.

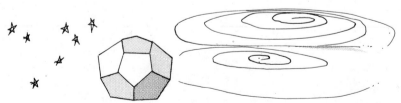

Polyhedra

The five shapes on the previous page belong to the general category of space figures called *polyhedra*. Just as a polygon separates a plane into three regions—interior, exterior, and the polygon itself—so a polyhedron separates space into three regions—interior, exterior, and the polyhedron itself. In this respect, it's much like a sphere or ball. Yet a polyhedron has one other requirement separating it from space figures with "curved surfaces": a polyhedron has a surface made completely of polygonal regions.

A polyhedron is a three-dimensional shape that is made entirely of polygonal regions and divides space into three regions – the interior, the exterior, and the polyhedron itself.

Some of the figures below are polyhedra, and some are not. Can you tell the difference?

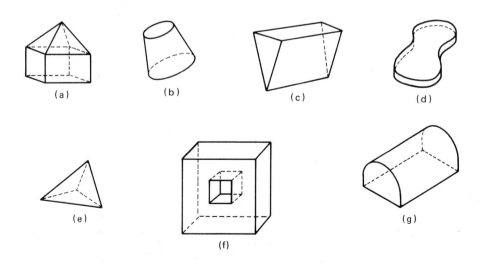

If you selected figures (a), (c), and (e) on the previous page, you are correct; if not, you need to look at the shapes again until you see why the ones you missed do, or do not, fit the definition of polyhedra.

There are similarities—other than the way they "separate" the dimension in which they lie—between polygons and polyhedra. Polygons have names like "octagon" and "decagon" that describe the number of basic shapes (line segments) of which they are composed. Polyhedra are named in a similar fashion: an "octahedron" has 8 polygonal regions (faces), and a "decahedron" has 10. Use the table of names listed below to name each of the polyhedra pictured:

Name	Number of faces
Tetrahedron	4
Pentahedron	5
Hexahedron	6
Heptahedron	7
Octahedron	8
Decahedron	10
Dodecahedron	12
Icosahedron	20

The respective names of the figures above are: hexahedron, tetrahedron, heptahedron, pentahedron, octahedron, and decahedron.

The *edges* of a polyhedron are just what you would guess they'd be—the line segments that determine the polygonal faces. The *vertices* are again what you'd imagine they would be—the points on the polyhedra where at least three edges meet. An interesting relationship between the number of faces, edges, and vertices has been known for quite some time, and will be explored in the next problem set.

Look back for a moment at the five polyhedra the Greeks knew were "special." Do you see anything that distinguishes them from ordinary polyhedra? It's hard to tell from these two-dimensional drawings of the three-dimensional shapes, but the faces are all congruent on each figure. Any polyhedron like this is said to be *regular* (just as any polygon with congruent sides is said to be a "regular polygon"). Thus we have pictured a *regular tetrahedron*, a *regular hexahedron*, a *regular octahedron*, a *regular dodecahedron*, and a *regular icosahedron*.

The amazing thing is that, unlike the limitless number of regular polygons, there are only a limited number of regular polyhedra. As a matter of fact, these five are the only ones. No wonder the Greeks thought that these five basic three-dimensional shapes were in some way related to the concept of the elemental substances that structure the universe!

The regular polyhedra have been studied since ancient times for their symmetry and construction properties. When the microscope allowed us to view the previously unknown "small worlds" around us, it was discovered that all five of the regular polyhedra exist naturally in nature. Contemporary interest in these solids include their use in architecture and the creative arts.

SET 8

1. What is the smallest number of faces possible in forming a polyhedron? What types of polygonal surfaces are used in forming such a polyhedron?

2. Fill in the chart below for each of the figures pictured:

Fig.	Name	No. of Faces	No. of Vertices	No. of Edges
(a)				
(b)				
(c)				
(d)				
(e)				
(f)				
(g)				
(h)				
(i)				

(a)

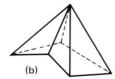
(b)

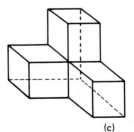
(c)

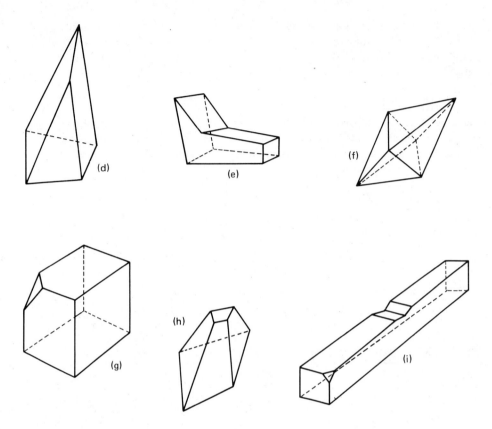

3. Look back at the chart for Problem 2. See if you can find the numerical relationship between the edges, faces, and vertices of a polyhedron. Once you have a conjecture, test it out on the polyhedra drawn before this problem set. If you do find a relationship, what type of rational thought process have you employed successfully?

4. Suppose you wanted to color some polyhedra so that no two adjoining faces would be the same. What's the smallest number of colors you could use?

5. Examine the following shapes. Note, of course, that these three-dimensional shapes are represented in a two-dimensional plane. Now for a "strange" question: Can any of them really exist as polyhedra in our three-dimensional world? Determine if any of them actually can, by using whatever method you can imagine.

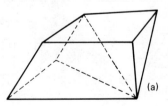
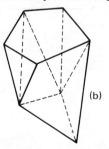

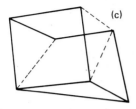 (c)

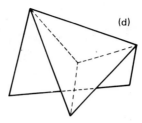 (d)

6. Would it make sense to talk about a polyhedron "tessellating space"? If so, what would this mean intuitively? Can you think of any polyhedra that would "tessellate space"?

REFERENCES

Ernst, B. *The Magic Mirror of M. C. Escher.* New York: Ballantine Books, 1976.

Hofstadter, D. R. *Gödel, Escher, Bach: An Eternal Golden Braid.* New York: Vintage Books, 1979.

Olson, A. T. *Mathematics Through Paper Folding.* Reston, Va.: National Council of Teachers of Mathematics, 1975.

Walter, M. I. *Boxes, Squares and Other Things – A Teachers's Guide for a Unit on Informal Geometry.* Reston, Va.: National Council of Teachers of Mathematics, 1970.

4

ANSWERS

Since the New Math era of the early sixties, mathematics educators have been among those debating the merits of process- and product-oriented learning. The argument has not been limited to any one subject area. The pure sciences, the social sciences, and even the arts and humanities have become embroiled in the controversy. All these areas have problems that need to be studied—problems for which reasonable approaches of both types would be beneficial.

A "product orientation" implies that the would-be problem solver is concentrating almost totally on the answer that would resolve a dilemma. This certainly doesn't sound all that bad, since without some prior thought about what constitutes a solution, we wouldn't know when the problem had been solved and when we still had more work to do. Any elementary school teacher will verify that a very real situation exists here. Many young students not only can't solve a given problem, but also can't judge when others have obtained a solution! Let's not kid ourselves; correct answers are very important. Buildings have caved in, ships have sunk, and planes have fallen from the sky because someone failed to supply a correct answer. It's quite natural that some educators and scientists fear that relinquishing the emphasis on correct answers in schools will tend to perpetuate this undesirable attitude in later life.

In the past we undoubtedly placed too much emphasis on a correct response and not enough on the footprints that lead to an answer. Certain negative feelings toward arithmetic might ultimately be traced back to an environment in which a numerical answer was either right or wrong, and the total amount of work done was judged by this one criterion. The starkness of such a system did not make it a popular one.

Furthermore, critics of product-oriented approaches to learning are quick to point out the accompanying problem of "fixation." Concentrating so hard on obtaining the answer may actually inhibit true problem-solving ability in that it causes us to oversimplify, underestimate, or in some way underanalyze the total situation. Once we see how to obtain the answer to a problem, it's more difficult to see that we may have misinterpreted a factor or that a shorter path may lead to the same result.

Most of the problems we face in life do not require hard-and-fast decisions; we have time to organize our thinking about various ways to approach the dilemma. A "process orientation" does not focus immediately on the nature of the answer, but on different ways of thinking about the problem and what the various factors mean, or appear to mean. The process orientation withholds judgments and decisions about the solution until time has been spent analyzing as many aspects of the problem as possible. Of course, carrying the analysis of a problem to the extreme may itself interfere with obtaining a good solution. We can become "fixated" on the analysis, just as on the answer.

Fortunately, the gap between products and processes of thought need not be so wide as people sometimes make it out to be. A most reasonable approach to solving the vast majority of problems would be to use some combination of the product and process orientations. And in situations where time and expense are factors, a realistic emphasis on one or the other would be considered appropriate by anyone's standards.

In view of this opening discussion, here are some general guidelines that might contribute positively to your personal "answering style."

> **If a probable solution occurs to you early in your analysis of a problem, test your answer against all conditions of the problem to see if it "holds up."**
>
> **If a reasonable answer has been tested against all conditions, and meets all requirements, attempt to find other solutions. There may be more than one correct answer.**
>
> **After obtaining an answer, search back through the problem for places where you may have made an interpretation different from someone else's. The "correctness" of an answer many times depends on interpretation.**
>
> **If you have tried and failed to find a solution to the problem, return and analyze each part and the relationship of the parts, and again search for misinterpretations.**
>
> **If you feel you've exhausted all possible ways of interpreting the problem to no avail, determine the form an answer would take and test several by guessing.**
>
> **If you're unable to find a solution that meets all conditions of the dilemma, perhaps you can offer an answer that satisfies some of the conditions. This sometimes provides further insight on a valid solution.**

Forms for Answers

Solutions to mathematical problems can take a variety of forms. The particular form for a solution depends not only on the kind of problem, but also on the information present and, frequently, the problem solver's own style. Many problems encountered in formal schooling are of the "one correct numerical answer" style, and these are certainly important, but a teacher truly interested in "personalizing the learning" of students will search out and make frequent use of other types of problems, those in which the answer may take on more than one form.

Some problems, especially geometrical ones, may call for a figure or description of a figure as its solution. You have already had the opportunity to solve some of these kinds of problems in Chapters 1 and 3.

Other problems may call for a rational argument as the solution. An answer of this sort consists of a logical sequence of statements describing relationships, assumptions, and a final judgment about the problem. Recall the experiences you may have encountered in a high school geometry course in which "proofs" played a major role.

Still other problems call not for a numerical answer, or a figure, or even a logical argument as the form of a solution, but for a conclusion. The conclusion may be as simple as a yes or no answer, or it may require listing several options and choosing the one that best fits someone's needs. Drawing inferences from empirical data is another important example of this kind.

Finally there are the problems in which the proper form of an answer is a "process" itself. You may be asked to find a way to do a certain set of tasks — for example, erect a perpendicular to any line segment or determine a set of steps that will make a quadrilateral tessellate the plane. In such problems, an underlying assumption is that a "good" answer is one that does the task, and a "better" answer is one that does it more efficiently.

The forms mentioned above certainly do not exhaust the possibilities for types of answers; they are merely some of the more common ones. Many problems require combinations of these forms with the most important one dictated by the questions asked. Certainly it doesn't surprise you to hear that "different strokes for different folks" applies here also. Each individual seems to identify with and enjoy solving problems of one type more than any other. Of those types discussed above, which do you prefer?

ANSWERS IN NUMERICAL FORM

Certainly a problem like $84 \div 4$ has only one correct answer in everyday arithmetic, and nothing further needs to be added in describing the solution. Many problems in mathematics are of this nature, and those that aren't frequently use such answers as part of the process of arriving at another type of solution.

For a previous problem in Chapter 1, you were asked to determine the number of rabbits and the number of chickens a farmer had. The solution to this problem is "The number of rabbits is 25, while the number of chickens is 13." Notice two points concerning this answer: it has two parts to it, but there is again only one correct answer.

Problems that require numerical answers dominate the textbooks in grades K-6, and are themselves mostly of the "one number, one interpretation" variety. So whether we agree or disagree with this emphasis in the curriculum, it behooves us to spend a special amount of time giving thought to helping students feel successful in such situations. We will take an in-depth look at gaining appropriate "numerical answers" later in this chapter.

Many problems ask for a solution in terms of a drawing. In a previous chapter, one of the problems was for you to create your own tessellation. In a later chapter you'll be asked to make graphs to illustrate certain points about a set of data. These are specific instances that illustrate our reliance on visual representations as correct ways to solve problems.

ANSWERS IN FIGURAL FORM

In some cases, a figure is not actually called for as a part of the solution, but is implied by the nature of the situation. You were asked previously to rearrange a set of toothpicks to form a different set. The problem did not ask for a diagram per se, but it would be difficult to demonstrate your solution without either a drawing or a physical model itself. In such cases, a diagram is appropriate as a solution to the problem.

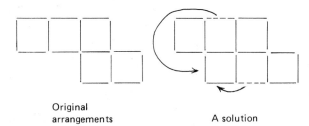

Original arrangements   A solution

A sequence of statements that follow from a given premise to a desired result constitutes another type of answer seen frequently in mathematics. Consider the following answer to the problem "Show there is no largest multiple of three":

ANSWERS AS LOGICAL ARGUMENTS

> Assume that a largest multiple of three exists, and is labeled M. Then M is 3 · n for some whole number n by the definition of "multiple of three." Next, let's add 3 to M, and we'll have that M + 3 = 3 · n + 3.
>
> The distributive property allows us to rename the right-hand side of this expression as M + 3 = 3 · (n + 1). But then we know that M + 3 is also a "multiple of three" by definition, and is certainly larger than M. But this contradicts our initial assumption.
>
> The only conclusion possible, then, is that there is no largest multiple of three.

The example above is probably in the "middle class" of logical arguments that constitute solutions. In many situations, it would not be considered rigorous enough, yet in others it would seem a little too formal.

We ask elementary students to give us such answers quite often in mathematics. For example, we might want a student to "explain how you borrowed to solve 453 - 278," or "show me why you think that 7 x 6 is 48." The logical argument, like other forms of solutions, depends on the audience and what will be understood. This characteristic of solutions is true as much for teachers and parents dealing with very young children as it is for adults who are trying to communicate their solutions and understandings to other adults. But the "logical argument" form of answer is used at all levels of mathematics, not just those in which formal, deductive proofs are required.

ANSWERS AS CONCLUSIONS OR INFERENCES

In real life, we don't solve math problems for the fun of it but to analyze and control our environment. The computations we perform generally lead us to make certain inferences or decisions based at least partially on the mathematical results obtained.

Suppose that you had a fixed income this year of $500 per month, and you wanted to plan ahead somewhat. You might begin this problem by considering alternatives like these:

Expenses	Solution 1	Solution 2	Solution 3	Solution 4
Savings	$ 50	$ 25	$100	$ 0
Gasoline	50	50	40	0
Rent	110	110	110	110
Food	200	200	250	0
Entertainment	90	115	0	390
TOTAL	500	500	500	500

The table above does not answer the question for you, but the information might be used to reach a meaningful personal conclusion (answer) about how to monitor your income.

In Chapter 2, several exercises might have enabled you to make some decisions on a personal basis. In one such problem, you were asked to determine your total net worth, and in another you tested yourself to see what kind of shape your body was in. Even though the problems did not call for conclusions on your part, you may have made some anyway. Did you decide to do something about your financial situation because of that problem? Did you decide that you were in good enough health that you didn't need to exercise? In any case, the problems are examples of situations where a worthwhile answer is not the number calculated but the personal inferences that can be drawn.

ANSWERS AS PROCESSES

Some of the more important situations we encounter, both in and out of school, require that a method be developed for attacking an entire set of problems with similar characteristics. In such situations, we answer the problem not with a specific number, figure, conclusion, or argument, but with a set of steps that can be applied to give the desired result any time we choose. For example, as a school teacher you'll need to be able to calculate the average of a set of numbers. So an appropriate problem for you to solve sometime before you have to perform that chore "for real" is to learn a process for "finding the average."

A surprising number of people have never developed legitimate processes for solving such problems as:

How does one find the miles per gallon that a car gets in going on a trip?

How much money is it going to cost me if I make a downpayment of x, and finance the rest for four years making payments of z per month, as opposed to buying the car with cash?

How can an angle be measured with a protractor?

All of the situations above call for answers in the form of processes, not specific to any particular calculations.

Elementary students are concerned with giving answers as processes also. When we ask such questions as "how do you solve a long division problem?" or "how can you change kilometers into meters?" we're asking them legitimate mathematical problems for which the appropriate form of answer is a process.

SUMMARY

There are many different forms for answers to problems in mathematics. All are appropriate and should be considered at different times. The labels attached to one form or another are not very important; it *is* important to expose yourself (and your future students) to the different types of solutions that may be required by "real life" problems.

SET 1

1. Look back through the first three chapters of this book, and consider each problem given in the exercise sets. Use the tally system to record the number of each type of answer called for, using the categories below:

    Numerical answer
    Figure answer
    Logical argument
    Conclusion answer
    Processes answer
    Other

   Organize the data you get, and see if you can draw a meaningful conclusion about the emphasis the authors feel is important for a course like this one.

2. What form of answer is called for in Problem 1 above?

Reasonable Numerical Answers

In the last section, we mentioned that by far the largest percentage of answers mathematics requires of us are those of the "one number, one correct answer" variety. This form of answer, then, deserves special attention. Keep in mind in the discussion to follow that our immediate concern is not how to find an answer, per se, but the form of the solution.

Two primary concerns about such answers are that they be judged "reasonable," whether correct or not, and that they be easily understood. In this section we'll concern outselves with the "reasonableness" of answers, saving the second concern (that of communicating our answer) for the next section.

In the cartoon below, Marcie is amazed at Peppermint Patty's answer. Even though Marcie doesn't know the answer herself, she recognizes that "green" is a ridiculous response to the problem $15 \times 7 = ?$, and is justifiably shocked. This is an exaggeration of a very real situation in which a student gives an answer that others consider totally unreasonable. Frequently such a verbal response so startles the teacher and other students that laughter is encouraged, resulting in "math scars" for the guilty party. Such situations should therefore be avoided if at all possible. Fortunately, there are some practical techniques employed by good problem solvers (almost subconsciously) that greatly reduce the chance of giving an unreasonable answer. No one knows whether these techniques come naturally to some people and not to others, or whether people become good problem solvers because somewhere along the line these skills were encouraged by a thoughtful teacher at just the right time. Until firm evidence to the contrary is available, it behooves us to assume the latter.

Source: © 1979 United Feature Syndicate, Inc.

ESTIMATING One popular technique for eliminating unreasonable responses is estimating. Two methods are used to approximate the answer to a problem, the preferred method (when it applies) being to use one's own *knowledge of the factors involved* in the situation to gain an approximate number. For example, if you encountered a problem in which the final answer would be the price of a college textbook for a basic psychology course, from your own experience you could probably guess that the answer would be somewhere between $5 and $25. If you had other clues, such as the text being hardbound or having about 300 pages, you could get even closer than that. Notice that you would not even have to know how to solve the problem to use this estimating procedure; you simply have to be familiar with the situation. Your estimate in this case might turn out to be a few dollars off the correct answer, but it would certainly prevent your being satisfied with an unreasonable answer.

This method of estimating is used by most of us very informally, but probably more often than we realize. From past experience, most of us could at least give "ball park" guesses for each of these numbers:

The price of a six-pack of cokes

The distance from New York to Los Angeles

The weight of a football player

The tuition for a semester at a state university

The cost of a new, subcompact automobile

Even if your estimate for each of these situations was not very close, you could certainly eliminate most unreasonable answers just from your own knowledge of the situations involved. And that's really all we're after with this form of estimation.

The value of this method of estimating answers to problems comes from the fact that you don't have to know how to solve the problem to gain the estimate. Unfortunately, many problems that confront us involve unfamiliar situations or outdated information. In such cases, this method is not much help, and we are left with the only other way to estimate an answer. We can *round off* the numbers involved so that they're easy to work with, and follow the steps to be used in calculating the correct answer, but with these "easy numbers."

Consider the problem to the right. Probably you're not familiar enough with this situation to give an estimate using personal experience, and so would not be able to use the previous method of estimating. But assuming you knew how to Sound travels about 1,000 feet per second. There are 5,280 feet in a mile. If you saw a lightning bolt, and then heard the thunder 9 seconds later, how far away, in miles, was the lightning?
proceed to compute the answer accurately, you could round off 5,280 feet to 5,000 feet, and 9 seconds to 10 seconds, and perform the calculations in your head. Can you make an estimate for this problem before going ahead?

Note that this method of estimating answers requires two skills that the first procedure did not: you must be able to round off accurately, and you must also know the steps involved in computing the final answer. These limitations force this method of estimating to take a "back seat" to the first, if both apply.

Whichever procedure is selected to estimate a numerical answer, the estimation should always *precede* the actual calculation. The temptation to be influenced by whatever has already been obtained, or merely to check the problem by redoing the same steps using the given numbers, frequently prevents an unbiased, ad hoc estimation.

Some people are quite fond of blending several different kinds of coffee beans together for their morning wake-up brew. If a mixture of Colombian beans ($4.30/kg.) and French Roast beans ($4.90/kg.) cost $4.75/kg., what was the percentage of Colombian beans?

Can you find a reasonable estimate for the answer above? Would the blend be more or less than 50 percent Colombian? Would it be as little as 10 percent, or as much as 90 percent? How can you tell?

ROUNDING OFF   The minimal competency tests presently in vogue in public education (grades K-12) have revealed several startling facts, not the least of which is that rounding off numbers is a stumbling block for many elementary students. And yet it's a crucial topic in the curriculum for several reasons, one of which was mentioned above: rounding off is an essential part of estimating answers.

A second reason for rounding off numbers is that it helps in communicating information; we will discuss this in more detail in the next section. Still a third justification for this important skill is that the concept being described may be so variable that accuracy is actually misleading. Population figures, distances in the universe, and the time of day are all ever-changing concepts, and hence are best described by approximate numbers.

The World population is 4,563,293,479.

The distance from the earth to the sun is 93,473,568 miles.

For a variety of reasons, then, we need to have at our fingertips the ability to replace a number with one that is less accurate, but more appropriate for our immediate purpose. We need to have mastered the skill of rounding off.

How do you respond to a simple question like, "How old are you?" Most of us would answer with a whole number of years, like 23, instead of with more accurate numbers like 23-4/12 years or even 23.25 years! Furthermore, we almost always "round down" our age; i.e., even if closer to 24 whole years than 23, it is the custom to report the smaller number. This is a common example of a real-life situation in which rounding down is the rule, rather than the exception.

The opposite rounding procedure is also encountered. If three cans of beans are on sale for $1.00, you should be able to purchase one can for 33-1/3¢, or 33¢ rounded to the nearest whole penny. Yet you must pay 34¢ for it; merchants always "round up," because it's in their favor and province to do so. Here is another real-life example in which the rule for "rounding off" depends on custom and who's doing the rounding!

For some reason, the curriculum tends to ignore situations like those above in teaching students to round off. Their total exposure comes in the form of approximating to the *nearest* number accurate to the desired degree. Certainly this is important, and in essence involves replacing the given number with the one closest to it, but accurate only to the specified degree. Consider a distance like:

2 yards, 1 foot, 7½ inches.

Past experience would have us arrive at an answer of "3 yards" when we round off to the nearest whole yard, or "2 yards, 2 feet" to the nearest whole foot. And in rounding to the nearest whole inch, we'd have the choice of going either with "2 yards, 1 foot, 7 inches" or "2 yards, 1 foot, 8 inches" since 7½ inches is equally close to seven inches and eight inches. Note that we have rejected the arbitrary rule that "when the number is halfway between two choices, always round up."

Approximating a base-ten numeral like 4,173,845.69 to a desired place value is handled in a similar fashion as the example above. Rounded to the nearest ten thousand, this number would be replaced by 4,170,000; to the nearest tenth, the answer would be 4,173,845.7; to the nearest million, we'd have 4,000,000. Note that the digit in the "hot spot," i.e., the one in the place value we're rounding to, is either left as is or replaced by the next largest digit, and in either case is followed by zeros. The decision on rounding up or rounding down is made by inspecting the digits to the right of the "hot spot" The only problem encountered is when we're asked to round off a number that is equally close to the choices available. Do we round up or down? Casting aside convention again (in light of what happens in real life), our suggestion is to make a choice based on the other factors involved in the problem.

Frequently we have an entire series of calculations to perform, and we gain the estimate by rounding off each number involved and going through the calculations with these easy numbers. In such situations, if we don't apply some intuition about rounding up about as often as we round down, we run the risk of being pretty far off in our estimate. The degree of accuracy of the final answer depends on the extent of errors in rounding for each individual number, so consistency becomes a factor. Accountants and others who deal with numbers in their professional lives have developed a variety of rounding rules for a series of calculations. Some of these will be explored in the next problem set.

SUMMARY

To avoid coming up with unreasonable answers for problems that require a numerical response, an estimate of the result should precede the actual computation. If possible, the estimate should be based on your own familiarity with the factors involved; if not, you can estimate by rounding off the numbers involved and proceeding through the computation with the approximate numbers.

The rules for rounding numbers are varied, and the one chosen for a particular situation should depend on an analysis of the problem itself.

SET 2

1. Estimate the answers to the problems below, using either your own knowledge of the factors involved, or rounding off the numbers. Indicate which method you use.

   a. You get on a very crowded elevator, and notice the sign to the right. You count and find there are 19 people in the elevator, but the cable didn't snap. What is the maximum average weight of the 19 passengers for this to be a safe trip?

   CAPACITY: 3000 lb. or 20 people

   b. The national debt is now slightly over $800 billion, and the population is 229 million. If the national debt were divided evenly among the entire population, how much would each person have to pay?

   c. A 1979 VW Rabbit (gasoline engine) was driven on a 268-mile trip and used 6½ gallons of gasoline. What miles/gallon figure could the owner record for this trip?

   d. A college football team had these tackles to choose from:

Jones, 245 lb.	Nathan, 270 lb.	Black, 249 lb.
Smith, 268 lb.	Martin, 252 lb.	Clark, 231 lb.
White, 253 lb.	Schult, 231 lb.	Brown, ?

   What would Brown have to weigh for the tackles at this school to boast that their average weight is 250 lb.?

   e. Joe stopped at a gasoline station when the gas gauge on his MG Midget showed a quarter of a tank left. He told the attendant to "fill 'er up." The attendant put in 8.8 gallons. How much gasoline does Joe's car hold when full?

2. Return to each of the five problems above and calculate accurately an answer, using the numbers given.

3. According to the 1978 *Guinness Book of World Records,* the longest word in the *Oxford English Dictionary* is *floccipaucinihilipilification*, which means "the action of estimating as worthless." This twenty-nine letter word might describe the results of your attempts to estimate the answers to certain problems. There are some problems for which it doesn't make sense to estimate the answer before working them. Find at least one problem in prior sets for which you would be unable to get an answer easily before working the problem.

4. Round off the numbers below as requested:

   a. 1 yard, 2 feet, 9¼ inches to the nearest foot

   b. 3,499 to the nearest hundred

   c. 3,499 to the nearest thousand

d. 300,852 to the nearest ten thousand

e. 0.453 to the nearest whole number

5. Estimate the *product* of (d) and (e) above using the rounded-off numbers you obtained. What do you get for this estimate? Why is this so far off the true answer for 300,852 x 0.453?

6. Role-play each situation below for a moment, and decide how you would round off the information asked for (up, down, or nearest).

> *You're a used car salesperson, and the interested party wants to know how many miles on the clunker.*

> *You're a college student, and your parents want to know your grade-point average from last term.*

> *You need a personal loan, and the credit union wants to know the value of your stereo system to use as collateral.*

> *You're describing to a possible future companion how many miles you backpacked last year.*

7. Estimate the sum of the set of numbers in the column to the right by rounding off the numbers in three different ways. Circle the estimate you feel is the most accurate.

8. Advertisers often list their prices as $.99, or $4,995.99, or in similar fashion (instead of as $1.00 or $5,000.00). Why do advertisers do this? What does this say about the potential buyer's tendencies in interpreting prices. Does this imply anything about our processes of rounding off?

9. Estimate your expenses for next term.

```
 93
 47
 35
 29
 75
 33
 89
 89
 45
 10
 83
 15
 64
 54
 92
 80
 11
 83
 65
 98
 37
 20
 28
 39
 46
 95
 34
 28
 39
 47
 56
 21
 39
 73
 24
 47
 77
 36
 48
 90
 76
 23
 45
 65
 68
 76
  8
+19
```

Communicating Answers

"Although humans make sounds with their mouths and occasionally look at each other, there is no solid evidence that they actually communicate among themselves."

Source: Courtesy of Sidney Harris and *American Scientist* Magazine.

Many ideas worth being understood and appreciated by people are sometimes lost or spoiled by inadequate communication. So it is with problem solving and answers in mathematics. Appropriate communication is an important part of all problem solving and involves more than just the solver. This section will examine the issue of effective communication of numerical information.

Having others understand our solution to a problem is sometimes imperative, particularly if decisions are going to be made based on the time and energy we've invested in the problem. Notice that we're leaving the arena of "correct answers" and entering the realm of "human experience" when we mention this aspect of problem solving. We must ensure that the answers we provide are, as much as possible, within the understanding of our audience.

PRECISION The accuracy of an answer is a quality associated with both the numerical computations performed and the instruments involved, if there are any. The final result must be reported in a manner consistent with the precision of the original numbers involved and any other factors inherent in the problem. For example, a square with *measured* dimensions of 4.3 cm by 6.4 cm, to the nearest tenth cm, has an area computed by 4.3 x 6.4. But even though we might arrive at an answer of 27.52 cm^2, to report the answer to the hundredth place is not warranted by the measuring device used. Another common example is that of calculating gas mileage. If a car travels 200 miles on 6.1 gallons of gasoline, reporting its gas mileage as 32.7869 miles per gallon is misleading in its precision. A general rule of thumb for such decisions is to be only as precise in your answer as the least accurate number in the original problem.

On the other end of the spectrum are the times when we can be extremely accurate in giving an answer, yet to do so would make us guilty of "information overkill"! The new digital watches have the potential for trapping us in this sort of situation. There's absolutely no reason in our normal routines for answering one of everybody's favorite questions in the manner shown to the right.

The minicalculator, microcomputer, and quartz watches and the many fantastically precise instruments that will undoubtedly be developed in the future will allow us to report extremely accurate numbers. But for communication purposes, we will need to decide whether we really need to do that. It is hoped we will make reasonable judgments about the accuracy necessary to convey the answer without being distractingly precise.

Rounding off is a crucial part of turning an extremely exact number into one that is less accurate but more meaningful. This basic skill, then, is important not just as a tool to help estimate answers, but as an integral part of the process of communicating answers.

LARGE AND SMALL NUMBERS

Believe it or not, the doctor below is actually trying to communicate! The patient would not have understood a statement like "One reason this condition is so hard to cure is that it's caused by a virus only 0.000037 inches in diameter!"

Reporting answers with numbers not in the experience range of the intended audience is quite common, yet for the most part people attach no meaning to large and small numbers beyond their own personal experience. Many people have no concept of how large one million is, for example, so any information containing that large a number cannot be communicated clearly to them. And likewise a number smaller than a thousandth is beyond the comprehension of the man in the street. So we're left with an interesting dilemma: how can an answer be transmitted with understanding if it involves large or small numbers?

Source: Courtesy of Sidney Harris and *American Scientist* Magazine.

The doctor in the cartoon has the right idea; try to *interpret* the answer for the listener! Sometimes this is fairly easy, and at other times it may prove impossible; yet the attempt should always be made. Consider the examples below:

Answer: *The national debt is $800 billion.*

Interpretation: *If every man, woman, and child would contribute only $3,500, the United States could pay its debts.*

> Answer: *The closest star to earth—not counting the sun—is Sirius. It's 51,666,655,200,000 miles away.*
>
> Interpretation: *If we could travel at the speed of light (a mere 186,000 miles per second) we'd reach Sirius in about 4½ years.*

> Answer: *A microscopic animal is probably 0.0004 inches in diameter, on the average.*
>
> Interpretation: *About 1200 microscopic critters could be lined up side by side across the tip of your little finger!*

Comprehension of numbers is different, of course, from person to person, as is a meaningful interpretation. Children are not likely to attach as much meaning to "100 miles" as adults would. Therefore, whether a number is "large" or "small," how to interpret it in a meaningful manner depends at least in part on the listener.

APPROPRIATE UNITS WITH NUMBERS

Answers may fail to communicate effectively when the unit used, together with the number, is not within the audience's experience range. This can happen at times when both the number and the unit, individually, are familiar to the audience. Consider for a moment an answer like "23,485 inches." How far is that?

The number part of the answer — 23,485 — is not that unusual. We frequently bump into terms like "$23,485" and "23,485 people" and have no problem grasping the essence of the situation. Likewise, answers like "23 inches" and "½ inch" are within our normal exposure. So both the number and the unit can be familiar to us, but used together they don't communicate effectively!

The solution is to interpret the answer differently, if possible. In the example of "23,485 inches," we could first change it into feet and see if that helps. Since 23,485 inches is 1,957 feet, we're perhaps a little better off but not much. So we'd try again, this time changing feet to yards. We're finally getting somewhere! Six hundred and fifty yards is about 1½ laps around a jogging track, or 6½ city blocks. How does this compare to your first guess about how long "23,485 inches" would be?

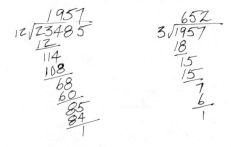

The concern at this point is not just with whether the numbers used in a numerical answer are generally understood, but with whether they're internalized in the context of the total answer. Again, the audience makes quite a difference; an answer like "0.000036 gm" might be understood by chemists but not cooks. On the other hand, an answer like "3½ tb" might communicate to a cook, but not to a chemist.

SUMMARY

Some of the most interesting and useful information reported with numbers is lost on the intended audience because the "human element" is ignored. Failing to communicate numerical answers effectively starts when elementary students solve "story problems" from the textbook, and the teacher neglects to emphasize this important aspect of problem solving. It's extremely important that students learn at an early age to interpret answers in meaningful terms, for anyone listening. The next problem set should provide you enough practice in this important skill that you will continue to consider this aspect of solving math problems until it becomes second nature to you.

SET 3

1. Interpret each of the answers below so that it would communicate effectively to the person who sits next to you in class.
   a. I baked a turkey yesterday for 18,000 seconds.
   b. Gold can be hammered into sheets as thin as 0.000004 inches.
   c. On the average, a female will blink her eyes about 1.9 billion times in her life.
   d. One loaf of bread costs 43¢, so the 100 loaves we purchased at Christmas for the needy folks cost 4,300¢.
   e. An adult male has around 100,000 miles of blood vessels in his body.
   f. A computer takes about 0.000125 seconds to perform a calculation.
   g. In the 1979 Great Britain Mileage Marathon, a mini Mercedes-Benz proved the eventual winner at 1,904 miles per gallon.
   h. An extremely fast sports car might have a top speed of 1/3,600,000 the speed of light (186,282.4 miles/sec.)

2. Give a reason to doubt the precision shown by each answer below, and rewrite the statement with something more meaningful.
   a. At rest, the subatomic particle we call an electron has a mass of 0.00000000000000000000000000910632 grams.
   b. By my calculations, so far in my life my heart has pumped 37,485,551 gallons of blood through my body.
   c. Last year my total assets amounted to $43,975.68.
   d. The world population at 12:00 p.m., December 31, 1980, was 4,280,413,612.

3. Solve the problems below and communicate your answer in written form. Practice estimating as your first step.
   a. An average American uses around 58 gallons of water each day at home. How much water did you use last year, if you were an average American?

b. A fast typist can type a letter of the alphabet in about 1/10 of a second. An average word is 5.9 letters in length. How many words per minute can such a typist type?

c. Americans own approximately 100 million cars. The average life span of a car is about 12 years, so we dispose of around 1/12 of the cars each year. How many cars gave up the ghost this year?

d. The sun is about 93 million miles away from the earth. Assume that light will travel through space at the rate given in Problem 1 (h) of this set, and find out how long it takes the light from the sun to get to the earth.

e. It takes a six-foot man about 1/60 of a second to realize he's dropped a brick and it's heading for his toe, and another sixtieth of a second to tell the foot to get out of the way. How fast do human responses travel?

Practicing

This last section is for the purpose of practicing your answering skills. The problems involve mass and temperature using metric units. You're probably vaguely familiar with these units because of the recent discussion about adopting metric units in this country. But if you're like most of us, you haven't yet felt a true identity with this system. This should represent the feelings that your future elementary students will have when you're teaching them arithmetic. They'll be somewhat familiar with the material but will also feel somewhat unsure of themselves. When you begin teaching, remember the feelings you had while doing these problems.

Most problem solving requires organizing your thoughts about the dilemma and having the necessary computational skills and memory to put a solution in motion. You may find that studying a problem with paper and pencil near you invites exploration, which can get you started toward finding an answer. If you have not developed the work habit of studying mathematics with scratch paper and pencil at hand, you might give it a try as you work with the problems in this section.

MASS Begin by making a small box out of paper, using the pattern to the right. This paper cube is a centimeter on each edge; hence it is called a *cubic centimeter*. After the glue has dried, fill it with cold tap water; it'll leak eventually, but you should be able to get a feel for its mass. By definition, a *cubic centimeter of water (at 4^oC) has a mass of one gram (1 g)*.

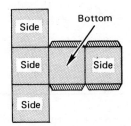

A gram isn't much, is it? This mass is used quite a bit in packaging medicines and other things that come in small quantities.

Familiar Objects with Small Masses			
Rubber band	½ g	Sheet of paper	3 g
Paper clip	1 g	Nickel	5 g
Dollar bill	1 g	New pencil	5 g
Penny	3 g	Pocket comb	6 g

A larger unit is needed, since we commonly measure the mass of things much larger than those above. A cube that is 10 centimeters (or, 1 decimeter) on each edge would hold 1,000 of the little cubic centimeters mentioned on the previous page. If such a cube were filled with cold tap water, its mass would be 1,000 g, or 1 kilogram (kg). To get a feel for this common unit of mass, take a waxed-paper, half-gallon milk carton and cut it so that the sides are about as long as the bottom edges. When filled with cold water, this container approximates one kilogram. The masses of some objects frequently measured in kilograms are provided below as "benchmarks" for your thinking.

Familiar Objects with Larger Masses			
Telephone	2 kg	Medium-sized dog	20 kg
This book	1 kg	Average female	55 kg
Gallon of milk	4 kg	Football player	100 kg
Turkey	7 kg	Small automobile	1,000 kg

An even larger unit—the metric ton—is used to report the mass of very heavy objects, such as large automobiles. The metric ton is equal to 1,000 kg, so the mass of the small automobile above could also be reported as 1 metric ton.

To keep a sense of perspective about reporting the mass of objects using the metric system, keep in mind that a kilogram is about 2.2 pounds. This might prove beneficial in estimating the mass of objects using metric terminology, keeping in mind that the preferred method of estimating comes from personal knowledge of the situation.

TEMPERATURE

Since water is the most common element on earth familiar to us, the points at which it changes from one state (solid, liquid, or gas) to another are used as the "critical" points on the Celsius scale for measuring temperature. Water's freezing point is labeled 0°C, and its boiling point is labeled 100°C.* It seems quite natural to break such a scale down into 100 equal intervals—for this reason, many people refer to this scale as

---

*At sea level, since these things vary quite a bit as the altitude, and thus the air pressure, changes.

the "Centigrade scale." A few of the more common temperatures are shown on the scale below. Try to remember these as "check points" for estimating the temperatures of other things in the Celsius scale.

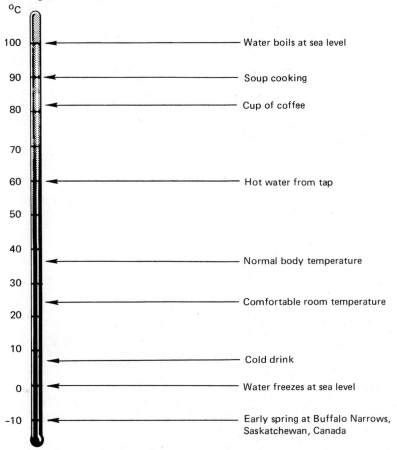

Notice an immediate advantage of the Celsius system over the Fahrenheit scale for measuring temperature: it's much easier to remember the two critical points in the Celsius scale (0°C and 100°C, rather than 32°F and 212°F).

Before you begin the next set of problems, take a few minutes and look back over the previous two sections to refresh your memory on the "answering skills" you need to be practicing. Make an obvious attempt to use the suggestions, where applicable, in solving these problems.

SET 4

1. From the two lists of masses of familiar objects in this section, you can find both the total mass of this book and the mass of a sheet of paper. These numbers are approximations, of course, but assume for a moment that they're quite accurate. Use this information to find the mass of the hard cover of the book.

2. According to the 1978 *Guinness Book of World Records*, the highest officially recorded shade temperature on earth is 56.7°C in Death Valley, California. The lowest temperature is –88.3°C in Vostok, Antarctica. What's the temperature range between the highest and lowest naturally occurring temperatures on earth?

3. Assume your empty billfold is an average one having a mass of 50g. You begin the day by putting in four dollar bills and four pennies (this is all the money you have to spend for that day) and putting it in your empty pocket. Later in the day you buy a pocket comb and pay $1.04 for it. You then put the comb in the same pocket as the billfold. How much mass are you carrying around in that pocket the rest of the day?

4. Consider the cartoon below:

Source: B.C. by permission of Johnny Hart and Field Enterprises, Inc., 1975.

Use your vast knowledge of cricket behavior to determine whether the formula is referring to Fahrenheit or Celsius scale. (Hint: Crickets die in cold weather, and chirp a lot during the summer.) Report your answer using a logical argument.

5. A can of Coca-Cola has a volume printed on the outside of 354 ml. Assume that coke has about the same mass as water, so 354 ml would have a mass of 354 g. How many cans of coke would it take to have the same mass as a gallon of milk?

6. The proper way to take a sauna bath is to go back and forth from the sauna to a pool of very cold water. This supposedly opens the pores, and if you don't have a heart attack, the shock of the varying temperatures produces a soothing effect. The temperature inside a sauna is quite bearable at 140°C, and very cold water might have a temperature as low as 10°C. If you take a sauna in the prescribed manner, what temperature range does your skin experience?

7. As mentioned in the text, a kilogram is about 2.2 pounds. Convert your present weight from pounds to kilograms, and then do the same but for your own "ideal weight." How many kilograms do you need to gain or lose to reach your goal?

8. Remember the critical points on the Fahrenheit and Celsius scales? The freezing point of water (sea level) is 32°F or 0°C; the boiling point (sea level) is 212°F or 100°C. Keep these numbers in mind while you carefully study the mathematical manipulations at the top of the next page.

$$100 = 5 \times 20 = 5 \times \frac{180}{9} = \frac{5}{9} \times 180 = \frac{5}{9} \times (212 - 32)$$

$$0 = 5 \times 0 = 5 \times \frac{0}{9} = \frac{5}{9} \times 0 = \frac{5}{9} \times (32 - 32)$$

Examine carefully the left- and right-hand ends of each expression above, and the critical numbers, and see if you can develop a process for changing a Fahrenheit reading to Celsius, or vice versa. Check your method using normal body temperature (98.6°F converts to 37.0°C).

REFERENCES

Stein, S. K. *Mathematics, the Man-made Universe.* San Francisco: W. H. Freeman, 1963.

5

CHANCE

> The night is almost gone, the bet is yours, and you can wager any amount that the next card turned up from the deck will be numerically between the two shown above. The two "up cards" are the only ones missing from the well-shuffled deck. How much would you be willing to risk? What are your chances of winning? What are your chances of losing?

Many people spend a good portion of their lives playing games like the one above without ever analyzing—even intuitively—their chances of winning or losing. Going with "seat-of-the-pants" emotion and blind hope, they tend to lose many more times than they win.

Would your chances of winning be at least as good as losing in the situation above? A cursory glance reveals that the only winning cards would be 6, 7, 8, 9, or 10, while losers would be 2, 3, 4, 5, Jack, Queen, King, or Ace. Since there are more losers than winners, a wise person would shy away from any heavy betting. This particular card game is called *in between*, and we will make use of it later in the exercises.

"Chance" is the underlying structure not only for gambling, but for many everyday decisions we make. The notion of "most likely" or "best chance" develops early in our learning. Parents and primary-grade teachers often see evidence of chance concepts emerging in five and six year olds. The games children play themselves, and the game shows they watch on television promote these initial ideas of "probability." Early learning is at an intuitive level, of course, and a formal development of the concept is not likely to be understood until somewhere between the ages of 12 and 15.

Many people depend on probabilistic information daily. Stockbrokers take chances usually with their investors' money, by purchasing stock in companies whose profits are rising and falling each day. In essence, stockbrokers study the market and gather as much information as possible, to help them predict which stocks will rise. Of course there's no guarantee that their final decision will be a profitable one; they only know that their money should go into investment programs with the highest likelihood of success.

Banks and loan companies make mortgage decisions only after they carefully measure the applicant's ability to pay back the money. The criteria the loan applicant must satisfy establish a high likelihood that principle and interest can be repaid. Among the items to which the applicant must respond are such things as:

Monthly income: _____

Rent or house payment: _____

Long-term monthly payments: _____

Place of employment: _____

   How long? _____

   Position: _____

   References: _____

                _____

                 _____

Using this information, the loan company rates a person as a good or bad risk. They're in the business of taking chances with their money based on hard evidence, not emotion.

Airline companies furnish another example of "big business" reliance on probability. Some airlines sell more tickets for a given flight than there are seats on the plane. After studying their flight records, the companies try to predict the number of ticket buyers who will cancel their plans or be "no shows." Through the years, for example, one airline determined that their St. Louis-to-Dallas flight would average 6 empty seats even though they had sold enough tickets for a full plane. This average eventually prompted the airline to "overbook" about 10 seats. Of course, many travelers on such overbooked flights will attest to the fact that this prediction is not borne out on some occasions. But over the long haul, the airline companies have been able to maximize their profits by using this chance-based procedure, even considering the penalty they pay when too many passengers arrive with tickets.

In weather reports, it has become customary to use probability statements to predict sunshine, rain, snow, hail, or other environmental events. The prediction is usually given in the form of a percentage. For example, we might hear "The chance of rain tomorrow is 70 percent." This prediction implies that the chance of rain in any given location is 70/100, or 7/10. Shortly we will use both percentages and fractions in describing the chance, or probability, of an event.

Chance is somehow involved in planning or carrying out each of the situations below. In which would you give at least an intuitive thought to probability before making a final decision?

CHANCE   179

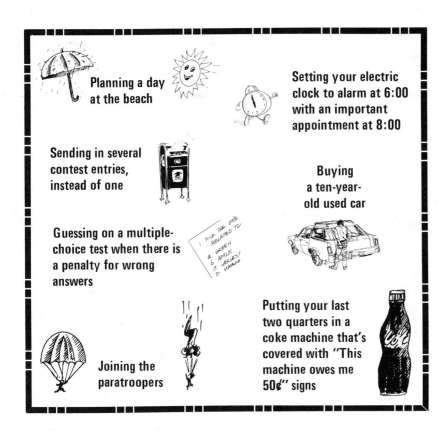

Can you think of other real-life situations in which either formal or informal thought is given to the chances involved, before a course of action is determined?

Consider the photograph below, which shows a chart on the wall of an elementary classroom. The teacher rewarded the students for solving extra problems by giving them stars. He then used these stars as the basis for a raffle, giving away such prizes as a calculator, a radio, etc. Each student's name went in the hat one time for each star. Can you tell from looking which student had the best chance of winning the first prize? Which had the worst chance? Did the student with the best chance to win actually have a good chance to win?

The chance that an event will occur is described as the *probability* of the event. Intuitively, we could say that the probability of an event happening is a comparison between the number of ways the event could occur and the total number of things that could happen. Usually this is put in the form of a fraction, and a formal definition given, along the lines of:

> **The probability of an event happening is**
>
> $$\frac{\text{number of favorable ways}}{\text{total number of ways}}$$
>
> **assuming that each favorable and each unfavorable way is equally likely to occur.**

Notice that the ratio above might appear as a fraction, or a decimal, or a percent. The basic meaning is the same in either case.

The "Star Wars" chart above can be analyzed in this fashion. Altogether, 140 stars were awarded. Randy's chances of winning on the first draw would be 16/140. Hazel's chance could be written as 8/140 since she had 8 favorable outcomes in the hat. Reversing for a moment what we consider a "favorable outcome," Randy's chance of losing on the first draw would be 124/140, while Hazel's would be 132/140.

Returning to the *in between* example that opened this discussion on chance, the total number of cards left in the deck—once the 5 and Jack pictured have been removed—is 50. The total number of winners left in the deck is 20 (four 6s, four 7s, etc.). So the exact probability of the next card being between the two "up cards" is 20/50, or 2/5. There are 30 losers left in the deck (four 2s, 3s, 4s, Queens, Kings, and Aces, and three 5s and Jacks), so the chance of *losing* on the next draw would have to be 30/50, or 3/5.

COMPLE-MENTARY PROBABILITIES

The examples above point out something that may seem intuitively obvious to you, but it bears mentioning anyway. When an event is certain to occur, its probability is given as a number equivalent to 1 (maybe 52/52, or 100 percent, or 1.00). So the largest number you'll ever see listed for a probability is 1, and that's only for events that are bound to occur. The smallest probability you'll ever see is a number equivalent to 0 (perhaps 0/52, or 0 percent, or 0.00), and this is reserved for events that absolutely cannot occur.

In this day of advanced technology, when nothing seems impossible anymore, probabilities of 0 and 1 are not nearly so common as probabilities close to zero or close to one. Consider these statements:

You'll live to be 125 years old.

A steak will cost more next year than this year.

You'll visit the moon sometime in the future.

A problem set is soon to follow.

Although we commonly think of the events above as certain to happen or not happen, the opposite is within the realm of possibility and hence requires that we list the chances as *close to zero* or *close to one*.

Any two events with probabilities that add up to 1 are said to be *complementary* events. Considering again the "Star Wars" example, two complementary events would be "Randy winning" and "Randy losing," since the respective probabilities for those two events are 16/140 and 124/140, and the sum of these two numbers is 140/140, or 1. For the "in between" example above, two complementary events would be "the next card being a winner" and "the next card being a loser," since these probabilities also yield *one* when added.

Generally speaking, given an event that has a chance statement attached to it, we also can easily find the chance of the "opposite" or complementary event happening. We merely subtract the known probability from one. If you figure that your chances of going directly from this paragraph to the next problem set, without getting something to eat or drink, as 75 percent, then your chances of getting something to eat or drink before attacking the next set of problems would be 100 percent – 75 percent, or only 25 percent.

SET 1

1. The chart below should help you in a deeper analysis of "in between." The only change is that an ace that's "up" can either be called "high" or "low." The cards across the top represent the right-hand "up card," while the vertical list represents the left-hand card. If you have trouble visualizing this, find a deck of cards and use it for the problems.

    Put the chance of the next up card being a winner in each box; a few have been done for you. Use any patterns and short cuts you can in filling out the table, and then answer the questions below:

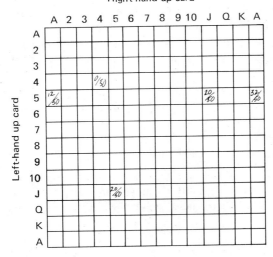

a. If the two up cards were an ace and an 8, should you call the ace "high" or "low" if you want to bet?

b. What should be the span of the two up cards to give you a probability of at least 50% of winning?

c. Which pairs of up cards will yield a probability of at least 75 percent of winning?

d. Work out a reasonable system of betting—using $10 to start with, and each bet being limited only by how much you have at that time—based on the probability table above.

2. In the chart below, list the total number of stars owned by each individual from the "Star Wars" chart shown on page 179. Also give the probability of each person winning the first prize. Leon's has been done for you.

Name	No. of Stars	Chance of Winning 1st	Name	No. of Stars	Chance of Winning 1st
Leon	8	8/140	Charles		
Leslee			Linda		
Judy			Diane		
Randy			Dean		
Chris			Nancy		
Hazel			Jane		
Kate			Kandi		
Sheretta			Kathy		
			Colette		

3. The questions below refer to the previous problem:

a. The person with the best chance of winning is Leslee. What is the probability that she will win the first prize? Do you think she would probably win?

b. Is it more likely that Leslee will win, or that one of the students who have fewer than half as many stars as she has?

c. A student can win only one prize, so if his or her name is drawn, the stars are discounted for the rest of the drawing, just as if they weren't there. Suppose that Kate wins the first prize—how many total stars are left for the second draw?

d. What would Colette's chance be of winning the second prize, assuming Kate won the first prize? What would Linda's chances be?

e. In the actual drawing for the group above, Linda won the first prize, Leslee the second, Diane the third, and Hazel the fourth. If there had been a fifth prize, what would Charles' chances have been of winning it?

4. Assume for a moment that you are taking a multiple-choice test that has five choices for each test question. If you are completely stumped on one of the questions and decide to guess, what is the likelihood that you'll randomly select the correct choice?

5. The dart board to the right has been used by some instructors to determine final grades. Assume for a moment that the thrower is unskilled and is therefore just as likely to hit one spot on the board as any other. What measurement would you have to know about the total target, and each piece, to be able to find the probability associated with each letter grade?

6. A famous time killer is throwing cards at a hat, seeing how many you can get in out of a typical 52-card deck. For purposes of standardizing the data, try the game yourself sitting on a sofa with the hat or its equivalent exactly two meters from the front edge of the sofa. Use the data you collect to estimate, as best you can, the likelihood that you'll get any particular throw in the hat.

7. A well-shuffled deck is placed in front of you, face down, and you are to select a card at random. What are the chances your card will be a 5? What are the chances your card will be red? What are the chances you'll get a red 5?

8. List two events that would have been given a probability of zero by the "average" person 50 years ago, but have been shown to be within the realm of possibility since then.

9. From your own feelings, make probability statements about the events below:
   a. Earth has at some point in time been visited by life forms from some other planet.
   b. There is a "Heaven" and a "Hell."
   c. A cure for lung cancer will be found in the next 10 years.

10. Return to the three situations above, and make three statements regarding the chances of the *complementary* events occurring.

11. Dave Logothetti uses a drawing similar to the one below to demonstrate to his students an event with probability one. He claims that *the tin can has probability one.* Can you figure out why, using the discussion of such events from the text?

12. The number of different, 5-card poker hands that can be dealt from a 52-card deck is 2,598,960. The number of ways each of the types of hands described below can occur is given in parentheses. Find the probability associated with each hand, first as a fraction, and then as a decimal. Once this has been done, arrange the hands from "easiest-to-get" to "hardest-to-get."

> **FLUSH:** Five cards of the same suit, not in sequence   (5,108)
> **STRAIGHT:** Five cards in sequence, not in same suit   (10,200)
> **FOUR OF A KIND:** Four cards of one face value   (624)
> **ONE PAIR:** Two cards with same value, other three match nothing   (1,098,240)
> **TWO PAIRS:** Two with same value, another two with same value, but different from first pair, last matches nothing   (123,552)
> **STRAIGHT FLUSH:** Five cards in sequence, all same suit   (40)
> **THREE OF A KIND:** Three cards with same face value, other two match nothing   (54,912)
> **ROYAL FLUSH:** 10, Jack, Queen, King, and Ace, all in the same suit   (4)
> **FULL HOUSE:** Three cards of one face value and two cards of another face value   (3,744)

Counting the Possibilities

Finding the probability of an event (in which any outcome is just as likely as any other) involves counting two things: the number of ways that "success" can occur, and the total number of outcomes possible. The probability is then the ratio of these two numbers.

Consider the rather "unusual" experiment of tossing simultaneously a fair coin and a fair die. For the coin, there are two equally likely outcomes, a head and a tail. For the die, there are six equally likely outcomes, since each of the six faces of the cube has the same chance of coming "up" on a given roll.

In tossing both the coin and the die simultaneously, there are a variety of combinations possible. Can you think of any possibilities other than these?

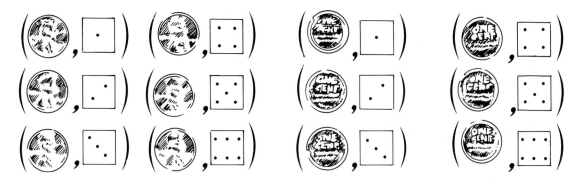

Give up? Good. There are only 12 possibilities, and all are shown above. In this case, counting the total number of things that can happen has been handled by actually listing all the possibilities, and counting. To find the probability of an event such as "getting a tail on the coin, and an even number on the die," we would only have to count the number of successes above, and form a fraction using this number as the numerator, and 12 as the denominator.

Could we have determined that there would be 12 outcomes without having to list them all? Consider this line of reasoning:

> **With each of the two outcomes for the coin, there are six outcomes for the die. So there are two times six combinations possible for a coin and die.**

The value of having a procedure for counting the number of possibilities becomes obvious when we get to situations in which it is impractical to try to list all the possibilities and count them one at a time.

If you were interested in getting a rough estimate of the maximum number of automobiles registered in a particular state, you might consider counting the number of tags that could be distributed before duplications would be required. The photograph to the left shows a Florida tag. Notice that the tag consists of three alphabetical letters followed by three single digits.

The sketch to the right is similar to the license plate above, but the six positions have been replaced by boxes with numbers in them. The first three boxes have "26" in

them to represent the number of possibilities for each position. The last three boxes have "10" in them, to represent that any of the 10 single digits are possibilities.

To find the total number of license plates possible for Florida's automobiles, we could either list all of them and start the boring, impossible task of counting them one by one; or, we could apply a little reasoning to cut short our task! Since each of the first 26 possibilities can be matched with 26 possibilities in both the second and third positions, there would be 26 x 26 x 26 ways that the first three positions could be filled with letters of the alphabet. In a similar fashion, for each of these 26 x 26 x 26 ways to start off a license plate with letters, there would be 10 x 10 x 10 possibilities for the three-digit numeral to follow. Altogether, the number of license plates possible would be given by 26 x 26 x 26 x 10 x 10 x 10 (which turns out to be 17,576,000). Aren't you glad we didn't try to count them all, one by one?

To formalize this method for avoiding counting individually the total number of possibilities, we might conclude:

When an experiment can be thought of as a sequence of unrelated events, the total number of possibilities is the product of the number of individual possibilities.

A few more examples should help internalize this procedure, which has traditionally been called *The Fundamental Counting Principle*.

COMBINATION LOCKS

The lock drawn to the right is typical of those found in public schools. Perhaps you've at some time wondered how safe those locks really were. One measure of this would be the chance that a thief could open the lock by trying random combinations. What would the probability be that any given random combination (R, L, R) would be correct?

The dial has 40 numbers on it (0–39), and theoretically each number has an equal chance of appearing as any of the three numbers used in a combination. The three boxes to the left are filled with the number of possibilities for each of the three positions. To find the total number of possible combinations, we could apply the fundamental counting principle and obtain 40 x 40 x 40 ways for the thief to try before duplicating his efforts. And 40 x 40 x 40 = 64,000.

The chance that any given combination would open the safe would then be given by the fraction 1/64,000. Better that the thief bring a pair of "lock cutters" than try to guess the combination!

TELEPHONE NUMBERS

Perhaps you've wondered how many phones there are that can be reached by our "direct dial" system. If so, we might again analyze the situation by finding the maximum number of phone numbers possible, using what we know about area codes and the seven-digit numerals that follow.

An area code can begin with any digit except zero, leaving nine possibilities for the first position. The second digit is always a zero or one—two possibilities. And the last digit is like the first, with only zero being excluded as a possibility. So the total number of area codes possible is given by 9 · 2 · 9, or 162.

The only restriction on the seven-digit phone number is that the first digit can't be zero. So there are ten possibilities for the remaining six positions, but only nine for the first. The number of seven-digit phone numbers would be given by 9 · 10 · 10 · 10 · 10 · 10 · 10, which turns out to be 9 million!

The total number of possible telephone numbers, under these restrictions, would be given by the fundamental counting principle. What is this number?

The problems in the next set will give you more experience in using the fundamental counting principle. Even though it might not specifically ask you to calculate a probability after finding the total number of ways an event can occur, keep in mind that this is the reason behind introducing this principle. Perhaps you can develop a probability question yourself for some of the situations.

SET 2

1. Calculate the maximum number of automobiles registered in the state where you reside, using the "license plate" method described in the text.

2. There are about 100 million personal cars in the United States. Design a license plate that could be used to register each automobile nationally, instead of by individual state, for the next few years to come.

3. The license plate to the right was on a car that almost forced you into a collision with oncoming traffic. Over your C.B. radio, you report the incident to a highway patrol car in the next town. How would you verbally describe the license to the authorities?

4. You have two dice, a red one and a white one. How many unique combinations of the two "up faces" are possible when this pair is rolled together?

5. In the chapter on geometry, the five *regular polyhedra* were discussed. These are shown again below, but each one has been turned into a die by numbering the faces consecutively, starting with "1."

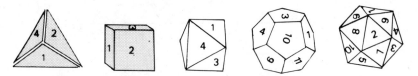

Suppose we agree to roll all five of the dice at one time, and record the numeral that comes down on each one, according to this scheme: tetrahedron, hexahedron, octahedron, dodecahedron, icosahedron. How many different possibilities are there for this experiment?

6. How many "pairs" are possible for the two "up cards" in *in between?*

7. The crossword puzzle below is typical of those you might see in a commercial sweepstakes contest, where a lot of money is at stake. You are generally allowed to enter as many times as you wish, which attracts a good many contestants. The puzzle is very difficult to solve correctly, however, because each word has at least two letters of the alphabet that will make sense in the blank space. The clues given are such that they would fit either word.

The clue for 1 across, for example, might be "A baseball pitcher does this to batters." "Walks," "talks," and even "balks" would all fit and make sense. For purposes of this problem, assume that each blank space has only two appropriate letters. How many unique possibilities would there be for this puzzle, with each possibly fitting all the clues given?

There are times when college students find themselves forced into guessing on an exam due to unforeseen complications the previous night. Maybe it was worth it, maybe not, but the price will have to be paid in either case—unless it's a multiple-choice test, and you're a lucky guesser! Suppose you find yourself in this predicament. You look the test over and notice that each question has five choices, so your chances of guessing the correct answer on any question is 1/5. The test consists of 10 such questions—what are your chances of guessing all 10 of the correct answers?

If we knew how many possible ways there were to answer a 10-question test, with five choices for each item, we could write the probability immediately as a fraction with numerator "1" (since there's only one way to successfully guess all ten) and the calculated number as the denominator. The fundamental counting principle can be applied here:

1 has 5 possible ways	6 has 5 possible ways
2 has 5 possible ways	7 has 5 possible ways
3 has 5 possible ways	8 has 5 possible ways
4 has 5 possible ways	9 has 5 possible ways
5 has 5 possible ways	10 has 5 possible ways

So there are 5 x 5 x 5 x 5 x 5 x 5 x 5 x 5 x 5 x 5 or 9,765,625 possible ways to answer a ten-item, five-choice, multiple-guess exam! Your chances are about one-in-ten million!

You may be wondering whether there's a more direct way to find the probability above than figuring the total number of possibilities. In fact there is. Notice that the probability of each answer being correct above is 1/5, and that

$$\frac{1}{5} \times \frac{1}{5} \times \frac{1}{5} \times \frac{1}{5} \times \frac{1}{5} \times \frac{1}{5} \times \frac{1}{5} \times \frac{1}{5} \times \frac{1}{5} \times \frac{1}{5} = \frac{1}{9,765,625}$$

At least in this case the probability of a total sequence of events happening together can be found by multiplying the chance that each will occur separately!

Consider another example. How likely is it that you could flip a coin three times, and get three tails? The chance of getting a tail on any one of the flips is certainly ½, so if the procedure noticed above happens to apply in this case also, the probability of obtaining three tails in a row should be ½ x ½ x ½, or 1/8. But is it really?

To check out this contention, we have two choices—either list all of the possibilities and start counting, or use the fundamental counting principle to determine the number of possible ways three coins can land. Since the latter is the easier, we'll do the former. In the box below are the different ways that three coins can land, when flipped:

```
(heads, heads, heads)      (tails, heads, heads)
(heads, heads, tails)      (tails, heads, tails)
(heads, tails, heads)      (tails, tails, heads)
(heads, tails, tails)      (tails, tails, tails)
```

Notice there are eight possibilities, only one of which would be considered "successful." So we can verify that the probability produced above is exactly the same—1/8—as that obtained by multiplying the probabilities of the individual events.

Are we on to something? Yes, we are, but we must be fairly careful in stating what it is. The general conclusion reached below has been around for quite some time, and is extremely useful in general probability theory—for lack of a better name, we call it "the chain law":

 For a series of unrelated events, the chance that they will occur together is given by the product of the individual probabilities.

The part above that needs your careful attention is where it says that the chain law applies if the individual events are *unrelated*. A few more examples might prove beneficial at this point.

Russian roulette. Four people find themselves with nothing better to do on a Friday night, so they decide to play "Russian roulette." This game starts by placing only one bullet in the barrel of a six-shooter, and then spinning the barrel so that, at least theoretically, the chance that the bullet will be in the firing position is 1/6. The gun is placed to the head of a participant, and the trigger pulled. The person then has a probability of 5/6 of living.

Each of the four people had a gun, and they decided to "fire" at the same time so no one could back out. What were the chances all four would survive?

Analysis. This is a series of unrelated, or independent, events since each person's actions have no influence over any other's. So we can apply the chain law, and find the chance that all will survive by the product that each individually will make it.

$$\frac{5}{6} \times \frac{5}{6} \times \frac{5}{6} \times \frac{5}{6} = \frac{625}{1296}$$

The chance that all will live to tell the tale is a little less than ½.

> **Basketball.** A few years ago, the finals of the NCAA basketball tournament came to a tense conclusion. With no time showing on the clock, a player was fouled and was given a "one-on-one" chance to win the game (that is, if he made a first foul shot, he'd get another). Since his team was behind by one point, he could win the game by hitting two straight foul shots.
>
> His season percentage on foul shots was 80, and he didn't seem to be a "streak shooter," implying that his success on the first shot would have no bearing on his success on the second shot. What was the probability that he'd win the game outright by hitting both free throws?
>
> **Analysis.** The chain law applied in this case, since his chances on each shot were unrelated. The probability of his winning the game, then, would be given by .80 x .80, or .64. So the likelihood that he'd go home the hero was .64.

RELATED VERSUS UNRELATED EVENTS

If a coin is flipped eight times, the chance that eight heads would appear is ½ x ½ x ½ x ½ x ½ x ½ x ½ x ½, or 1/256. Suppose for a moment that you've already flipped the coin seven times and have seven heads—what is the probability that the next flip will yield a head?

Freddy the Lush figures that he has a likelihood of .80 of taking a drink on his way home from work each Friday. What is the chance that he'll have *two* drinks this Friday? *Three* drinks?

The two examples above illustrate the difference in *related* (or dependent) and *unrelated* (or independent) events. In flipping a coin, the results of past flips have absolutely nothing to do with upcoming flips. Even if seven heads have appeared, the chance that the next flip will be a head is still ½, same as always. Freddy's in a different situation, however, since he's a known lush. Having one drink increases the chance he'll have a second, which will increase the chances that he'll have a third! The "coin flips" example shows a series of *independent* experiments; past results have nothing at all to do with the chances of future events. But Freddy's drinking history would reveal a set of *dependent* events; each drink would change the probability that he'd have another drink.

Failure to make this distinction between dependent and independent events has caused more than one gambler to lose his shirt. One famous example involves a roulette wheel, which is designed so that an ivory ball will fall at random into one of three different-colored compartments. The chance that it will fall into a black area is 47 percent, the same as the chance that it'll go into a red area. The green has only a 6 percent chance of getting the ball.

A roulette wheel yielded 26 straight "blacks" in a Monte Carlo casino in 1913. Who cleaned up financially from this long run of blacks? The casino did, because after the first ten or so black rolls, the crowd began to bet heavily on red, thinking that so many blacks were bound to be followed by a red on the next roll. The casino kept its cool, knowing that each roll was independent of the previous ones and that the chances of a black on the next roll would be the same as on previous ones!

Being aware of the distinction between dependent and independent events is very important when considering a sequence of events in which the chain law might apply. As stated previously, if events are *unrelated,* the chance that they'll occur together is the product of the individual probabilities. But what about a sequence of *related* events? Is there a chain law for events that change the chance that another event will occur? Fortunately, the answer is "yes!"

HISTORY AND THE CHAIN LAW

A modified form of the chain law can be used when past history affects future events, even though in such a case the sequence of events is obviously not an independent one. We can still find the probability of the sequence of events occurring together by finding the product of individual probabilities, but we must be sure that the individual probabilities used reflect that previous events have occurred. Consider this example.

A flush is five cards from a deck, all in the same suit. A "club flush," then, is a five-card hand in which each card is a club. The chance that a hand drawn at random would be all clubs is *not* given by:

$$\frac{13}{52} \times \frac{13}{52} \times \frac{13}{52} \times \frac{13}{52} \times \frac{13}{52}$$

since the chance of drawing each succeeding club changes as you go. The chance that you'll select a club for the first card is 13/52. But if you are successful, the chance that you'll get one the next time is 12/51, since there would then be 12 clubs in the 51 cards left. The chance of getting a third club, if you've gotten a club on the first two draws, is 11/50, and the chances for the fourth and fifth clubs, if you've been successful on previous attempts, are 10/49 and 9/48, respectively. So the chance that you'll draw five clubs in a row would be given by:

$$\frac{13}{52} \times \frac{12}{51} \times \frac{11}{50} \times \frac{10}{49} \times \frac{9}{48}$$

This is an instance in which the *modified chain law* can be applied. We know the chance that each individual event will occur, given that prior ones in the sequence have occurred. So the chance of the entire sequence happening is the product of these individual events happening.

Reconsider for a moment Freddy the Lush, and his 80-percent probability of having a drink on Friday after work. Suppose we knew that his chance of having a second drink, given that he'd had the first, was 90 percent. The modified chain law could be applied to this sequence of dependent events, yielding a probability of .80 x .90 (or .72) that, on any given Friday, Freddy would have at least two drinks.

This new rule can be stated somewhat formally in the following manner:

 For a sequence of related events, the chance that the entire sequence will occur is given by the product of each individual probability, given that prior events in the sequence have occurred.

The next set of problems will give you a chance to examine your own thinking about these two chain laws.

SET 3

1. Look back for a moment at the example in the text of flipping three coins or one coin three times. What is the likelihood that you'd get two heads and a tail, in any sequence, in this experiment?

2. Suppose the "Russian roulette" game were played by three people, but each person put two bullets at random in the gun's chamber. What are the chances that all three would live?

3. The standard slot machine pictured below has three windows. The dials underneath the windows have 20 symbols as shown in the table below; supposedly each dial is independent of the other two, and the symbols on each dial come up randomly.

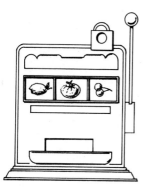

Symbol	Dial 1	Dial 2	Dial 3
Bar	1	3	1
Bell	1	3	3
Plum	5	1	5
Orange	3	6	7
Cherry	7	7	0
Lemon	3	0	4

On one pull of the handle, what is the probability of getting three plums? three bars? three lemons? For which of the six symbols would there be the biggest payoff if three of them showed up on the dials? (Keep in mind that the size of the payoff is *inversely* related to the chances.)

4. Consider two friends of yours. What are the chances they were both born on a Wednesday? What are the chances they were both born in the morning hours? Finally, what are the chances they were both born on a Wednesday morning?

5. Do you have ESP? To test yourself, try this experiment. Mark one of 32 three-by-five cards, and mix them all up with the mark "face down" so it can't be seen (or use 32 cards from a standard playing deck, and just select one of them as the marked card). Divide the cards into two equal stacks, and use your ESP to choose the pile your card is in. Discard the other pile, and redivide your stack in half again, and again choose which pile your marked card is in. Continue this process until you are down to only two cards; select one and look to see if you're correct.

   What is the likelihood that you would wind up with the marked card at the end, assuming you had no ESP and were guessing each time you had two stacks to choose from?

6. Look back at the "Star Wars" chart used in the opening unit of this chapter. What is the probability that this sequence of events would occur?

   Leslee wins the first prize.
   Hazel wins the second prize.
   Linda wins the third prize.
   Judy wins the fourth prize.

7. Insurance companies base their rates on statistical evidence, and generally the rates increase with each accident a person has. Knowing this, would you assume that their statistics show that the probability of a person's having a second accident is greater than that of having the first, and that the chance of having the third is greater than that of the second, etc.? Does this have implications about the dependence or independence of such events?

8. A hamburger chain recently offered to let customers who purchased a large coke keep the glass, which had one of eight Disney characters embossed on it. What are the chances that you'd get all eight of the Disney characters, buying only eight cokes?

9. A chain smoker tries to quit, and actually makes it through the first week without a cigarette. But at a weekend party, temptation rears its ugly head. Statistics reveal that, in such a situation, the former smoker has a likelihood of .88 of yielding to temptation and having a smoke. Estimate the chance that such a person would smoke a second cigarette and a third. What is the overall probability of such a smoker having at least three cigarettes at the party?

10. Toss the cards toward the hat again, as in the problem in the first set of exercises, but this time keep track of each throw in order. Would you say that you have streaks of luck, or do your hits really appear unconnected to the other hits? Can you make a conclusion about the dependence or independence of your throws?

11. A well-shuffled deck is placed in front of you. What is the likelihood that you'll draw three aces in a row, if you do not replace a card after you draw it?

    Assume now that you have performed this unlikely feat, and the person next to you wants to bet that you can't draw the other ace on your next try. How much should the person be willing to risk against your $1 for the bet to be fair?

12. Back in the early sixties (before the day of the metal detectors at major terminals) hijackers of airplanes frequently used homemade bombs to threaten the safety of the airplane. The chance of being on a flight with such a hijacker was estimated at about one in a million. The probability of being on a plane with *two* such hijackers, working independently of each other, would then be one in a trillion. The working theory at the time was that you should carry a bomb with you when making a flight, reducing the chance that a real hijacker would be on board!

    Does this argument make sense to you? If so, why? If not, where's the loophole?

Through the Back Door

Did you ever see a house with no back door? Probably not: the chance of finding such a thing is close to zero. They're simply too useful for home builders even to consider saving the few extra dollars it costs to install them. The safety factor alone—being able to go in or out the back door if the front door is blocked—makes them worth putting in any building.

Probability theory has its own "back door" method of calculating the chance that an experimental event will occur. This method is useful when the front door is blocked, i.e., when it seems impractical to calculate *directly* the likelihood that an event will occur. At such times, this alternate procedure merely gives another way to approach the problem. Sometimes it works, and sometimes it's no better than a direct approach, but it is nice to have around.

The back door approach to finding the chance that an event will occur rests on the fact that the event has a *complementary event*. Knowing the probability associated with either an event or its complement gives us the other probability quite readily, since we merely have to subtract the known probability from 1 to get the other. And surprisingly, at times it's much easier to find the probability of the complementary event directly, than it is to find the chance of the event itself. Sometimes the complementary event is much simpler in concept.

The famous parlor trick—finding the probability that, among 30 people, at least two will have the same birthday—is a prime example of using the back door method. Finding this probability "head on" is difficult, and requires procedures beyond the scope of this text. However, we can perhaps "sneak through the back door" by finding the probability of the complementary event. That is, maybe we can calculate the chance that, among 30 people, *none* have the same birthday. If we're successful, we can subtract the resulting number from 1 to find the answer to the original query.

> **Object:** Calculate the probability that among 30 people, at least 2 will have the same birthday.
>
> **Method:** Find the chance that among 30 people no 2 will have the same birthday. Then subtract this number from 1 to answer the question above.
>
> **Calculations:** The modified chain law applies here. Assume for convenience that the people are numbered from 1 to 30. Start with 1 and ask his birthday, then 2, then 3, etc., on to 30. It doesn't matter what 1 says, in terms of finding the chances.
>
> Probability 2 doesn't match 1 is $\frac{364}{365}$. (Why?)
>
> Probability 3 doesn't match 1 or 2, given 2 doesn't match 1, is $\frac{363}{365}$.
>
> Probability 4 doesn't match 1, 2, or 3, given no prior matches, is $\frac{362}{365}$.
>
> $$\vdots$$
>
> Probability 30 doesn't match any, given no prior matches, is $\frac{336}{365}$.
>
> Therefore the chance that there are no matches in the room is given by:
>
> $$\frac{364}{365} \times \frac{363}{365} \times \frac{362}{365} \times \frac{361}{365} \times \ldots \times \frac{336}{365}$$
>
> Multiplying all this out on a calculator, we arrive at a chance of 0.294 that there are no matches among the 30 people. So the chance that there is at least one match is 1 - 0.294, or 0.706 (about 71 percent). Surprised?

One note of caution is in order. When trying the back door approach to calculating probabilities, take care in describing the complementary event. Above, the opposite of *at least one* match is *no* matches. But the opposite of having *exactly one* match would open up a real can of worms, since having no matches, or 2 matches, or 3, or 29 would all qualify as part of the complementary event!

Another example is in order. Do you remember the tale of the pauper who was in love with the king's daughter, and she with him? The pauper was thrown in jail until the king could decide his fate. Eventually the king said he would arrange a fair way to let the gods decide. He would put 50 black beads and 50 white beads in a jar, mix them, and allow the blindfolded pauper to select one at random. A white bead and the princess was his; a black bead and the lions got a free meal.

The king's daughter agreed to this, but only on the condition that *two* jars be used to hold the beads, and that she could place them in the jars as she desired before they were mixed thoroughly. Not yet having had this course, the king didn't realize he was being taken, and conceded. The princess then put *1 white bead in a jar, and the other 99 in the other jar.* She thus "adjusted" the chances of her true love's getting a white bead to well above ½. Exactly what were the chances of the pauper selecting a white bead at random?

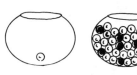

The chance of selecting the jar with the lone white bead is ½. This case will automatically give a white bead since it's the only type there. The complication is that half the time the other jar will be chosen, and the chance of a white bead in this case is 49/99. So *directly* finding the chance of obtaining a white bead on the draw is fairly complex, in that two cases must be considered and the results somehow combined. Let's try the back door, and see what happens. The complement of selecting a white bead is choosing a black bead.

Object: Calculate the probability that a white bead will be selected at random, given a jar containing 1 white bead and another jar containing 49 white and 50 black beads.

Method: Calculate the chance that a black bead will be drawn. Then subtract this answer from 1 to find the answer to the original question.

Calculations: The chance that the jar with some black beads in it will be chosen is ½. Given that this has occurred, the chance that a black bead will be chosen is 50/99. So the modified chain law can be applied, yielding:

$$\frac{1}{2} \times \frac{50}{99} = \frac{50}{198}$$

Converting to a decimal, the chance of getting a black bead is about 0.25, so the chance of getting a white bead must be 0.75. The pauper had a 75-percent chance of winning the princess!

Again, using this back door approach to finding the probability of an event is not always a wise choice, as it's frequently just as difficult to calculate as a direct attack on the problem. But it does offer another option, and at times comes in quite handy, as in some of the problem situations to follow.

SET 4

1. Return for a moment to the example of the princess and the pauper in the text. Extend the solution the princess arrived at by considering questions such as these:

   a. If she had used three jars instead of two, and arranged them to the best advantage, what chance would the pauper have had?

   b. Same as above, but with four jars.

   c. What number of jars would give her lover the best possible chance? What would that chance be?

2. How many students are in your math class this term? Suppose the instructor decided to post final grades by using the last two digits of each person's Social Security number. What would be the chances that the instructor would have to modify this system, because there'd be at least one match in the class?

3. In modifying the system above, the instructor decided to go with the last three digits of the social security number. How likely is it now that no matches will appear in the class?

4. A married couple is ready to start having children but find that they disagree on the final result desired. The man wants only to have at least one son and doesn't care about the total number of children. The woman wants to have three children and doesn't care whether they're males or females. What is the likelihood that, if they do have three children, there will be at least one boy?

5. Sheila and Shirley, sorority sisters, had blind dates with Jack and Jerry, fraternity brothers. Sheila was paired with Jack, but later found Jerry more to her liking. When she secretly mentioned it to him, she found that he felt the same way. The problem they faced was how to get together and make it appear accidental.

   Their plan was for each to go alone to the local pub the following night, sometime between 8:00 and 9:00 p.m. They agreed to wait for the other person for only 15 minutes. If their appearances overlapped, they would take it as a sign to continue seeing each other.

   The picture to the right can be used to figure the chance that they met, and also the probability that they didn't meet. Study the picture carefully and you'll notice that any point in the *shaded area* represents an overlap of their arrival times. A point *outside the shaded area* represents that they missed each other. The various probabilities can be found by finding the area of the total graph, then the areas of the shaded and unshaded regions.

   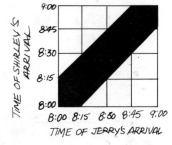

   Calculate the chance that fate would approve a match of Sheila and Jerry.

6. The patient at the top of the next page happens to be a friend of yours. Explain to him the way you feel about the situation.

The Way Things Are

The introduction to this chapter mentioned just a few of the big businesses (banks, airlines, insurance companies, and brokerage firms) that rely extensively on chance-based procedures. In most cases, the probabilities they use come from examining the real-world evidence they've collected. Their statistics reflect "the way things are," regardless of either supporting or contradictory theoretical arguments.

The data in the chart to the right came from calculating and categorizing information from the 1960 United States Census. The decimals can be interpreted as probabilities that a newborn infant in a particular category would live at least to the ages shown in the left-hand column. For example, a newborn white male had about a 26-percent chance to live to be 80 years old, while a newborn, nonwhite female had a likelihood of almost 29 percent of living to that same age. Does this seem "unfair" to you? Should life insurance companies be allowed to base their rates totally on this view of "the way things are"?

	Chances of survival of newborn infants in the United States			
Age	White Male	White Female	Nonwhite Male	Nonwhite Female
1	.974	.980	.951	.961
10	.967	.975	.939	.952
20	.959	.971	.920	.945
30	.944	.965	.897	.928
40	.924	.953	.849	.892
50	.874	.925	.766	.828
60	.755	.863	.626	.721
70	.538	.725	.418	.539
80	.261	.448	.189	.288

A less dramatic example involves our use of various letters of the alphabet. The percentages shown below came from researchers (or their graduate students) who looked at a variety of manuscripts and actually counted the number of times each letter appeared. Again, the information can be interpreted somewhat loosely in terms of probability—e.g., the chance that a randomly selected letter from a manuscript will be P is 2.2 percent. Since the percent frequency at which a V appears is listed as 1 percent, the likelihood that a randomly selected letter will be V is 1 percent. This is again an example of basing probability statements on "the way things are" as opposed to a theoretical argument.

Percentage frequency of letters of the alphabet							
Letter	%f	Letter	%f	Letter	%f	Letter	%f
A	8.1	H	5.5	O	8.0	U	3.0
B	1.4	I	7.0	P	2.2	V	1.0
C	3.1	J	0.2	Q	0.2	W	1.7
D	3.8	K	0.5	R	6.3	X	0.3
E	12.6	L	3.7	S	6.6	Y	1.6
F	2.6	M	2.5	T	9.3	Z	0.1
G	1.5	N	7.2				

There are inherent dangers in drawing conclusions from statistical evidence. The conditions might have changed since the data were collected, or the data may not have been an accurate measure of the true conditions. Yet at times this is the most practical solution available, and so experimenters proceed cautiously to gather unbiased, accurate reflections of the way things appear in the real world.

Collecting the desired data for each of the two examples above—the survival chances of newborn infants, and the percentage frequency of appearance of the alphabetical letters—can be thought of as an "experiment." Someone had to decide exactly what specific information was desired, and then how to gather accurate information of the right sort. Another type of experiment, which somehow seems more personal, is described below. Notice that the eventual outcome is similar to the previous two examples: the ultimate goal is to base probability statements on real-world evidence rather than theory.

PITCHING PENNIES AT THE FAIR Did you ever waste money at a fair, trying to pitch pennies so they would land on a saucer? If you were lucky—or perhaps unlucky, depending on the interpretation of the prize—you won the animal (a chick, or goldfish, or parakeet) underneath the saucer.

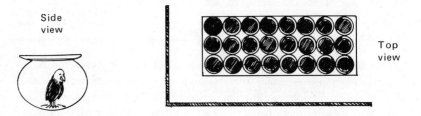

Side view

Top view

Using the top view shown above, it would be fairly easy to develop a mathematical model for the probability that your penny would land on a saucer. Find the total area of the saucers, and the total area of the rectangle surrounding them. The ratio of the two would be the theoretical chance of landing on a saucer. However, penny pitching being the inexact art that it is, this model would probably not be worth very much. When a penny lands on a smooth surface, it's quite likely to roll or slide.

The best way to verify or reject the theoretical model for penny pitching would be to set up some saucers in the position shown, and actually throw pennies and record data. If the data seem to support the theoretical model that doesn't consider rolls or slides of the coin after it lands, then the model could be accepted. However, if the data do not support the model, you would be better off to cast aside the theory in favor of the real-world chances based on the data alone. Try this before your next visit to a fair, and see how accurately your probability of winning can be predicted beforehand!

The example above shows how to collect data through a *simulation* of a problem situation. You try to represent the conditions in the booth at the fair as accurately as possible, using the saucers you can get your hands on at home. Hopefully your simulation would be close enough to the actual situation so the evidence you collect would transfer to the real world. The example to follow shows still another way of representing "the way things are" while you sit at home, collecting data for later probability statements, in the comfort of your own world.

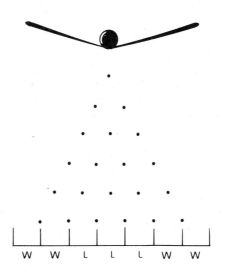

AN EARLY PINBALL MACHINE

One of the first gambling devices is shown to the right. The ball is released when the two metal bars open; the contraption is designed so that the ball has a 50-50 chance of bouncing either to the right or left when it hits each nail. The ball will hit exactly six nails on its trip down to one of the bottom slots, one nail at each horizontal level. Since there are seven slots where the ball can wind up, and four of the seven are marked with a W for a "win," it seems that on any given drop of the ball, the chances of winning would be 4/7. The three middle slots are labeled L to represent a "loss," and the temptation is to think that your chance of losing on any given drop would be 3/7.

But you might be suspicious of this line of reasoning, since the device is intended for gambling, and so automatically conjures up feelings of distrust! And perhaps your reservations would be justified. But how can these probabilities be put to the test? You wouldn't normally have access to such a device, so collecting data by actually running through the experiment a large number of times is out of the question. Can you simulate the experiment and collect the data more easily?

The path that a ball might take on its descent can be modeled by using six coins, one for each level that the ball must strike. A tail on a coin means "to the left," while a head means "to the right." The sketches below represent three paths for the ball to take. The two on the top have only partial paths shown, while the lower picture shows a complete descent.

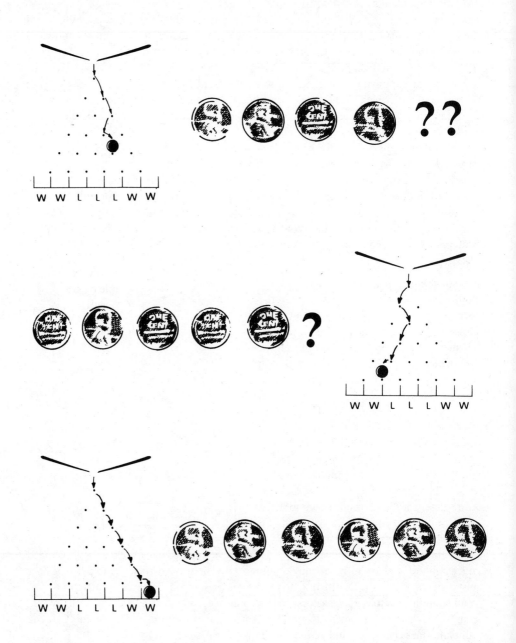

Notice that the coins shown beside each path above could be used to simulate the path of the ball. Just to test your understanding of this process, find the path that corresponds to "heads, tails, tails, heads, heads, heads."

Did you get the path shown to the right? If so, you could then proceed with the experiment. You'd want to take six coins, mark them somehow so you could keep them lined up in order, and then flip the group of coins a large number of times. After each turn, you'd mark where the ball wound up, and keep a tally of the results.

We tried this experiment, flipping the group of coins 100 times. Our result is shown below. Yours would differ somewhat, but would probably be fairly close. If the entire class put its results together, the data would be even more realistic of what could be

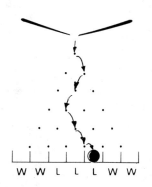

| 9 | 12 | 24 | 23 | 2 | 1 | 15 | 6 |
 W W L   L   L W W

expected. An approximate probability statement of the chances of winning could then be made, based on this evidence. Once again, you would have an example of basing a chance statement on "the way things are" rather than on a theoretical analysis.

The problems in the next set will give you a chance to explore the examples already given even further. You'll also have the chance to design a couple of your own experiments, with an eye to making a probability statement with your results.

SET 5

1. Look back at the chart labeled "Chances of survival of newborn infants in the United States." For your own category (white male, white female, nonwhite male, or nonwhite female):

   a. Write in a complete sentence the chance that a newborn infant will live to retirement age (60).

   b. Write a complementary statement for that in (a) above.

   c. Decide in which 10-year span a newborn infant is most likely to die.

2. Pictured on the next page is part of the game board for SCRABBLE. Notice the "letter distribution" chart provided to the left of the game board.
   Does this distribution appear to be based on the same information as that given earlier in the text, labeled "Percentage frequency of letters of the alphabet"? Explain how you came to your conclusion.

3. Consider for a moment the gambling device in the section immediately before this set of problems. Do two things with this problem situation:

   a. Use the data given for 100 trials to write a probability statement for the chance of winning on any given drop of the ball.

   b. Investigate the possibility that the *order* of getting heads or tails on the six coins might not be that important (i.e., perhaps four heads and two tails would have the same final result, no matter in what sequence the heads and tails appeared).

4. Each of the three situations below is meant to give you a chance to make an *initial prediction* based on a theoretical understanding of the problem, and then to *design an experiment* to gather evidence on "the way things are." Both of these parts are crucial for each of the three situations. Choose one of the three to study in depth.

   a. A well-known football coach told his team's captain always to let the captain of the other team call the flip of the coin at the start of each game, even when he was the visiting captain. The coach's reasoning was:

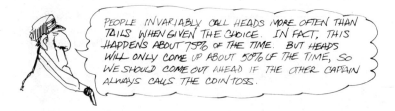

   Does this sound reasonable to you? Design an experiment to test your conjecture.

b. In 1979, McDonald's had a large advertisement campaign for their "Happy Meals." The big attraction for kids was the Space Raider that came "free" with each Happy Meal. Assuming that the Space Raiders were randomly distributed, about how many Happy Meals would a child have to purchase to have the entire collection of eight different creatures?

c. A magician had three cards, identical in physical appearance except for the coloring. One card was black on both sides, one card was white on both sides, while the third was white on one side and black on the other side. The magician would draw a card at random from a hat, lay it on a table without seeing the other side, say some magic words, and then turn the card over. He claimed he had magical powers, since about two-thirds of the time the underside matched the top in color, even though this should logically happen only about ½ the time.

His argument that it should happen only half the time went something like this:

Did the magician really have magical powers, or was he doing nothing out of the ordinary, relying only on an apparent paradox?

REFERENCES

Mullet, G. M. "Watch the red, not the black." *The Mathematics Teacher* 73 (1980): 349–53.

Shulte, A. P., ed. *Teaching Statistics and Probability.* 1981 Yearbook. Reston, Va.: National Council of Teachers of Mathematics, 1981.

6

THERE MUST BE A BETTER WAY

Source: Courtesy of Mathematical Alternatives, Inc.

Mankind has been calling on manipulative devices to aid in arithmetic computation for thousands of years. However, advances made in this area over the past 25 to 30 years are truly astounding. Computations are now performed literally at the speed of light on hand-held calculators by children so young they don't even know what the speed of light is!

The abacus is probably the best known of the early devices for calculating with numbers and is still used today in some cultures. Commercial calculators have been around since the early 1900s, but they were mostly heavy, cumbersome instruments with electric motors turning the synchronized wheels, doing the work that fingers had done previously. A hand-powered portable model (the slide rule) became the identification mark of the scientist in the 1950s. While all these tools did make "number crunching" a lot easier, they are eons behind what is on the market today for less than $50.

In the late 1940s, electronic technology had advanced to the stage where the first "modern" computer could be built. Although extremely large and slow, compared to what has since evolved, this computer and its relatives spawned mankind's first successful attempts to escape the earth's atmosphere. For the first time, the millions of instantaneous calculations required to launch and control a massive rocket in flight were in our grasp.

As mankind began to explore beyond the confines of earth, the space program began to reciprocate the favors and patience bestowed on it. The practical concern for making small, lightweight components that were reliable under extreme conditions of heat and pressure opened up the world of the "minis," producing common everyday items used around the house. Besides calculators and computers, such things as surgical instruments, communication devices, frying pans, clothing, and even automobile ignition systems, received a "new look" from these efforts. The technological boom created by the money pumped into the space program in the sixties has profited us immensely.

Not only has the size of calculators and computers diminished in the past 20 years, but so has the price, which is significant in light of the rise of inflation in recent years. And technology is still advancing in electronic wizardry, so much so that no one can accurately predict if and when it might "peak out." One fact is clear, however, the minicalculator and microcomputer will play increasingly important roles in our lives!

Calculator versus Computer

Generally speaking, a calculator is capable of performing the four basic arithmetic operations with decimals—addition, subtraction, multiplication, and division—as fast as the operator can punch in the numbers. Usually when a computer is mentioned, we are thinking of a device capable of thousands, even millions of computations per second, all performed on data that has been stored in the memory of the computer, with the computations being directed by a sequence of instructions also stored in the computer. Compared to this standard, a hand-held calculator is rather primitive.

In the case of most hand-held calculators, the operating rules represent a relatively simple language. To have a calculator find the answer to "3 + 7," we merely push certain buttons: $\boxed{3}$ then $\boxed{+}$, then $\boxed{7}$, and then $\boxed{=}$. The "language" represented here is one with which most people have been familiar since elementary school. The "grammer" of the language establishes the proper order of pushing the various buttons. The buttons complete electrical circuits inside the calculator, and in so doing instruct the calculator to do such things as:

$\boxed{3}$ means "enter 3"

$\boxed{+}$ means "add the number to follow'

$\boxed{7}$ means "enter 7"

$\boxed{=}$ means "give the answer"

So in a real sense, you already know a primitive computer language. Hand-held calculators typically respond only to variations of this simple language.

More sophisticated computers have the capacity to "read" and follow instructions expressed in many different languages. A few of the more widely known computer languages are FORTRAN (from FORmula TRANslation), COBOL (COmmon Business Oriented Language), APL (A Programming Language), and BASIC (Beginner's All-purpose Symbolic Instructional Code). Most of the several hundred computer languages were developed to handle a large class of problems related to a particular segment of society (e.g., scientific or business concerns). BASIC was derived from FORTRAN as a simpler language for the beginner who merely wants to dabble in the field. An introduction to this language is included in this chapter.

Use a hand-held calculator as you work the next problem set. Some of the exercises are meant to provide you with practice in using this tool, while others are problem solving in nature.

SET 1

1. Time yourself doing the problem below by hand:

    $$2347 + 618 + 7251 - 536 + 1809 - 921 = ?$$

    Now do the same problem again, but on a calculator. Which takes less time? If you had obtained two different answers, which one would you have more faith in?

2. Can you do the problem $8 \overline{)16553847713}$ on your calculator in one step? If not, can you devise a way to do it on your calculator in several steps?

3. Try the problem "$45 \div 0$" on your calculator. What happens?

4. The $\boxed{8}$ on your calculator is broken. Describe a way to do the problem 38476 − 247 on the calculator in its state of disrepair.

5. The $\boxed{\times}$ on your calculator is not working correctly. How can you proceed to do the problem 675 × 237 anyway?

6. The $\boxed{\div}$ key on your calculator isn't working—describe a method of solving 876984 ÷ 239 using your calculator "as is."

7. Write down a method of obtaining 1000 in the display of your calculator, using only the $\boxed{3}$, $\boxed{+}$, $\boxed{-}$, $\boxed{\times}$, $\boxed{\div}$, and $\boxed{=}$ keys.

8. Consider the "cross-number" puzzle below. Do the calculations on a calculator and record the digits *precisely* as they appear on the display of the calculator. When finished, turn the puzzle upside down and you should have a cross*word* puzzle completed also!

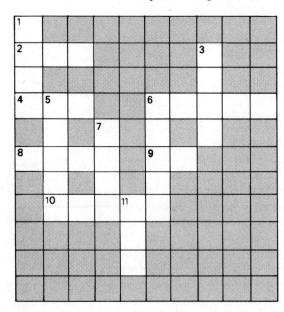

DOWN:
(1) (28 × 126) − 21 = ?
(3) ? + 716 = 4220
(5) 6521 + 9963 − 12321 + 42896 + 30286 = ?
(6) (−364 × −265) + −41282 = ?
(7) Average of 4728, 9630, 7465, & 725
(11) $\sqrt{100489}$ = ?

ACROSS:
(2) 6000 − ? = 5486
(4) 280644 ÷ (300 + 64) = ?
(6) $35^3 + 100^2 + 170$ = ?
(8) $3 \times 10^3 + 3 \times 10^2 + 7 \times 10^1 + 6 \times 10^0$ = ?
(9) Age the second year as a teenager
(10) $\{[(238 \div 14) + 20] \times 1560\} + 18$ = ?

9. Practice the following computations on your calculator, paying particular attention to the paired expressions and comparing answers when through. Correct answers are indicated to the right of each exercise.

   a. (82 ÷ 4) x 12   (ans: 246)
      82 ÷ (4 x 12)   (ans: $1.708\overline{3}$)
   b. (193 - 73) ÷ 2   (ans: 60)
      193 - (73 ÷ 2)   (ans: 156.5)
   c. (7 x 8) - (4.2 x 8) + (17 ÷ 3)   (ans: $28.0\overline{6}$)
      {7 x (8 - 4.2) x 8 + 17} ÷ 3   (ans: 76.6)

10. First do the two problems below using paper and pencil.

    a. (1 ÷ 3) x 3 = ?

    b. $12\frac{2}{3}$ x $3\frac{2}{5}$ = ?

    Now do them on your calculator. Do the answers agree?

11. Old books frequently offer pleasant excursions into the "dream world" of lower prices and obsolete expressions. Solve the two problems below (Winslow, 1901, pp. 110, 175):

    a. William Brown, in account with Edward Davis, owes a balance of $12.42, according to a statement that has been sent him. Dec. 8, 1899, he buys on credit 12 yd. cloth @ 42½¢ and 4 pr. socks @ 22¢. On Dec. 17 he buys 1 suit clothes @ $14.50. On Jan. 1, 1900, he pays the amount in full. Make out a proper bill and receipt it.

    b. One tenth of a cent is called a mill and is written in the third place from the decimal point.

    > $ .003 is three mills
    > .015 is one cent, five mills, or fifteen mills
    > .234 is twenty-three cents 4 mills

    What is $.043?

12. Your great grandfather invested $805.437 in the bank in 1890, drawing 8% interest per year. After ten years, if the interest had not been removed at any time, how much would be in the account? How much did the 7-mill portion add to the savings?

13. Someone has suggested that manufacturers begin putting a red light on their calculators that would come on when the batteries were discharged. This would alert you to buy some new ones before you needed the calculator late one night when the stores were closed. What do you think of this idea?

Flowcharts

Neophytes in mathematics sometimes have the mistaken impression that calculators and computers can solve problems. Of course, neither a calculator nor a computer can think; they merely do the arithmetic computations we direct them to do. So the decision of what steps to take to resolve a problem situation is still ours. *We* must decide what to do before we begin pushing buttons, if we expect to obtain a correct numerical answer.

In a previous chapter, mention was made of the fact that for some mathematics problems, the appropriate form for an answer is a *process,* not a number. This is quite often the case in working with calculators and computers; the answer to a problem might be a set of steps to follow to solve a host of problems, all of a similar nature. Just as it was appropriate to spend some time previously discussing meaningful ways to report numerical results, it would now behoove us to consider methods of communicating processes to other interested parties.

Flowcharting is a useful technique for reporting an answer that is a set of instructions. It is particularly useful when the solution requires a large number of steps and has some "loose ends" that need to be kept separated. A flowchart is a chart (and thus reaches visual learners) that "flows" from one step to the next by arrows. In this section, we will view flowcharting as a method of organizing and communicating special kinds of problems.

The two examples below show simple flowcharts. The first shows how to find the future price of a $100 item in two years, at an inflation rate of 9 percent. The second involves a nonmathematical situation that you'll undoubtedly meet at some inconvenient point in life:

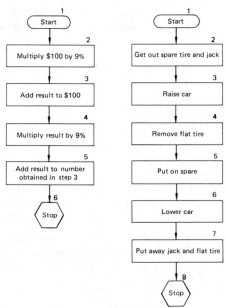

The second example gives an oversimplified version of how to change a flat tire. Notice the symbolism used in these two flowcharts: "Start" is enclosed in an ellipse, directives or commands in rectangles, and "stop" in a hexagon. Although this symbolism will be standard throughout this chapter, you can expect to see variations from place to place. Note also that each step is numbered for easy reference.

Unfortunately, changing a flat tire is not quite as simple a chore as the flowchart indicates. For one thing, "raising the car" might not get it quite high enough the first time around, so we'd have to raise it some more. This typifies the fact that resolving real-life problems frequently involves making decisions as we proceed along the solution path.

DECISIONS, DECISIONS

For simplicity, only questions that have yes or no answers will be considered. These decisions can be included in flowcharts as long as we have *two* arrows leaving the symbol: an arrow is needed to tell what to do if the answer is "yes," and the same if the answer is "no." The traditional flowchart symbol for a decision is a diamond, as shown to the right. We'll go along with this also, merely pointing out that a triangle would probably be more appropriate.

A decision has been inserted into the "flat tire" situation in the example below to accommodate this new factor. Notice that here we're allowing three tries to get the car "high enough." In most cases, this would be adequate.

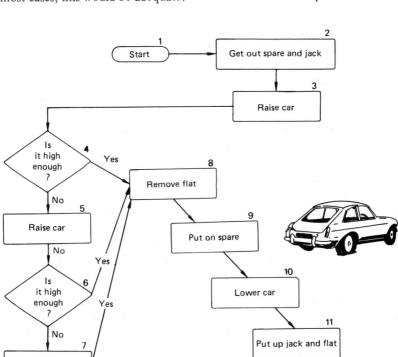

Notice that, with the way this flowchart has been numbered, we proceed with each step to the next numbered step unless we've answered "yes" to a decision.

Since automobiles cause about as much grief as anything else we come in daily contact with, another example from this area might not be out of place. Have you ever had the following experience? You got in your car and turned on the ignition, and absolutely nothing happened? If so, you might have found directions like these in your owner's manual:

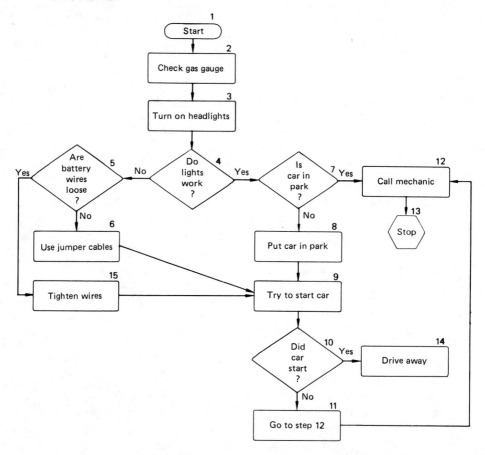

Even more painful than our experiences with automobiles are the once-a-year trials and tribulations associated with filing an income tax return with the Internal Revenue Service. Perhaps if the government would put the instructions into a flowchart, we could follow them much more easily. A simplified version, yet typical through the years, for the calculations of income tax for a single person, is illustrated in the flowchart to follow:

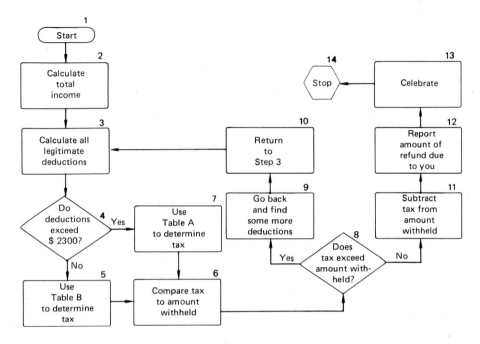

The flowchart above is only slightly facetious: many people do keep trying to find more deductions until they finally have enough to get a refund.

Illustrated in the example above is an extremely important concept in flowcharting, that of returning to a previous step in the instructions. This process is so vital that it warrants a special definition:

> Going back to a previous step in a flowchart is called "looping."

You can think of a "loop" in a flowchart as a circular path: the only way to get out of the loop above (theoretically) is to be able to answer "no" at step 8.

If you've got some time on your hands, and the facilities and ingredients, try making some brownies according to the flowchart on the next page. If you don't have time, at least look closely at the *loop* in the set of directions. Notice that steps 6, 7, 8, and 9 constitute the loop; the only way out of this circular path is when the oven is at least 350°F.

Before starting the next problem set, look back over the flowcharts used as examples in this section and observe a very important point: in some cases, the order of some of the steps could be switched without affecting the final outcome (above, steps 2, 3, and 4 could be reordered). Yet in other instances, the steps must be in the order shown if the solution is to be correct. The implication of this variable is that some times there's only one way to get the correct path, and in other cases there may be several alternate routes from which to choose.

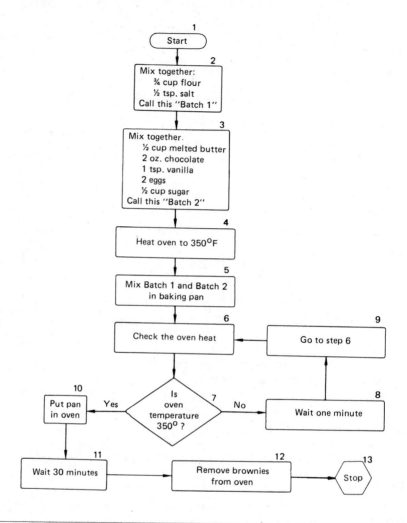

SET 2 1. Write a flowchart using the commands below, after first placing them in the correct order. The procedure is for reporting the letter grade earned from five test scores.

> Record on the report card the letter grade corresponding to the average of the scores.
>
> Divide the sum by five.
>
> Find the letter grade that corresponds to the average.
>
> Add the five test scores of the student.
>
> Decide what numerical range will correspond to the grades A, B, C, D, F.
>
> Stop.
>
> Start.

2. Construct a flowchart for buying a $50 pair of shoes, using the set of directions listed below. You will have one "decision" in it.

   Stop.

   Buy the shoes.

   Wait till next month.

   Calculate all essential expenses for this month.

   Start.

   Find current balance of your checkbook.

   Do you have at least $50 to spare?

   Subtract essential expenses from balance.

3. There are several things wrong with the flowchart below, written to plan a two-week trip to the beach. Rewrite the entire flowchart in a reasonable sequence of steps, adding any you feel necessary for the logical flow of events.

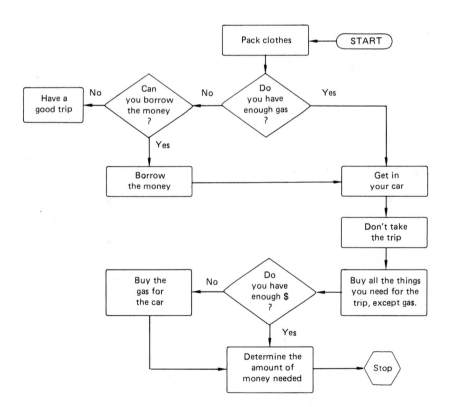

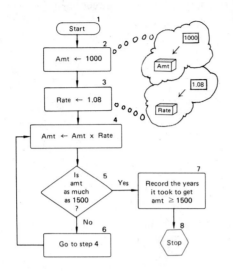

4. The flowchart to the left was designed to help you find the number of years you must leave $1000 in a savings account, at 8% simple interest, until you have $1500 in the account. The interest is added back into the account each time it's calculated at the end of the year.

Steps 2, 3, and 4 can be interpreted as "store the number to the right of the arrow, in the box to the left of the arrow." There are two boxes mentioned, one labeled "Amt" and the other labeled "Rate." The number in "Rate" stays the same, but the number in "Amt" changes each time you get back to step 4.

Go through this flowchart step by step on your calculator, and record your results in the table below.

Value in Amt.	Number of Years	Times Through Loop
1000	0	0
1080	1	1
1164.40	2	2
———	—	—
———	—	—
.	.	.
.	.	.
.	.	.
1586.87	?	?

5. Return to the "flat-tire" example of a flowchart included in the text preceding this problem set. Rewrite the flowchart, using a loop, so that you could check on the height of the car as many times as necessary to get it high enough off the ground. That is, include a "loop" to check this.

6. Write your own flowchart for one of the activities below. Practice using loops, but try to keep your chart relatively simple for this exercise—perhaps 15 to 25 steps in length.

You're about to leave for a job interview. The competition will be tough, but you really want the job. Prepare for it.

Bake a lemon meringue pie.

Plan a way to meet that stranger in class that has been catching your eye.

*"There's one advantage in our inability to make decisions—
we never make the wrong ones."*

Source: Courtesy of Holden-Day.

Flowcharts that Count

Look back for a moment at the flowchart for Problem 4 of Set 2. As you set about finding the answer to this problem, you probably had to run through the steps several times before you realized that the flowchart was missing something needed to find the final answer. As it stands now, the flowchart doesn't keep up with the number of years automatically; that part is left up to the individual. It would be beneficial if this chore could somehow be included within the flowchart, so that anyone following the flowchart verbatim would have the correct answer without having to do any work other than directed.

But how can we do this? We could start by labeling a box "years," the same as we did with "Amt" and "Rate." The essence of the problem is that each time another year passes, the amount in the bank is increased. Or, every time we get a new number in the box labeled "Amt," we'd want to increase the number in "years" by one. So we could try putting these two steps close together in the flowchart, so that each time one of them changed, the other would be right behind.

The flowchart below is a modified version of the one from Problem 4. Carefully go through the flowchart and step-by-step explanation offered to the right:

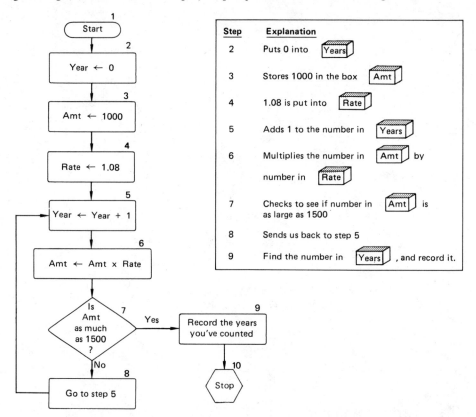

Notice that zero was initially put into the "years" container, since we need that box to be zero at the same time that the money in the account is $1000.

The flowchart above is said to have a "counter" in it, because it counts each time through the loop. The position of the counter is not sacred, of course, and the number we put into it initially, and how we change that number each time through the loop are all variables that must be made to fit the problem being described with the flowchart.

Before going ahead, try your hand at writing a flowchart with a counter in it for this problem:

> *The world population increases each year about 2 percent over what it was the year before. Assume the world population is 4½ billion. What will it be when you reach 60 years of age?*

Important questions to consider are such things as: (a) what will the counter represent, (b) what number will go into the counter first, (c) how will the counter be changed each time through the loop, (d) how will the world population be increased each year mathematically over the previous year, and (e) when should the flowchart get out of the loop? After you've given the exercise a good chance, compare your flowchart to the one below:

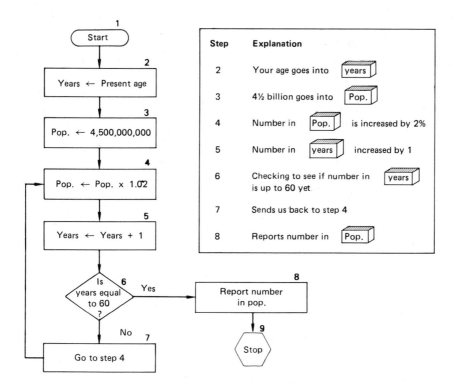

Notice several differences between this flowchart and the previous one, even though both have counters:

The number in the counter initially is different.

Changing the value in the counter comes at slightly different points in the set of steps.

In the first example, the loop is halted and then the counter checked for the answer to the problem, while in the second, the loop is halted when the counter reaches a certain point, and the answer is found in another box.

Variations such as these are common, and represent the essence of the difference in the problems given to be described with the flowchart. There are no "standard rules" for constructing such flowcharts; each must be done thoughtfully with respect to the problem at hand.

Given that you are not yet very close to 60, you would have quite a task following the last flowchart through to its conclusion, even with a calculator. And before you started on such a chore, you would probably want to ensure that, in fact, your flowchart would give you the correct answer in the end, assuming you followed all the steps correctly. So the question arises—how can a flowchart be "checked out" to ensure that it's a valid procedure?

TRACING A technique found very useful in checking out flowcharts is called "tracing," and involves nothing more than running through the steps of the flowchart with simple numbers. The method behind this madness is that you can determine what the end result should be for these simple numbers without the flowchart. If the steps in the flowchart give you the answer you know is correct for the simplified version, you assume that the logic of the flowchart is valid.

In the previous example involving world population, we could certainly calculate what the population should be in two years, under the conditions of the problem, without looking at the flowchart: Population in 2 years would be 4½ billion x 1.02 x 1.02 = 4.68 billion.

To *trace* the flowchart, then, we could start "years" off at zero, and stop it when the number in "years" is at two. If the logic is valid, the number in "Pop" at step 9 should be 4.68 billion. Complete the table below, which is an example of a *trace* for this particular flowchart:

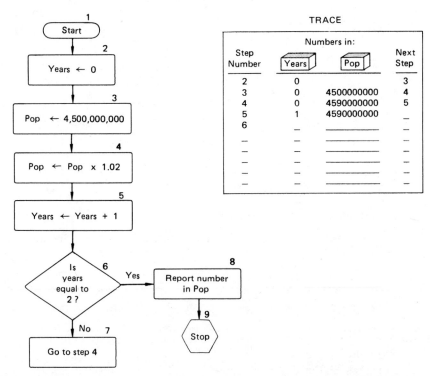

What did you get in "Pop" at the last step above? Do you think the original flowchart is a correct set of instructions for finding the world population at age 60?

Whether or not to use a *trace* to check the logic in a flowchart depends, of course, on how complex the flowchart itself is. Generally speaking, a trace is necessary only when there are several variables that we have to keep track of simultaneously, and when there's a decision step involved that might lead us astray. Certainly the trace doesn't have to take the form of the one above to be legitimate; an informal "running through the steps" might suffice for many situations. Each step of the flowchart doesn't have to be recorded in a trace either, even though they were above: only the ones in which the numbers inside the boxes change are essential for checking out the steps of the flowchart.

SUMMARY

In this section, we've covered two main ideas in flowcharting: *counting* the number of times through a loop, and *tracing* a flowchart to ensure that the logic is correct and will solve the problem at hand. In the set of problems to follow, you'll want to do both of these things at times, but at other times you'll probably reject them as inappropriate. Remember that the flowchart must be constructed to fit the problem; there are no hard and fast rules that will ensure success for all problem situations.

SET 3

1. Take your pulse when you're at rest, and assume that this is the average number of times your heart has beat each minute you've been alive. Plug this number and your age into the appropriate place below, and then use a calculator to go through the flowchart to determine about how many times your heart had contracted by your last birthday.

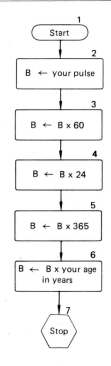

2. The following flowchart provides a routine for converting temperatures from Fahrenheit (F) into Celsius (C). Use the flowchart to convert each of the following Fahrenheit readings into Celsius: 32°F, 100°F, 212°F, 131°F, 293°F, −15°F.

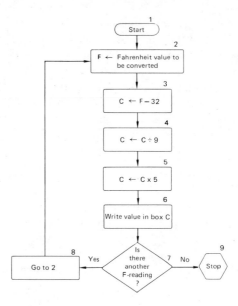

3. Construct a flowchart that gives instructions for converting Celsius readings to Fahrenheit numbers. Follow the pattern for the previous flowchart, but note that the arithmetic will be different in the assignment steps. Use your flowchart to change each of the following Celsius readings to corresponding Fahrenheit values: 20°C, 10°C, 0°C, –10°C, –40°C.

4. Write flowcharts, and then follow them on your calculator, to solve each of the problems below. If appropriate, use a loop or a trace to check out the logic.

   a. You borrow a nickel from a friend and say you'll pay her back tomorrow, with a penny interest. She jokingly says, "Okay, but each day the interest doubles." You forget to pay for a whole month; how much money do you owe your friend?

   b. You want to have $50,000 in a savings account when you retire at age 65. You know of a bank that will give you 8% interest on a long-term savings account. How much do you need to invest now for this to come true?

   c. Estimate your starting salary as a teacher, and how many months you'll be employed (9½ or 10 usually). Assume that, as is the case in many districts, you're paid only during the months you work. For most teachers, this means a lean summer without proper planning. Devise a plan so that you'll have the same amount of money during the summer months as during the months you receive a paycheck.

   d. There are approximately 330 million cubic miles of ocean water, each cubic mile containing about 38 pounds of gold. From current money exchange quotations in a newspaper, find out the going rate for an ounce of gold, and use this to find the total worth of all the gold in the oceans.

Introduction to Programming

When people use a computer to help them solve a problem, they typically follow a sequence of steps that has proved to be both effective and efficient. The first of these steps is certainly the hardest, because that's where the real problem solving comes in. But luckily, that's the step we've already discussed—analyzing a problem and writing a flowchart to resolve it. These steps are given to you below:

   Analyzing the problem, writing and checking the logic of a flowchart

   Translating the flowchart into a language the computer understands

   Typing in the program and data so the computer will have both the set of directions and the necessary numbers

   Having the computer run the program it's been given

   Accepting the answer the computer gives

The second step above is the one we'll concentrate on for the moment. The language we'll learn is BASIC, and we'll use the philosophy that the best way to learn a foreign language is with only a few instructions and a lot of practice and experimentation. BASIC is a straightforward computer language in that its special codes are very intuitive. However, this does not lessen the requirement for precision. For the computer to understand even this simple language, you must be sure to translate your flowchart into BASIC in a very exacting manner.

Before getting into BASIC, however, we might get a "total view" of the five steps mentioned previously, particularly because of their relationship to this thing we call a "computer." The first two steps can be done sitting in the student union while you enjoy a cup of coffee; they do not require direct interaction with a computer. But the last three are not like this; they come directly from the basic nature of a computer itself.

A computer has three main components:

The "input" component refers to putting into the computer both the set of directions for the computer to follow and any data necessary for it to perform its job. This can be done in several ways—using prepunched cards, or calling up a program and data

stored in the computer's memory bank, or sitting down and typing in the program and data by hand. This last method is the one preferred for this course due to its simplicity. Inputting the program and data is handled (usually) at a *terminal*, which is connected electronically with the computer itself.

"Processing" refers to the computer running through the program, using the data it has. This is all handled electronically (the program you typed in and the data become little spurts of electricity) and so is almost error free. The last component—output— involves getting the computer's answer back out so you can understand it. This last part, which is also handled at the *terminal,* is where the electrical spurts are changed back into words and numbers.

Of the five-step process on the previous page, the first step—writing a flowchart for the problem after deciding what to do—is the most difficult. Coding the flowchart into BASIC becomes almost routine after you practice a little, and typing the BASIC program into the computer is not hard if you're careful and not easily frustrated by events beyond your control (like the computer "going down" for maintenance while you're hard at work). Having the computer run the program and give you the answer involves no work at all on your part, so you're essentially through when you've finished the third step; you can sit back and wait for the computer to do the last two!

BASIC   As you will soon see, most of the symbolism we've been using with flowcharts translates very easily into BASIC. The memory units inside the computer can still be thought of as boxes that hold numbers, and these are labeled with letters of the alphabet instead of word names.

The example below has the computer print the numbers from 7 to 11. Study all three of the parts very carefully:

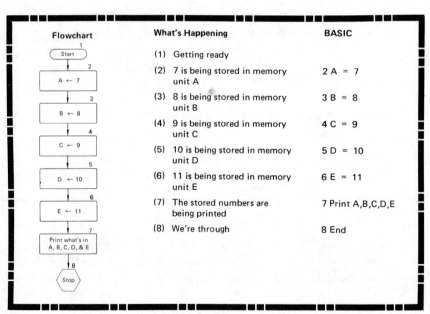

Notice that there is no BASIC statement for step 1 in the flowchart (which corresponds to "logging on" at the terminal), but there *is* a BASIC statement that must be used to STOP the program. You'll also notice that each BASIC step must be numbered, with no period after the number.

The PRINT statement is interesting in that slight modifications produce different answers. For example, if the statement above had read: 7 Print "A, B, C, D, E," those five alphabetical letters would have been printed on the screen as the answer. Or, to have the program print letters of the alphabet, or words or numbers that you happen to want on the screen, merely use the PRINT statement with quotation marks around what you want to see.

This program has been slightly modified below, to give the *sum* of these five numbers. Note that an extra step has been added to store the sum, and that the PRINT statement has been written so that the answer will appear in a complete sentence.

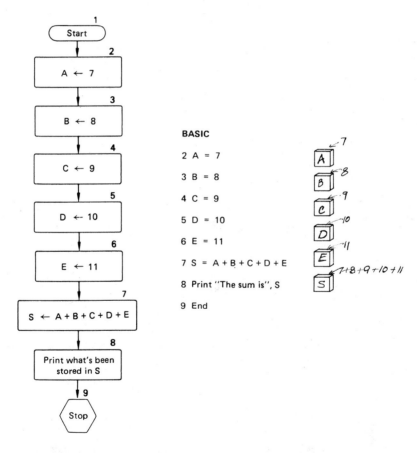

Still another slight modification would give the *average* of these five numbers. One way to handle this is shown below:

226 MATHEMATICS FOR TEACHERS

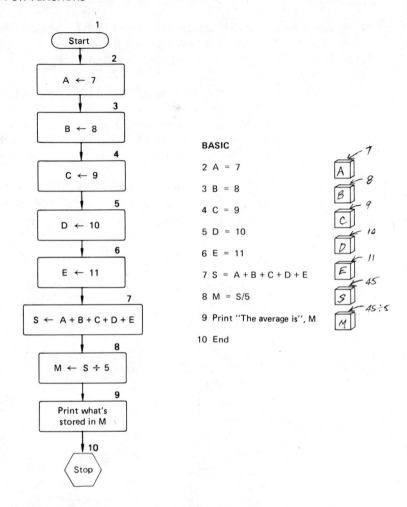

As you can see from the examples above, BASIC is a very straightforward, intuitive language. The major symbols that you'll need in writing programs from this text are given below.

BASIC	Meaning	BASIC	Meaning
+	add	<	less than
−	subtract	>	greater than
*	multiplied by	< >	is not equal to
/	divided by	< =	less than or equal to
**	exponent (to the power of)	> =	greater than or equal to

THE LAST THREE STEPS

Your instructor will have to give you the specific instructions for "logging on" to the computer you'll be using; they vary somewhat from one another. Once you've finished that minor chore, however, you're ready to start keying in your program. A few hints might be in order.

Some of the BASIC statements you've seen in the examples must be typed in exactly as you've seen them, even down to such minute details as having the proper spacing between symbols. For other commands, this doesn't seem to matter on some computers. The watchword here is "experimentation" to see what will and will not work on the computer you're using. If you type in a line and hit the return key, and the computer prints back something like "illegal statement," your line of the program has not been accepted for some reason—either electrical interference or failure to type it in exactly as it should be done. Correcting an error is easy; you just retype the line, wiping out anything that was on that line of the program previously. If you try to get a line in unsuccessfully a few times, don't get frustrated—get help! We all need it at times.

Once you've finished keying the program into the computer, it's a good idea to get the computer to show you what it has accepted as your program. Type in LIST (with no line number, since it's not a part of your actual program), and the computer will reprint your program with all of the lines as they have been accepted. If you see no errors, you're ready to have the computer do its job. Type in RUN, again without a line number. You'll hear some electrons moving through the wires, and then the computer will give you the results of your program.

If the answer is a very large or small number, the computer will print it out using "scientific notation." You might see such things as $2.0976118 \text{ E}^{+}10$ or $5.210000 \text{ E}^{-}9$. The number after E tells you how many places to move the decimal point, and + means "move it to the right" while − means "move it to the left."

After you're through, you need to "log off" the computer to let it know that you have no more programs to run. Keying in "BYE" will accomplish this chore on most computers that accept BASIC, so you can wrap up your work in an intuitive fashion also.

In the next problem set, you'll get the chance to use all five of the previously mentioned steps. The number of problems has been kept small so that you'll be able to extend the concepts while you're working on the computer. Remember the philosophy of learning this foreign language called BASIC: a few instructions and a lot of practice will give you a "working knowledge" of the code.

SET 4

1. As part of a project in a science class, a sixth-grader found some information about how a hummingbird flies, and then made up the following problem. Some of his scratch work is shown below the problem, which tells how he thought the problem should be solved.

> A hummingbird beats its wings **60** times per second, and can fly a non-stop distance of **500** miles. Its top speed is **75** m.p.h. Using these figures, how many times would its wings beat on such a flight?
>
> FLYING 500 MILES at 75 m.p.h. MEANS FLYING FOR 6.67 HRS.
>
> ```
>      6.67
> 75 ) 500.0
>      450
>      500
>      450
>      ANSWER ⟶
> ```
>
> 60 BEATS/SECOND × 60 SECONDS/MIN × 60 MIN./HR × 6.67 HRS

Decide first if you agree with his solution method. If so, finish the steps missing from the computer program to do the calculations. If you disagree, modify the program below to fit the way *you* think it should be solved. In either case, run the program you write and find the answer to the question asked.

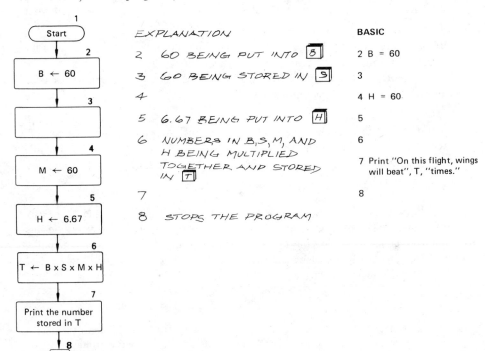

EXPLANATION

2 60 BEING PUT INTO [B]
3 60 BEING STORED IN [S]
4
5 6.67 BEING PUT INTO [H]
6 NUMBERS IN B, S, M, AND H BEING MULTIPLIED TOGETHER AND STORED IN [T]
7
8 STOPS THE PROGRAM

BASIC

2 B = 60
3
4 H = 60
5
6
7 Print "On this flight, wings will beat", T, "times."
8

2. The drawing to the right is the creation of Ms. Linda Hobbs, an elementary education major at the University of Wyoming in 1979. Her program consisted of 79 PRINT statements, and an END command.

Source: Courtesy Linda Hobbs, University of Wyoming, 1979.

Her 10th line looked like this:

    10 Print "XXXXXXX XX   XXX XXXXXXXXXXXXXXXXXXXXXXXX   XXXXX"

What did line 52 look like, in BASIC?

3. Write and then run programs to solve the problems below:
   a. Find the number of McDonald's hamburgers that have been sold, and measure or estimate the height of an average one. If all of the ones ever sold could be stacked up, how high would the stack be? (Be sure the computer gives a meaningful answer, i.e., the computer perhaps should convert inches to feet, feet to miles, etc.).

   b. About 135 billion stars make up our galaxy, the Milky Way. Each one is a potential sun like ours. A conservative estimate is that about 1 star in 200 would have an earthlike planet orbiting it; of these unique stars with earthlike planets, about 90% are located in the central cluster of the Milky Way and are thus subjected to such violent explosions and intense radiation levels that life could not get off to a decent start. Using the 10% left in the outskirts of the galaxy, calculate the number of planets in the Milky Way that *could* support life. (For your further speculation, scientists now feel that any planet that *can* support life, *will* do so at some time in its existence.)

   c. Modify your program above to calculate the number of planets in the known universe that could support life. Our galaxy is considered only an average one out of the 10 billion known galaxies!

Using a Loop

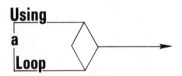

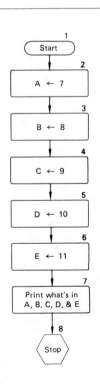

In the previous section, we considered several flowcharts like the one to the right. This flowchart was one method for getting the computer to print the whole numbers from 7 to 11. Notice that five different storage units (A, B, C, D, E) were used, one for each number we wanted to have printed out.

Possibly you wondered about the efficiency of such a process, and rightly so. Given a similar problem, but wanting *many* numbers printed, we'd have to use many storage spaces and spend a lot of time just typing the program in, unless we could come up with a better method. But one of the great things about computers is that we can turn over to them many of the routine tasks we have.

If we want to have the computer print the numbers from 7 to 100, we can have the computer follow the steps below (after the steps have been coded into BASIC). Look carefully at what is going on in this example.

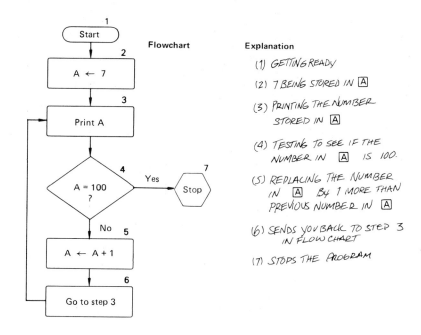

The essence of this flowchart is that 7 is entered into the computer originally, and printed. Then 1 is added, and the result printed. This process of adding 1, then printing, goes on until we finally have 100 in the memory unit, and then the program jumps over to the command to stop everything. The computer is now performing a first-grade task—counting and printing—using only one storage unit!

This concept of *looping* is a very important one in dealing with computers. It allows us to perform the same task over and over again till we get to a certain point and then stop automatically. Only two new BASIC commands need to be learned for performing a loop:

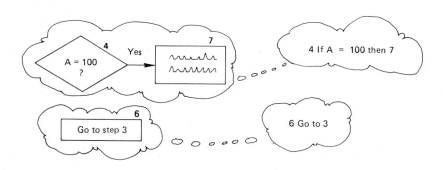

Notice again the intuitive nature of the BASIC statements. One point that should not be overlooked, however, is that the computer will automatically go from a *decision step* to the *next line* of the program if the decision is "no." If it is "yes," however, the flowchart sends us to another point in the sequence of steps.

Suppose we wanted to have the computer give us the *average* of those 94 numbers from 7 to 100. We'd need to insert another storage box (to hold the *sum* as the computer gets each new number). The flowchart to the right is an example of one such program—study the logic of the chart carefully, and see if you can code it into BASIC, just for practice.

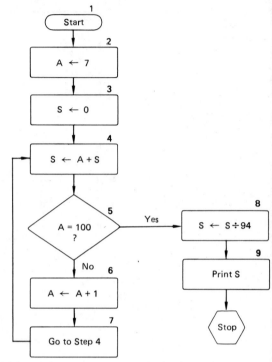

To satisfy ourselves that the flowchart above would really do its job, we could run a *trace* of the program. Instead of finding the average of the numbers from 7 to 100, though, let's modify the program so that it'll find the average of just the two numbers 7 and 8, which we know is 7½. So if the logic of the flowchart above is correct, the modified version should give us 7½ at step 9. Only two steps in the flowchart above need to be altered to run through the *trace*; can you find them for yourself, and then compare your decision to the version below:

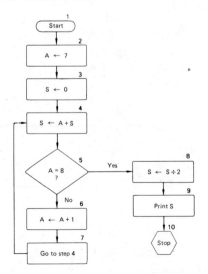

Step	A	S	Explanation
2	7	—	Loads 7 into A
3	7	0	Puts 0 into S
4	7	7	Loads 7 + 0 into S
5	7	7	Checks to see if A = 8; no
6	8	7	Adds 1 to number in A
7	8	7	Sends us back to step 4
4	8	15	Puts 8 + 7 into S
5	8	15	Checks to see if A = 8; yes
8	8	7½	Puts 15 ÷ 2 into S
9	8	7½	Prints what's in S (7½)
10	8	7½	Stops program

TRACE

Does the program work? Did the computer print out 7½ for the average of 7 and 8? If so, we can assume that the former, unmodified flowchart has the correct logic for finding the average of the numbers from 7 to 100. And certainly it's easy to imagine how to change the flowchart to perform such tasks as:

Print the *sum* of the first 1000 whole numbers.

Print the *average* of this sequence of numbers:
5, 10, 15, 20, ... , 635

Print the *integers* in this sequence:
-7, -49, -343, . . . , -5764801

You'll have a chance to do the modifications required for the problems above, in the next problem set.

THE INPUT COMMAND

One of the most useful BASIC statements comes into play in situations in which you have a good bit of data to get into the program, and you'd like to use only a few storage boxes so you don't have to type in such a long computer program. The INPUT statement gives this capability. Suppose, for example, that you wanted to write a program to find the average of any set of test scores you had, no matter how many there were. You could then save this program and use it over and over again, in any situation in which you needed the average.

Consider the flowchart and BASIC statements below. S is the place where we'll keep up with the *sum* of the numbers, C is a *counter* so the computer can keep up with how many numbers you put in, and X is the storage box that will hold the *test scores* as they come in.

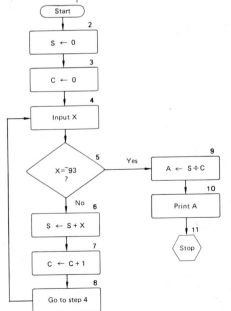
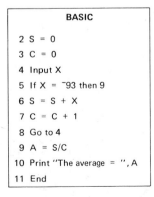

```
          BASIC
2  S = 0
3  C = 0
4  Input X
5  If X = -93 then 9
6  S = S + X
7  C = C + 1
8  Go to 4
9  A = S/C
10 Print "The average = ", A
11 End
```

When the computer is actually running this program, every time it gets to Step 4, a ? will appear on the screen, indicating the computer wants you to key in the next test score. When you have keyed in all the test scores, and ? appears, you type in the *dummy number* -93. This kicks the decision step to the "yes" answer, and finishes the calculations by dividing the *sum* by the number in the *counter,* and printing the result as the average.

Before leaving this example and going to the problem set below, you should take a group of four or five typical test scores and run through a *trace* of this program for finding the average. You would need four "boxes," labeled S, C, X, and A.

SUMMARY This section has been concerned with how to get the computer to run through a "looping procedure," a technique that takes advantage of the computer's capacity to perform routine functions tirelessly. Care must be taken when you use a loop inside a program however; it's quite easy to get the decision step in the wrong place, and have the computer go through the loop either one time too many or one time too few. If you're not extremely careful, the decision might never be "yes," in which case the computer goes into what's called an *infinite loop* and continues in the circular path until you realize something is wrong and stop the program manually. Don't let these factors discourage you. They merely indicate the need for a *trace* of your program almost every time you use a loop. And even if the computer does go through the loop a few times extra, or gets caught in an infinite loop, it won't tear up the machine (or, we hope, your confidence)!

SET 5

1. Modify the examples shown in the text, writing both a flowchart and BASIC, for the problems below:

    Print the *sum* of the first 1000 whole numbers.

    Print the *average* of this sequence: 5, 10, 15, 20, . . . , 635.

    Print the *integers* in this sequence: -7, -49, -343, . . . , -5764801.

    Run the programs.

2. This flowchart describes finding the sum of the first 99 unit fractions. A *unit fraction* is any number that can be represented as $\frac{1}{n}$ for any counting number n (for example, 1/2, 1/3, 1/4, etc.). Write the corresponding BASIC code for this problem.

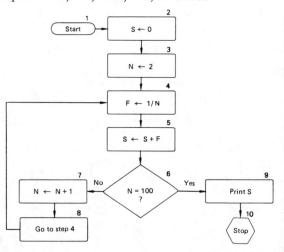

3. Examine the two BASIC programs below, and find the significant errors that would keep the program from running successfully.

```
1 A = ⁻2              1 Input A
2 B = 1               2 Input B
3 B = B + 1           3 C = A + B
4 C = 1/(A + B)       4 A = A + 1
5 If B = 10 then 7    5 If A = 10 then 7
6 Go to B             6 Print C
7 Print C             7 Go to 3
8 End                 8 End
```

4. A college professor graded his class by simply totaling each student's test results, the five tests being worth 87, 111, 102, 98, and 116 points respectively. He then found 90% of the total of 514 points, and that total score was an A. The typical grade scale of 80% (B), 70% (C), 60% (D) was employed to find out the other points at which to award various letter grades. He then wrote the flowchart on the right so that the computer could assign the letter grade for each score he fed into the computer.

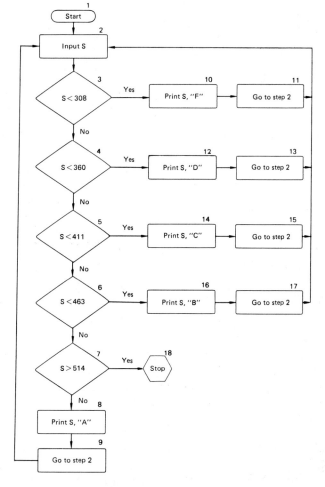

The first four numbers he typed in were 432, 506, 411, and 382. What would the computer print out for each of these scores?

What "dummy number" could he use to "shut down" the program after he was finished with all his test scores?

MATHEMATICS FOR TEACHERS

5. Write and then run programs to solve the problems below.

   a. A salmon starts on a 100-mile journey upstream to the placid lake where she was born. Each day she is able to swim three miles upstream, but each night while she rests she is pushed two miles back downstream. Exactly how long will it take her to reach the quiet spawning grounds?

   b. A few years back, a longitudinal study of the residents in Erie County, New York, produced some startling results: the number of people dying from respiratory diseases in this heavily polluted area was *doubling* every five years.

   There were 263 deaths attributed to respiratory ailments in 1950. If this rate continued, how many deaths of this sort could be expected in Erie County in 2000 A.D.?

   c. Jack got a pair of bunnies as an Easter present in 1979. This pair became a pair of "young rabbits" in May of that year, and grew into a pair of "adult rabbits" the following month. A pair of adult rabbits produces a pair of "bunnies" each month from then on, and this growth cycle continues.

   The number of pairs of rabbits of each type is provided below, for the first 8 months after Easter:

Pairs of:	Apr	May	Jun	Jul	Aug	Sep	Oct	Nov	. . . .
Bunnies	1	0	1*	1	2	3	5	8	. . . .
Young	0	1	0	1	1	2	3	5	. . . .
Adults	0	0	1	1	2	3	5	8	. . . .
Total	1	1	2	3	5	8	13	21	

*This pair of baby rabbits, remember, came from the pair of adult rabbits.

What is the maximum number of rabbits that Jack could have in two years under the condition that, of each pair born, one is a male and one a female, and none die?

d. A gifted fifth grader was frequently bored in class, and so got into trouble often. For her punishment, she had to write things like "I will not take Mary's notebook again" 100 times. The last time she was asked to do this, she accepted the teacher's challenge to get the microcomputer in the media center —the one she had used when she learned elementary programming—to do the work for her.

Can you write a program that will perform this chore? Can you get the computer to number the sentences, as she eventually did? What other variations can you make?

The Right Tool

Guess which modern tools bear the brunt of their mistakes. If you've ever heard someone on the other end of the telephone line say "I'm sorry, but our computer has been messing up lately," or read in the paper of an "expert" blaming students' lack of computational skill on the hand-held calculator, you are aware of the problem.

The hand-held calculator and microcomputer will play increasing roles in our lives. They are simply too valuable as tools to downplay their worth in education. We need to shift our attention from the argument about whether to use them, to the debate about how to teach students to use them wisely. Several factors are important in this discussion.

One consideration, of course, is the person's own skill in computation and whether he or she can expect to improve that skill by practice. If an older person has very weak arithmetic skills, that person should be using a calculator for most mathematical manipulations. On the other hand, those who are quick in the four operations with paper and pencil might quite often find that having a calculator handy is as much a nuisance as a benefit, as coming to rely on the device will undoubtedly slow down their ability to operate on their own. For these reasons, we should definitely analyze our own abilities so that we know in general the effect of using a computational aid.

Most adults, if they performed this self-analysis, would conclude that, in fact, they should not use these devices all the time. So for most of us, the decision about when to use either of these tools to help resolve a mathematical dilemma must be made on an individual basis. Inherent in such a decision would be two factors: the characteristics of both the problem situation and the two tools at our disposal. Assume for a moment that the problem is one that *will* involve a lot of computation, thus justifying one or the other of our choices.

If the problem is a one-shot deal, and is not complicated by a good many variables to keep up with, probably the calculator is the best bet. It's usually around somewhere, doesn't need to be programmed in a special language, and yields an answer very quickly.

On the other hand, if you might meet the problem situation again some time in the foreseeable future, or if the problem is a complex one with a good many "loose ends," you might choose to take the time and effort necessary to write a program that you could save. Or if you wanted to have a record of the numbers you used (to check for mistakes, or keep for later), or the answer to the problem, a computer could give you a "hard copy." And as a future classroom teacher, certainly you realize the potential of the computer for handling some of the routine teaching chores previously performed by hand, or ignored because of the work involved.

The purpose of the problems in this next set is to provide you with a variety of situations, and have you make an intelligent choice about using paper and pencil, a calculator, a computer, or a combination of tools. The most difficult part is still analyzing the problem itself, however, and writing a flowchart. Once this has been done, the choice of "the right tool" takes on more meaning.

SET 6

1. A popular upper-elementary science project is for the kids to grow one-celled animals, and then examine them through a microscope to watch their movement, their absorption of food, and even their method of reproduction. Once you have a bunch of

paramecia, for example, you're quite likely to see one splitting apart, each part eventually becoming a new animal. This method of reproducing is called "cell division."

It's easy to grow paramecia in the classroom: merely take a glass of tap water, add some hay or dead grass, and leave the concoction for a week or so. Cysts that happen to float by this attractive environment will just "drop in" and start growing (our apologies to biology purists). A healthy paramecium will go through cell division about five times a day.

If only one paramecium "drops by" and starts growing in the manner described above, how many should be in the glass in a week?

2. The bristleworm can reproduce by splitting into 24 segments, each of which then becomes a new bristleworm. Then these 24 new bristleworms continue the tradition and reproduce in the same fashion. Starting with only one of these creatures, how many would you have after 5 generations of splittings?

3. Fold a sheet of paper as many times as you can (i.e., fold it in half, then in half again, then again, etc.). What is the largest number of times you can fold it? Someone has claimed it is impossible to fold it more than 8 times, no matter what size sheet you start with!

But imagine for a moment that it *is* possible to fold it a large number of times. If you could fold it 50 times, how high would the stack be? (Hint—in a previous chapter, you found the thickness of a sheet of paper to be about .025 inches. Each time you fold it, the height or thickness doubles.)

4. You should have the records of your cumulative grade-point average for your college career so far, and you know how many total hours you've taken and how many you have left to take. Calculate what kind of grades you'll have to make for the remainder of your college career to wind up graduating with a B average.

5. Your rich uncle deposited $5000 in an account for your college education the day you were born. The account draws 5 percent simple interest, and the interest is added back into the account each year. How much would you have received if you had removed the money on the last birthday before you entered college?

6. Most of us never stop to think how much we pay for something bought "on time." Houses are good examples of this.

Find out from someone you know how much money they financed as the original mortgage when they purchased their house. Calculate how much they'll actually pay for that mortgage by using their monthly payments and the total length of the financing period. For a *real* scare, do this exercise with a brand-new house, using the interest rate presently in effect!

7. Obtain from a local school district a copy of the salary schedule. If theirs is like many districts, the salary is fairly competitive for a beginning teacher, but loses out in the long haul (that's one major reason that so many teachers leave the profession after a few years).

   Start out with a beginning teacher's salary, and figure what it should be from year-to-year merely to keep up with inflation, using today's inflation rate. Compare this to the actual salary scale.

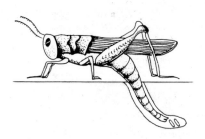

8. A female grasshopper has a structure at the tip of her tail called an *ovipositer*. With this she digs holes in the ground or a rotten log, and then lays her eggs, anywhere from 20 to 100.

   Assume for a moment that you are a scientist doing an experiment with grasshoppers over a ten-year period. You are applying for a grant to support your work, so you have to estimate how much you'll need for food, tags, etc. This boils down to a good estimate of the number of grasshoppers you'll have during that time period.

   Starting with just one pair—male and female—and knowing that grasshoppers only live one year (i.e., they lay eggs and then die), how many would you expect to come through your lab in the ten-year period? What assumptions must you make in doing this problem?

REFERENCES

Channell, D. E., et al. "Using calculators to fill your table." *The Mathematics Teacher* 74 (1981): 199–202.

Hay, L. "Using the computer to help prove theorems." *The Mathematics Teacher*, 74 (1981): 132–38.

Spencer, D. D. *The Illustrated Computer Dictionary.* Columbus, Ohio: C. E. Merrill Co., 1980.

Vannatta, G. D., and Hutton, L. A. "A case for the calculator." *The Arithmetic Teacher* 27 (1980): 30–31.

Winslow, I. O. *The Natural Arithmetic.* Book II. New York: American Book Co., 1901.

7

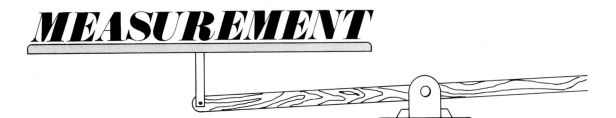

MEASUREMENT

Whether you realize it or not, you're already familiar with the concept of measurement. As a matter of fact, you've been "measured to death" already. You know the size shirts, shoes, rings, and hats to buy; your normal body temperature; your grade-point average; your height, weight, and age; and even where to place yourself on the scale below:

Source: B.C. by permission of Johnny Hart and Field Enterprises, Inc.

Ten, right?
    Human characteristics certainly aren't the only measurements we make. We hear statements like the ones below constantly:

> *In Palm Springs today, the temperature will hit $39^\circ C$, the probability of rain is 60 percent, the barometric pressure is 30.20 and rising.*

> *It'll take us about five hours and 1½ tanks of gas to make the 300-mile trip to Boston.*

> *Peanut butter has 135 calories per ounce, and costs $1.79 for a 1 lb., 12 oz. jar. So you get about 21 calories per penny with peanut butter!*

> *Plainville had 1752 residents at the last census, and an average family income of $13,540.*

To be sure, most of the numerical information we receive daily can legitimately be considered as a measurement of something.

In general, "measurement" means assigning numbers to a set of things, the numbers telling us how big the individuals are in relation to each other and the standard we choose as "one." Selecting the standard unit is done thoughtfully and with regard to the size of the resulting numbers for the set of things we wish to describe. Remember a point brought out in the unit on answers: extremely large or small numbers don't communicate well with most people. In choosing the unit for measuring several things, we should try to select a unit that will result in "familiar numbers."

Frequently the standard or scale we select is adequate for many things we want to measure, but has to be changed when we begin to judge things that are too large or small to communicate well using that standard. The metric system is becoming quite popular in this country primarily for this reason: adjusting the unit used to describe a set of measurements is a very systematic, simple process that can be handled with no computation (other than moving a decimal point). From the chapter on answers, you might recall that an answer such as "4500 g" can be easily changed into "4.5 kg" for communication purposes. Perhaps you remember these relationships from that chapter:

A kilogram is 1000 grams.

A milligram is $\frac{1}{1000}$ gram.

These examples illustrate the system for changing from one unit to a "derived unit" in the metric system: derived units are produced from the base merely by adding a prefix. The most common prefixes are shown below:

kilo	(1000)	deci	$\left(\frac{1}{10}\right)$
hecto	(100)	centi	$\left(\frac{1}{100}\right)$
deka	(10)	milli	$\left(\frac{1}{1000}\right)$

Since the mass of most familiar objects can readily be described using only the kilogram, gram, and milligram, these are all the "man in the street" needs to be concerned with. So these were the only units previously mentioned for measuring mass in the metric system, even though such derived units as "hectogram" and "decigram" could also be used.

The first four sections which follow will deal with some common measurement concepts that you'll be teaching in a few years. The emphasis is on "personalizing the experience" by selecting both base units and derived units as reasonable ones that you can internalize via physical and mental involvement. The last section deals with the most common measurements, other than the physical ones, used to describe the students in a classroom.

Measuring Length

When you measure your waist, the distance to the nearest ski slope or beach, or your height, you're essentially measuring in one dimension. You're assigning numbers to tell how far it is from one point to another point, along a certain path. Line segments (•————•) are the shapes chosen to define the standard unit, even though the path to be measured may be some sort of curve (•〜〜〜•).

The most commonly accepted standard for measuring length is now the *meter*. In case anyone asks you, you should always have a definition handy:

> *The meter is the length equal to 1,650,763.73 wavelengths in vacuum of the radiation corresponding to the transition between the levels $2p_{10}$ and $5d_5$ of the krypton-86 atom.*

Or about 39.37 inches.

Take a few minutes now to become familiar with this standard. If you don't have a krypton-86 atom handy, use a meterstick (or cut off a stick that's 39.37 inches) and make a meterstring in similar fashion. Find some common things that are about one meter in length. Measure parts of your body, things in a room, a "giant step," etc., until you have a "feel" for things that are about a meter in length. Don't just measure in a straight line; go around some curves!

SMALLER UNITS

In exploring the meter, you probably found some parts of your body that couldn't be measured very well with a meterstick or meterstring: like your nose or the distance around your ring finger. To measure smaller distances like this, the meter is divided into tenths, hundredths, and thousandths to produce these units:

decimeter $\left(\frac{1}{10} \text{ meter}\right)$ •—————————————————•

centimeter $\left(\frac{1}{100} \text{ meter}\right)$ •————•

millimeter $\left(\frac{1}{1000} \text{ meter}\right)$ •

Since the *decimeter* and *centimeter* are used quite often by most of us—or will be in the future—we'll concentrate on gaining a "feel" for both of these units too. In the cartoon Sally is about to develop a "feel" for one of these units.

Source: © 1974 United Feature Syndicate, Inc.

Make a stick and a string for both the decimeter and centimeter units, and repeat your process of finding common things around you to use as references for measuring things with these two units. This activity should be repeated until you're comfortable talking about centimeters and decimeters. Be sure to find at least one part of your body that's very close to each of these two units. You can then use it to measure other things approximately.

LARGER UNITS The meter is not very useful for measuring very long distances, such as the distance across the United States. The resulting number is simply too large to be meaningful, implying that a larger unit should be used so that the numerical descriptor will use a smaller number.

Ten meters is a *dekameter* (dam), one hundred meters is a *hectometer* (hm), and 1000 meters is a *kilometer* (km). Of these, the "man in the street" needs to have internalized about how far the kilometer is, so we'll concentrate on it for a moment.

Most people would walk a kilometer in about eight or nine minutes, ten at the most. If you can do so conveniently now, take such a walk and see about how far you go in that amount of time. If you're not up for such exercise right now, try gaining a personal feel for this unit by converting some familiar distances that you already know from miles into kilometers, using the approximation

1 mile is about 1.6 kilometers

to complete the chart below. As an example of this conversion factor, a distance of 5 miles would be (5)(1.6) km, or about 8.0 km.

Distance	In Miles	In Km
From where you are now to your hometown	?	?
Your car has travelled in its lifetime	?	?
Around the earth at the equator	25,000	?
From your college campus to a close recreation area	?	?
You can legally travel on an open highway in one hour	?	?
From the earth to the moon	240,000	?
You can jog in 30 minutes	?	?

Hopefully the distances above will give you an intuitive feel for the *kilometer*, and you can relate new distances to the ones above in gaining estimates.

SUMMARY

The basic unit chosen to measure in one dimension is the *meter*, which is a little longer (3 inches or so) than a yard. Derived units are obtained by attaching the previously mentioned prefixes to this basic unit, as summarized below:

$$\text{decimeter} = \frac{1}{10} \text{ meter} \qquad \text{dekameter} = 10 \text{ meters}$$

$$\text{centimeter} = \frac{1}{100} \text{ meter} \qquad \text{hectometer} = 100 \text{ meters}$$

$$\text{millimeter} = \frac{1}{1000} \text{ meter} \qquad \text{kilometer} = 1000 \text{ meters}$$

A measured length in the metric system could be renamed using another unit, merely by moving the decimal point in the number involved (for example, a ski that's 180 cm could also be called 1.8 m). This follows from noticing that moving from one derived unit to the next one merely involves multiplying or dividing by 10.

SET 1

1. Use the body measurement you found for a centimeter to make these measurements:
    a. The length of the pencil you're using
    b. The length of the diagonal of this rectangular sheet of paper
    c. The distance around your balled-up fist, at the knuckles

d. The length of an average hair from your head

e. The length around your ring finger

2. Practice going from one unit to another in the exercises below:

   a. 5.4 km = ? m
   b. 13 dm = ? mm
   c. .05 m = ? cm
   d. 2.47 km = ? dam

   e. 48 cm = ? m
   f. 328 mm = ? dm
   g. 100 m = ? cm
   h. .364 m = ? cm

   i. .021 dm = ? km
   j. 88 m   = ? km
   k. 1.926 m = ? cm
   l. 1 km   = ? mm

3. Draw new speed limit signs for those below, using kilometers per hour (kph) instead of miles per hour.

SPEED LIMIT 55 MPH	QUIET ZONE 15 MPH	SCHOOL ZONE 25 MPH	RESIDENTIAL ZONE 35 MPH

4. Is it true, as claimed, that the distance of your extended arms is the same as your height? Try it and see, making your measurements using the metric system.

5. Most of the athletic tracks that we use in this country were designed so that exactly four laps would constitute a mile. As a "mixed number," about how many laps would be one kilometer?

6. The wire used for coathangers is about 2 millimeters (mm) in diameter. A *micrometer* (μm) is one-millionth of a meter. A coat hanger is then about how many micrometers in diameter?

7. People have different "comfort distances" when carrying on normal conversations with others. When others get too close to us, we are uncomfortable. Some have suggested that this distance has cultural ties: Europeans, for example, seem to be more comfortable than Americans in the "close-up" range. What is your own comfort distance? Talk to some friends who aren't aware of what you're doing, and estimate your comfort range using metric units.

8. As you might imagine, kilometers are not convenient units for expressing distances once we get off the earth. The distance from the earth to the sun, for example, is about 158,000,000 km. Astronomers made up their own standard for measuring such distances in the solar system, and decided to let the mean distance from the sun to the earth be "one" Astronomer's Unit (AU). Then the distances of the other things in the solar system turn out to be less than one (if they're closer to the sun than the earth is) or greater than one (if they're farther away). The chart on the next page gives part of this information. Complete it!

	Distance from Sun (Km)	Distance from Sun (A Units)
Mercury	?	.4
Venus	?	.7
Earth	?	1.0
Mars	?	1.5
?	?	?
Jupiter	?	5.2
Saturn	?	9.5
Uranus	?	?
Neptune	?	30.1
Pluto	?	39.5

Hint: There's a relationship at work above. The pattern (called "Bode's law") is most obvious if you play around with the column above that lists distances in "A Units." Bode's law helped astronomers predict that a planet should lie between Mars and Jupiter before the asteroid belt was discovered there.

9. The sunlight that bathes our planet takes about 8½ minutes to get here. How long would it take sunlight to get to Pluto?

10. Look back at the sketch in Problem 8 for a moment. If this drawing were "to scale," the distance from the sun to Pluto would be about 40 times the distance from the sun to the earth. Using your familiar body parts to gain a metric estimate, how large a sheet of paper would be required to put this sketch "to scale"?

11. As a special challenge, you might consider that the greatest distance conceivable to man is the distance across the known universe—about 15,600 million light years. A light year is about 100 trillion km. So about how many kilometers is it across the known universe?

Measuring Area

In the last section, the concentration was on measuring in one dimension, or measuring *length*. When the two-dimensional aspects of a figure are of interest, we're thinking about the *area* of the figure. In choosing a shape as the basic one for measuring area, we need one that is itself two dimensional. And it would certainly be helpful if the selected shape "tessellates the plane"—that is, one that covers a two-dimensional figure without overlapping. As you can see from the pictures to the right, a circle wouldn't be an appropriate shape to use as the basic one for measuring area, because it doesn't tessellate the plane.

But this still leaves a wide choice. We could select any triangle, any quadrilateral, or even a regular hexagon, since they all tessellate. If we wanted to be truly creative, we could even select an "Escher-type" shape as the basic one for measuring area. However, the square has always been used as the basic shape for measuring area, probably because it's the simplest having both "right angles" and "parallel sides." These characteristics have endeared the square to architects down through the ages, because it enables them to construct sturdy floors and walls. So for the time being anyway, we'll go along with this "rational choice" of a square as the basic shape for measuring area.

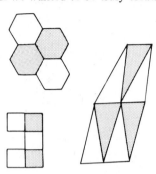

FINDING THE AREA BY COUNTING UNIT SQUARES

The basic concept of measuring the area of a geometric figure involves deciding on a basic square to use, imposing a tessellation of this unit square onto the figure, and counting to see how many unit squares are required to cover the figure. The areas of the figures to the left can be considered as representative of how to do this. Verify to yourself by counting unit squares that the area of figure A is 9, that of B is 6½, and the area of C is about 7 unit squares. This procedure of finding the area of a figure by counting unit squares should not be forgotten in a rush to generalize the counting process, producing "formulas" for finding the area. If the counting procedure is first internalized, the most familiar formulas will become obvious, based on a true understanding of the concept of area.

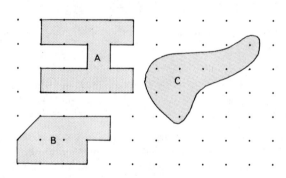

FINDING THE AREA VIA FORMULAS

If you count the area of enough very basic, very familiar figures, you'll gradually internalize some generalizations about their areas. One such figure is the rectangle, and certainly the formula below isn't new to you:

   Area of a rectangle = length times width

   (A = L x W)

Of course, the formula above holds true for fractions, decimals, and even irrational numbers, not just the whole number lengths and widths that are used in the initial counting experiences.

The area of a parallelogram comes intuitively, once the area of a rectangle is accepted to be "length times width." From observing the figure below, you can see that a piece of the parallelogram can be "cut off" and rearranged, resulting in a rectangle to which the formula above can be applied. Notice, however, that the area of a parallelogram is *not* given by the product of the length of the two sides, but rather by the product of a side and the "altitude" of that side.

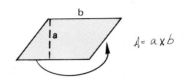

Extending things a little further, once the area for a parallelogram has been internalized, the area of a triangle seems to pop out of the woodwork. Given any triangle, you can make a parallelogram by joining it to another triangle congruent to the first, so each triangle has half the area of the parallelogram that could be made from it. Therefore, the area of a triangle is one-half the product of a side and the altitude drawn to that side.

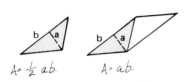

A familiar formula, if ever there was one, is that for finding the area of a circle. Note that this generalization does not follow very easily from counting unit squares—advanced mathematics is used to verify that the area of a circle is given by the product of π and the square of the radius.

STANDARD UNITS FOR MEASURING AREA

After deciding on a square as the basic shape for measuring area, we still need to settle on some common, standard sizes of squares to use as basic standard units. Metric measurements seem the most logical ones to describe the length of the sides of such basic squares, since the metric system was used in the previous section. In the activities to follow, the emphasis will be on gaining an intuitive feel for the standard units with which the "man on the street" should be familiar.

Take out your meterstick and meterstring, and lay them on the floor end to end, in a right angle. Form the rest of the square in your mind; the area of this standard unit is a *square meter* (m^2). A fairly large square, isn't it? This basic unit would be useful in measuring the area of a floor you were about to carpet, or a wall you were about to paint.

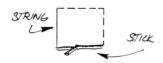

The square surrounding these words has sides that are one decimeter in length. Hence such a square is the standard unit *square decimeter* (dm^2). Such a unit might be used to measure the area of the desk you're writing on, or the size of an artist's canvas.

Cut out a square this size and measure, as best you can via counting, the area of your desk. Then think of several more items whose area would appropriately be measured in square decimeters.

Find a part of your body (your hand, maybe) that would be useful in estimating the area of figures using this basic unit.

The other common standard unit for measuring area is the *square centimeter* (cm^2). Such a square is shown in the bottom left corner of the larger square above. This standard unit might come in handy if you needed the area of a sheet of paper, or perhaps the sole of your shoe. Find a familiar body part that approximates in area a square centimeter.

LARGER AND SMALLER UNITS FOR MEASURING AREA

The everyday measurements for area have already been mentioned, but several others deserve at least a cursory glance. It's not that unusual to encounter area measurements using either larger or smaller units than square meters, decimeters, or centimeters.

The end of a pencil is shown to the right, and has a *square millimeter* (mm^2) tessellation covering it. Can you find its area by counting? The square millimeter is a common standard for reporting the area of small figures. The size of a microscopic plant or animal might be given in mm^2, for example.

If you purchase land in the future, you might find a need for a larger basic unit for area. The *aire* (or square dekameter) might prove useful in describing a lot for a single family dwelling in a residential district of town. The *hectaire* (or, square hectometer) is about the size of a football field, and would perhaps be useful in describing the land area of a small town. Even larger units are obtained from considering such things as a square that's a kilometer on each side: a *square kilometer* (km^2). The area of Pennsylvania would probably be expressed using this basic unit.

And, in truth, it'll be a while before we rid ourselves of the vestiges of the English system in measuring area. Square inches, square yards, square feet, and square miles will remain in our vocabulary for quite some time to come. For example, consider the photograph below of the wall of a pizza parlor in Laramie, Wyoming. The proprietor obviously prefers square inches to square centimeters, but his calculator and sense of humor will probably carry him over the rough spots!

Source: Photo by Jim McClurg.

SET 2

1. Use your meterstick and meterstring and your imagination to find the area of the following things in square meters by counting:
   a. The floor of your bedroom or dorm room
   b. Your bed
   c. The door into your room
   d. The windows in your room

2. Use your body approximation of square decimeter and square centimeter to estimate the areas below:
   a. The seat of your favorite chair
   b. The surface area of your skin
   c. A record album

3. Find the areas of the figures below by counting unit squares:

CAN BE DONE EXACTLY, BUT ITS TOUGH!

4. Take care in considering these "conversion problems," and check that you've done the first few correctly before you tackle the others:

   a. 1 dm^2 = ____ cm^2

   b. 1 m^2 = ____ dm^2

   c. 1 km^2 = ____ m^2

   d. 4.2 m^2 = ____ cm^2

   e. 2000 cm^2 = ____ dm^2

   f. 348569 m^2 = ____ km^2

5. Determine about how much time it would take an average adult to walk around a square flat piece of land with an area of 1 km^2.

6. Determine the dimensions of a rectangular field with area exactly 1 km^2, but in walking at the same rate as in the previous problem, you would need at least *twice* as much time to walk around the field.

7. Consider the tessellating figure to the left to have area 1. What, then, would be the area of the figure to the right?

8. Return for a moment to the photograph of the pizzapans, which immediately preceded this problem set. Determine the radius of each of the three sizes, in inches.

9. Scientists estimate that there are about 13,000,000,000 hectaires of habitable, arable land on this planet (actually, the figure is high; it includes inland water, but not Antarctica). Assume the population of the world is about 4½ billion. How big a piece of land is each person entitled to?

10. Devise a method for finding the area of polygonal figures like the one below, using only a ruler, compass, and one of the previously mentioned formulas.

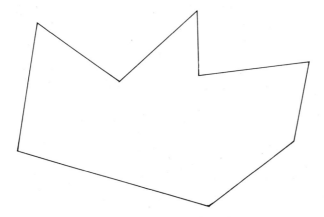

11. Using the map below, find the approximate area of the lake, and communicate your answer clearly.

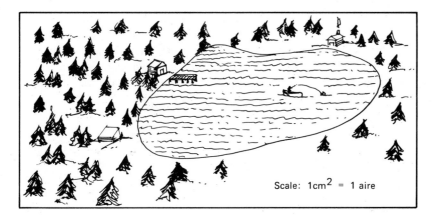

12. Estimate as best you can the size of each person's space in the cartoon on the following page. Use your personal body estimate to find the approximate length and width of the rectangles shown, in meters.

"Excuse me, sir. I am prepared to make you a rather attractive offer for your square."

Source: Drawing by Weber: Copyright 1971 The New Yorker Magazine, Inc.

Measuring Volume

Measuring volume or capacity requires a three-dimensional shape that will "fill" or tessellate space. This notion of measuring the space in which we actually live is merely an extension of two-dimensional space and the concept of area. Finding a three-dimensional shape that will do this and settling on appropriate dimensions for it, will provide us with a standard, basic unit for measuring volume.

Do any of the shapes below tessellate space?

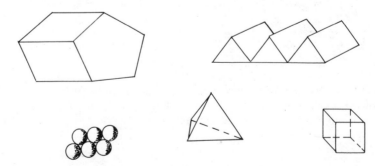

Because our decisions always seem to be dominated by rational thought we have chosen the cube as the unit for measuring volume. It certainly tessellates space, and is actually an extension of the shape used to measure area.

The basic concept involved in measuring the volume of a three-dimensional shape is to count the number of unit cubes it takes to "fill up" the shape. Verify by counting that the volume of figure A below is 4, figure B has volume 20, C has volume 18, while the volume of figure D can be estimated visually as about 51 unit cubes.

FINDING VOLUME BY COUNTING UNIT CUBES

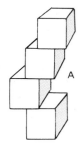
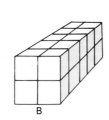
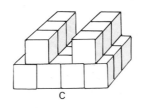
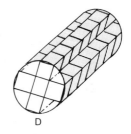

Again, after the basic idea of counting unit cubes to find the volume of a three-dimensional object has been internalized, this relatively inefficient procedure is replaced by generalized formulas for finding the volume of common shapes. Some of these are reviewed briefly below.

The most common three-dimensional shape for which we must obtain the volume is an ordinary box shape, as shown below. Look at figure B above and you'll see such a shape; notice that the volume (20) can be obtained by multiplying its length (5), width (2), and height (2) together. After counting a number of examples similar to this one, the general formula that the volume of a box is the product of its three one-dimensional measurements becomes obvious.

FINDING THE VOLUME BY FORMULA

Relying on concepts from advanced mathematics, we have learned that the volume of a sphere is given by $4/3\ \pi\ r^3$, where "r" is the radius of the sphere. For example, if we assume the earth is a sphere with radius 6,700 km, the volume of the earth could be obtained by finding the value of $(4/3)(\pi)(r^3)$, or, $(4/3)(\pi)(6{,}700\text{ km})(6{,}700\text{ km})(6{,}700\text{ km})$. The answer to such a calculation, without worrying about communicating in meaningful terms, would be $1{,}260{,}000{,}000{,}000\text{ km}^3$. That's about 300 km^3 for every man, woman, and child on the face of the earth! But what's a km^3, anyway?

STANDARD SIZED CUBES FOR MEASURING VOLUME

A km^3 is just what you'd expect it to be: the volume of a cube that's a kilometer on each edge. A pretty large box, isn't it! *Cubic kilometers* are needed only for measuring extremely large volumes, such as the amount of water in the oceans or the size of the Grand Canyon. Average people will probably never have to deal with such volumes in their lifetime.

But the *cubic meter* (m^3) is another story. This cube, which measures a meter on each edge, will be used quite often to measure the volume of everyday things that concern us. The size of a storage room, or a moving van, or a backyard swimming pool might be expressed using this standard unit. To gain a feel for this volume, take out your meterstick and use the corner of a room to mark off about what a cubic meter would be. Then curl up as tightly as you can in the corner, and have someone mark off as a cube about how much space you take up. Estimate what fraction of a cubic meter you measure.

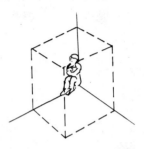

One of the handiest sizes for measuring things around the house is the *cubic decimeter* (dm^3), which is about the same size as a quart. Recall that a decimeter is 10 cm, or 1/10 m, and that you have a familiar body part to estimate this particular length. Find a one-pound coffee can, or cut off the top of a half-gallon milk carton 10 cm from the bottom all around, and you've found a nice approximation for this particular volume. This particular size has become very popular, and has thus been given its own name, *liter* (ℓ). So a liter is the capacity of a cube that's 10 cm on each side.*

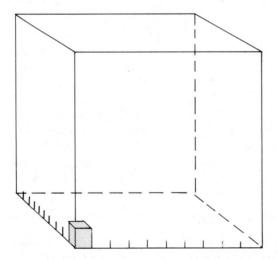

The two cubes to the left show the relationship of a cubic meter and a cubic decimeter. Can you tell from this drawing how many cubic decimeters would equal a cubic meter? That is, how many small cubes would fill up the larger one?

The same drawing could show the relationship of a cubic decimeter and the next smallest common standard, the *cubic centimeter*. Cubic centimeters would be useful in expressing the volume of a cup of coffee, or a dose of medicine. A good approximation for this capacity can be obtained by considering a sugar cube.

---

*It's interesting to note in passing that originally the liter was defined differently, and thus only approximated the volume of a dm^3. Originally, the liter was defined as "the volume of one kilogram of water at 4°C, at the standard atmospheric pressure of 760 mm of mercury." Hardly an "intuitive definition"!

MEASUREMENT   257

From the drawing above, we can judge that it would take 1000 cubic centimeters to fill up a liter (dm³). From this fact we can determine another name for a cubic centimeter: we can also call it a *milliliter*. In fact, "milliliter" is used much more often in reporting volume than "cubic centimeter."

SET 3

1. As accurately as you can by counting unit cubes (even if you have to use your imagination for parts of a cubic meter), find the volume of the following:
   a. A bathtub full of water
   b. A piece of fruit
   c. The inside of a small automobile
   d. An ice cube

2. Use a ruler marked off in the metric system, and a formula, to calculate the volume of the following:
   a. Your dorm room or bedroom
   b. A box of kleenex
   c. A ball of some sort (tennis, basketball, volleyball, etc.)

3. You need to find the volume of an irregularly-shaped, fairly small object, such as a light bulb. You can find a beaker large enough to hold the object, and the beaker is graduated in millimeters. You also have access to a water supply. How could you quite accurately find the volume of the object?

4. Your elementary science class has been studying the human body, and gets interested in how much air each person can hold in his/her lungs. As the teacher, could you suggest a way this could be measured?

5. Ocean A has a surface area of 10,000 km², and Ocean B has a surface area of 20,000 km². The volume of Ocean A is 20,000 km³, the same as that of Ocean B. Which of the two oceans would have more of those weird fish that live in the depths where sunlight can't reach them?

6. You need to design a box that has a volume of 1ℓ, but the height of the box can only be 5 cm. What can the length and width be to meet both of these requirements?

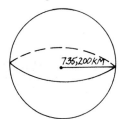

7. The radius of our sun, which is only an average size star, is about 735,200 km. Using the formula for a sphere, find its volume. Then calculate (using the information given in this section on the volume of earth) how many earths it would take to "fill up" the sun.

8. The VW van on the right is yours. After the beans are cleaned out, you're going to pack it as full as you can and head across the country to a new job. You need to know the inside volume of the van so you'll have a good idea of how much of your stuff will fit in, and how much will be left over for the traditional "garage sale."

   Devise a way to arrive at the inside volume of the van, using the beans.

Source: Courtesy of Volkswagen of America, Inc.

Measuring Angles

An angle is usually defined as "the union of two rays that have a common endpoint." It therefore seems likely that whatever standard shape we use to measure rays would be used again to measure angles. The problem with this line of reasoning is that rays are figures that are not measured in terms of size, since they extend infinitely in one direction. In the picture to the right, for example, $\vec{BA}$ and $\vec{BC}$ both extend forever in the direction of the arrow, so we certainly couldn't find the length of either one. Consequently, we cannot measure their union in terms of their length.

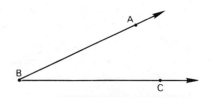

Technically, when we say we "measure an angle" we are describing the sharpness with which the two rays meet at the endpoint (the *vertex*). We assign a number to the angle, and the number tells us whether our angle is sharper or duller than our standard unit or another angle. We now have the problem of choosing the standard angle to compare all others to.

Each of the angles below would certainly need to be measured using our standard angle—each angle is "the union of two rays with a common endpoint." The "dullest"

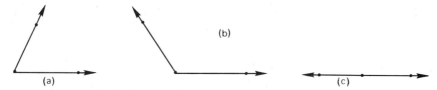

such angle would be of the kind in (c) above; what we therefore need is a standard angle that tessellates a semicircle, i.e., an angle that, when used continuously without overlapping, covers a semicircle exactly. Each of the angles below meets this criterion, and is certainly not unique in this regard. An infinite number tessellate a semicircle,

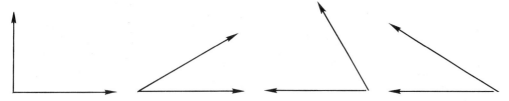

but most are too small or too large to be useful. Again, we want to select a standard unit that gives us accurate measurements but still results in understandable numbers for most of the angles with which we come in contact.

The angle to the right was selected as the "standard angle" in ancient times. It's small enough to measure accurately the angles we can draw with a pencil, and 180 of the things tessellate a semicircle. Hence any angle we measure can be assigned an understandable number between 1 and 180. This unit is called the "degree"

The counting concept also underlies this sort of measurement. The measure of an angle is the number of degrees it would take to "fill up" the angle. Each of the angles below has been conveniently "filled up" for you—about what is the measure of each?

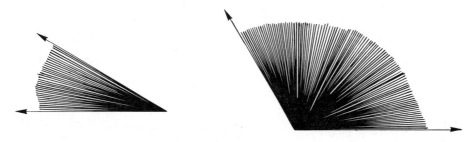

MATHEMATICS FOR TEACHERS

The "degree" is further partitioned into smaller units that are useful in such fields as meteorology, astronomy, and navigation.

one degree = 60 minutes = 360 seconds

One of the apparent ironies in this case is that the smaller subunits of a degree are primarily useful when studying large shapes and distances, such as with huge tracts of land, or interspace computations.

TEACHER, GIMME' A D

The only time students will ask for a D is when they want a tool that has been designed to measure an angle in degrees with a minimum of fuss and a maximum of accuracy. One such tool is shown below, the real name being "protractor." Note that in measuring angles with a D, the vertex of the angle is placed underneath the tip of the arrow, one of the angle's rays is lined up with an edge of the protractor (after twisting and turning, if necessary), and the measure is obtained by reading the number where the other ray (extended if necessary) meets the "rounded part" of the protractor.

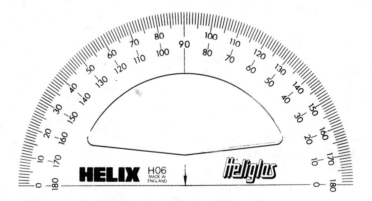

Measure each of the angles below using your own protractor and the procedure just mentioned. If you don't have a protractor yet, trace the angles on a thin sheet of paper and place them over the protractor shown above.

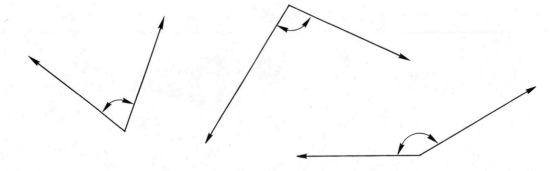

If you're doing things correctly, you should get 72°, 98°, and 150°, respectively, from left to right.

ANOTHER STANDARD FOR ANGLES

A completely different partitioning of a circle produces another basic unit for measuring angles: the radian. Rather than partitioning the circle into some arbitrary number of pieces (like 360), the radian uses the radius of the circle to decide how to divide it up.

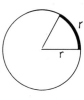

Imagine a flexible strip of metal exactly the length of the radius of a circle. This metal strip is bent to the exact curvature of the circle and placed "on top of" the circle as shown to the right. The angle with vertex at the center of the circle and rays intersecting the endpoints of the "bent" metal radius on the circle has a measure of *one radian*.

The radian is sometimes described as the naturally derived standard for measuring angles because of the use of the radius of a circle. In some areas of mathematics and science, although not in education yet, the *radian* is more widely used than the *degree* because of this "organic" relationship with the circle. It should be noted in passing that an angle measured to be *one radian* does not tessellate the circle, as does the degree. Can you determine the number of nonoverlapping angles of one radian each, sharing the center of a circle, that would fit into a semicircle?

Both of these standard units for measuring angles will be used in the next problem set.

SET 4

1. Suppose you decided to develop a "base-ten standard unit angle" to use in measuring all other angles. You'd want to divide a semicircle into 10 basic angles, and each one of those into 10 smaller angles, etc. How many degrees would the standard, base-ten angle be? How many degrees would the smaller angle be?

2. Politicians are often accused of "making a 180-degree turn" on a controversial issue. Exactly what would this mean?

3. Draw several large triangles on scratch paper, and shade the corners as shown to the right. Cut the triangles out, and then snip the corners off and line them up on your protractor. If you're careful, the three angles of each triangle will give you the same answer when lined up. What is it?

4. Does the same sort of thing work with polygons that have four angles (or equivalently, four sides)? Draw some quadrilaterals on scratch paper that are large

enough so you can measure the angles accurately with your protractor (or the one in the text.) Add the measures of the angles. What conclusion can you reach?

5. Can you extend the problem above to five-sided polygons? How about six-sided? Or seven-sided? Perhaps even an "n-sided" polygon!

6. Are you using rational, intuitive, or autistic thinking to solve the problem immediately above? Is this the same mental process you used for problems 3 and 4 above?

7. Our friend π is present again in an exploration of determining how many angles of one radian measure can be fitted into a semicircle, assuming that the angles share the center as a vertex and that they do not overlap. Study this question and try to draw a conclusion about π's relationship to the radian unit, and a circle. Recall that π is approximately 3.14.

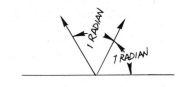

8. Which is larger, an angle of 180 degrees or an angle of 3 radians?

Measures Common to All Classrooms

Educators quite possibly have the highest calling of any of the professionals. We are constantly concerned with molding the intellect of young children and we hope our efforts will produce adults capable of contributing positively to society and eager to pursue intellectual growth.

Measuring the outcomes of education is a problem that has plagued researchers for quite a while, particularly of late when the cry from society has been "accountability." As George Miller succinctly states in the report of a conference on research on testing (1978),*

> The theory that has, explicitly or implicitly, dominated most thinking about tests is that they provide measures, analogous to length and weight, that can be used to arrange children and what they have learned, in a simple linear order. This theory supported initial research on testing, but it is obviously inadequate to our present conceptions of the complex nature of human knowledge.

We can easily measure the physical attributes of a child but beyond that we're on tenuous ground.

---

*The quote is from a paper presented at the August 17–26, 1978, Conference on Research and Testing. This meeting was sponsored by National Institute of Education and the U.S. Dept. of Health, Education, and Welfare. The paper was titled "The Issues Concerning Tests" and is found on pgs. 5-9 of *Testing, Teaching, and Learning: Chairmen's Report of a Conference on Research on Testing*, US Gov't Printing Office, Washington, DC 20402. The publication date is Oct., 1979.

But proceed we must. The need to ascertain what a child knows, or is capable of learning, is one that won't go away. Parents are no longer satisfied with statements like "Johnny reads well" or "Suzy isn't doing well in arithmetic." They want numbers, like grade levels, IQs, percentiles, stanines, etc. It behooves us to spend some time talking about some of the measurements commonly used in education.

IQ, MA, CA

Theorists differ greatly on what "intelligence" means and, consequently, on how to test for it. The most prevalent definition of intelligence these days depends on the test chosen to measure it; it says that "intelligence is whatever is measured on this particular test."

Regardless of the definition, the *mental age* (MA) of a person is determined by finding the age at which the raw score would be an average one. For example, if a raw score of 53 on the intelligence test corresponds to a mental age of 6 years, 11 months, then 53 is presumably the average score that would be made by a random sample of children that were 6 years, 11 months old. *Chronological age* (CA) is simply the age of the person since birth.

An indication of how bright the child is would then be found by comparing MA and CA. Intuitively, if the MA is far ahead of the CA, the child would be bright, and vice-versa. The traditional version of the *intelligence quotient* (IQ) is then defined to give us this type of measurement:

$$IQ = \frac{MA}{CA} \times 100$$

IQ scores have met with a good deal of criticism in the last few years in terms of cultural and socioeconomic biases reportedly imbedded in the tests themselves. They should certainly be viewed with caution by anyone concerned with the educational process.

While IQ is a rather static measure applied to a student, MA is certainly not and is quite a bit more useful in planning instructional tasks. A student with an IQ of 120 has different MAs at 9 and 12, and should therefore be doing conceptually different types of learning tasks. Most learning theories are reported as related to MA rather than either IQ or CA.

"You did very well on your I.Q. test. You're a man of 49 with the intelligence of a man of 53."

Source: Courtesy of Sidney Harris and *American Scientist* Magazine.

GRADE-LEVEL SCORES

Standardized achievement tests measure not a student's native intelligence, but how well the student has mastered what the test maker feels schools are supposed to be teaching. The scores are frequently reported back as "grade-level scores," both *actual* and *anticipated*. The school year has 10 months, so the grade level score can be reported back as a number like 8.3 or 6.1. The digit after the decimal point indicates the number of months completed of that school year. A score of 6.1 means that the first month of the sixth grade has been completed, while 8.3 indicates that three months have passed in the eighth grade.

Grade levels that correspond to raw scores on the achievement test are considered the grade levels at which those scores would be average. If a raw score of 14 on a section of a standardized test corresponds to a grade level of 2.7, then 14 is presumably the average score that would be made by a random sample of children that had completed the seventh month of the second grade.

Anticipated grade level for a student is determined by the average score of the students who have the same IQ, CA, and grade level as the student in question. This is helpful in determining whether a given student is living up to potential (in terms of mastering the skills measured on that test).

Just because students are above or below grade level, we can't assume that their needs are, or are not, being met. The purpose of the anticipated grade-level score is to compare how well students are doing to how well they should be doing. Consider the scores reported below:

Reading Comprehension

	Grade Level	Anticipated Grade Level	Living Up?
Mike	4.7	6.9	no
Andy	3.6	2.5	yes
Frieda	5.0	5.2	yes

Regardless of what grade the students were in at the time this test was given, we'd have to conclude that Mike is not reaching his potential. But Andy might be labeled an "overachiever," and Freida is about where she should be.

Actual grade-level scores are useful as gross measures of how well a student is reading, or performing arithmetic skills, and as such enable the teacher to pick instructional materials that the student can handle without frustration. *Anticipated* grade-level scores give us some measure of where students should be were their needs being met. Used with caution, both measures can aid the teacher in planning instruction.

PERCENTILES AND STANINES

Other quite common ways of reporting the results of standardized testing are *percentiles* and *stanines*. Each of these compares a student's results with that of all the other students in the same grade at the time of the year when the test was taken.

When Suzy is reported as being in the "40th percentile" on a certain section of the test, she did better than the bottom 39 percent of all the students comparable to her in terms of grade and time of the year. If Jack's raw score on a section of the test was better than 79 percent of the students who took the test at the same grade and time of year, Jack would be in the 80th percentile.

"Stanine" comes from contracting the phrase "standard score with nine categories," and is just what the phrase implies. Comparable students (those who are in the same grade, and take the test at the same time of year) are divided into nine categories to match the percentages given below:

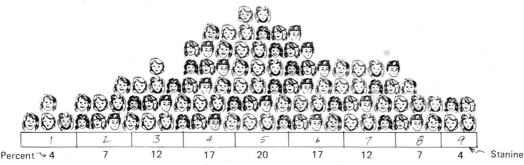

Notice that stanine 1 contains the lowest 4 percent, stanine 9 has the upper 4 percent, and stanine 5 contains the middle 20 percent of the students tested. This method of reporting achievement is becoming quite popular. It's easier for us to recall with understanding an intuitive interpretation of "stanine 4" than the meaning of "a percentile of 44."

As a teacher, you'll undoubtedly come into continuous contact with the measures mentioned so far in this section. Keep in mind the caution urged regarding placing too much faith in these descriptions of a young child's ability or achievement level. The same warning applies to the discussion to follow, that of measuring the "reading level" of a passage from printed material.

MEASURING THE READING LEVEL OF A PASSAGE

Knowing a student's reading level is not worth much to a teacher unless the teacher can also judge the reading level of instructional material to be used with the student. For example, to say that a social studies text is written "at a 5th grade level" may mean only that the *concepts* are appropriate for that level. The language used to describe those concepts may or may not be appropriate for a student on a fifth grade reading level. Attempts to measure the readability of a text have consequently been welcomed by the educational community.

One of the more popular methods used to measure the reading level of a written passage comes from Edward Fry, and is briefly described below:

> Count and mark the end of the first 100 words. Numbers (like 5,000) count as one word, as do the names of people and places like New York and Mammoth Mountain. Next go back and count the number of syllables for those 100 words —don't bother looking any up—taking the number as you pronounce the word is close enough. 5,000 has three syllables, while Mammoth Mountain has four.
>
> Then go back and count the number of sentences or thoughts that are grammatically independent of other sentences (or clauses). These independent

thoughts must end in a period, question mark, exclamation mark, semicolon or colon. Do not count "tags" such as "he said" or "she responded" as sentences by themselves.

Examples of:

    1 sentence . . .   He fought, but did not win.
                               The plan is: (a) timely, (b) comprehensive, and (c) eloquent.

    2 sentences . . .   Why did they leave? Embarrassment!
                               Result: He passed with flying colors.
                               The Lord is my shepherd; I shall not want.

    3 sentences . . .   There are two reasons not to go:
                               (1) It costs too much. (2) Bob will probably be there.

If the 100th word does not end a sentence or independent thought, count the last words as a fraction of a sentence.

Then the number of sentences and syllables can be used together to determine the approximate grade level at which that particular 100-word passage is written, using the chart below.

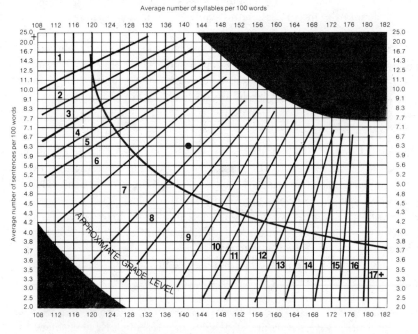

GRAPH FOR ESTIMATING READABILITY — EXTENDED

Source: Edward Fry, "Fry's Readability Graph: Clarifications, Validity, and Extension to Level 17." *Journal of Reading,* 21, 3 (Dec. 1977): 249. Used by permission of the author and the International Reading Association.

An example follows below. See if you can apply the rules above to the following passage to determine its approximate reading level.

> *The perspiration dripping from his entire body, George exerted himself completely. Then it was over once again. Quietly he withdrew into his own world, lighting a cigarette and slowly drawing its subdued irritation into his throat.*
>
> *"There now," Angela whispered quietly in George's ear, all the while rubbing his temple. "I'm glad you finally proved it to me. For a while I was beginning to doubt, but you've renewed my spirit."*
>
> *"Yeah, I'm happy too," said George. "But it seems to be more and more of an effort each time. For a minute there I thought I'd completely forgotten the proof that $\sqrt{2}$ is irrational!"*

Above, the 100th word is "the," there are about 150 syllables up to the 100th word, and the passage contains 8 and 10/15 sentences to that point. The chart tells us that this passage is written at about the 7th grade level.

Measuring the reading level of an entire text would require that we apply this procedure over and over again, collecting and analyzing the data. For the moment we'll content ourselves with using the method on individual passages from a text.

SUMMARY

This section has presented a different view of *measurement* than that presented in the first four sections of this chapter. We have briefly described some of the more common ways that numbers are assigned to measure nonphysical characteristics of familiar objects found in the typical classroom (children and books). We must take care not to place too much reliance on such measurements, since their fallibility is well known. Yet we must also realize that such measurements can aid in planning appropriate instruction for a student and, as such, deserve our attention.

SET 5

1. If a student's *mental age* were exactly the same as his *chronological age*, what would the student's IQ be?
2. Use Fry's method to determine the reading level of Henderson and Pingry's quote that opens the first chapter of this text.
3. Fill in the chart on the next page supplying the missing MA, CA, or IQ.

Student	MA	CA	IQ
Sam	10.2	8.5	
Mary	6.7	7.2	
Kim		11.8	119
Jan	13.7		88
Vera		8.6	131
Alan	15.6		107

4. Julia Raskin has just transferred into your fourth-grade class, and it's fairly early in the year. She was tested before entering, and the test results are below.

    Prepare a paragraph or two that analyzes her situation, so that you can inform Julia and her parents about what to expect during the remainder of the year.

Name: Julia Raskin   Grade: 4
Teacher: Paul Mayno   Date of Testing: 10/18/78
School: Aiken School   City: Milton   State: Ohio

Score Summary Box

Test	Number Possible	Number Right	Scaled Score	Grade Equivalent	Percentile Rank	Stanine	Instructional Reading Level
Reading	60	46	675	3.9	52	5	Grade 4
Mathematics	50	27	548	3.5	36	4	
Language	60	42	636	4.9	64	6	
Science	45	22	521	3.4	42	5	
Social Studies	45	24	559	3.8	50	5	
Basic Battery (R+M+L)	170	115	615	4.1	54	5	
Complete Battery (Basic+S+SS)	260	161	572	3.8	50	5	

Percentile Ranks and Stanines based on tables for Fall ☒ Spring ☐

Source: Reproduced by permission. From the Metropolitan Achievement Tests, Copyright © 1978 by Harcourt Brace Jovanovich, Inc. All rights reserved.

5. Below is a student profile for Charles Conrad. From the total information given,

    a. Find the grade-level score that would be reported (from looking at the grade he's in, and the testing date) if he were exactly "on grade level."

    b. Determine whether or not he's living up to his potential on each of these subtests:

    Reading vocabulary
    Reading comprehension
    Arithmetic reasoning
    Arithmetic fundamentals
    Mechanics of English
    Spelling

c. Determine what the *X*s mean on the graph. The circles? The straight line that goes all the way across the page?

d. Characterize Charles in terms of his intelligence (in very general terms, of course). Is he an over or under-achiever?

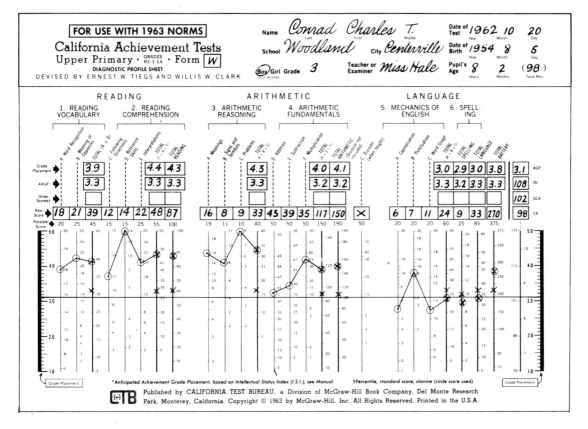

Source: Reproduced by permission of the publisher, CTB/McGraw-Hill, Del Monte Research Park, Monterey, CA 93940. Copyright © 1963 by McGraw-Hill, Inc. All rights reserved. Printed in the U.S.A.

REFERENCES

Baird, E. L., and Wyler, R. *Going metric the fun way.* New York: Doubleday & Co., 1980.

Gaughan, E. D., and Wisner, R. J. "A measured metric statement." *The Mathematics Teacher* 74 (1981): 262–265.

Miller, G. The Issues Concerning Tests. Testing and Learning. Washington, D.C.: U.S. Government Printing Office, 1979.

8

DESCRIPTIVE STATISTICS

For his own survival, a sailor on the open sea must learn to "read" a variety of clues regarding the weather. The direction of the wind, the height of the waves, the air temperature, the color and type of clouds, the barometric pressure, and many other bits of information must be integrated and interpreted in terms of both the present situation and the forecast for the future. It is the integration and organization of separate bits of information that is so important. No single piece of information would enable the sailor to make reasonable determinations of the course to be charted.

Probably most of the things we do require similar kinds of integration and organization of information. Our desire for sound judgments, and our past experience with similar situations, make us realize the importance of gathering and interpreting any available information that may affect the decision. Educators in particular seem concerned with gathering large amounts of numerical information for the purpose of identifying patterns and problems associated with both individual students and entire classes.  In order to diagnose deficiencies and evaluate learning, many different kinds of measurements must be made. And to comprehend what all these data mean, we resort to fairly standard, traditional ways of analyzing it. Today, the systematic and orderly treatment of numerical information (data) is achieved by the branch of mathematics known as *statistics*.

Of course, it sometimes seems that our world has gone overboard on the topic of statistics. It's a rare bird that hasn't at some time identified with the Charley Browns of the world:

DESCRIPTIVE STATISTICS   271

PEANUTS®   By Charles M. Schulz

Source: © 1960 United Feature Syndicate, Inc.

Apparently one of the keys to survival in the future is developing an intuitive, working knowledge of statistics. Our rapidly advancing technology, capable of storing thousands of bits of information on computer parts smaller than your fingernail, won't allow any other conclusion than "Statistics is here to stay!"

The two main branches of statistics can be intuitively described in terms of what each is intended to do:

> *To reduce large quantities of numerical data to an interpretable, understandable form which describes the situation at hand, and*

> *To enable us to draw meaningful conclusions and make sound predictions, inferring from the data we have at hand what to expect in other, similar situations.*

The second branch mentioned above is called "inferential statistics" while the first—the one we'll concentrate on in this text—is aptly called "descriptive statistics."

Descriptive statistics itself has two main purposes, the first being to organize the available data so its comprehendable and tells us something. The second concern is to communicate the information clearly to other interested parties. Since our first concern is to analyze the data so we understand them ourselves, the next few sections deal with methods for organizing and interpreting batches of numerical information.

Grouping and Classifying

There seems to be an uncanny desire in many people to categorize, group, and classify phenomena according to some numerical criteria. This really isn't a dehumanizing practice, even when applied to numerical descriptors of people. Usually it is an attempt to reduce a large amount of information to a more manageable form, so that any commonalities present can be found. There are two good reasons for this valuable practice in statistics.

For one thing, we often suspect that the data imposed on us seem more accurate than the instrument used to gather the data. IQ scores (discussed in the chapter on measurement) are perfect examples. In such cases, we tend to look at the score for an individual as a rough estimate of what we're measuring. Consequently some information is lost in grouping similar data together, but we feel much more comfortable speaking in generalities than about specific individuals.

A second reason for grouping data is that we can comprehend much more easily a few characteristics of classes of things than we can a lot of descriptions of a great many individuals. Grouping decisions can therefore be a great help in allowing us to overlook the trees and see the forest.

There are two common methods of grouping and classifying numerical information, each scheme being useful at different times. The first method is of the "hard-and-fast" variety; the groups are established (usually before even looking at the data) by setting the boundaries and letting the individual scores fall as they may into the appropriate slots. The second method involves looking at the data present, and from this examination determining if there are some "natural" groupings that suggest themselves. Each of these techniques deserves further examination.

PRESET CATEGORIES If there is a large amount of data present, and there is no reason to assume that some natural groupings will suggest themselves, presetting the categories is the most logical thing to do in most cases. Consider the data below:

Barrels of Motor Gasoline Consumed, Per Capita, United States, 1978
(1 barrel = 42 gallons)

State	Value	State	Value	State	Value
Alabama	13.9	Kentucky	13.2	No. Dak.	17.0
Alaska	12.4	La.	13.0	Ohio	12.2
Arizona	14.3	Maine	12.8	Okla.	16.6
Arkansas	14.5	Maryland	11.8	Oregon	13.7
Calif.	12.7	Mass.	10.1	Penn.	10.6
Colorado	14.3	Michigan	13.1	R. I.	10.1
Conn.	11.0	Minn.	13.7	So. Caro.	14.0
Delaware	12.8	Miss.	13.2	So. Dak.	17.1
D. of C.	8.0	Missouri	14.4	Tenn.	14.1
Florida	13.7	Montana	16.9	Texas	16.0
Georgia	14.6	Nebraska	14.6	Utah	13.9
Hawaii	8.9	Nevada	18.4	Vermont	13.2
Idaho	15.3	New Hamp.	12.5	Virginia	13.2
Illinois	12.4	New Jer.	11.4	Wash.	12.9
Indiana	13.3	New Mex.	16.1	W. Va.	11.6
Iowa	15.0	New York	8.3	Wisc.	12.4
Kansas	14.7	No. Caro.	13.9	Wyoming	21.9

Source: Federal Highway Administration.

This tells us that, on the average, each person in Alabama used 13.9 barrels of motor gasoline in 1978, while each person in Wyoming used 21.9 barrels.

Depending on what you wanted to determine from the data, you might try several different classification schemes. One might be "regions of the country"—you would put all states from the same region into a group, and then look for trends that differed from one region to another. Another method might be to use numerical categories such as:

less than 10 barrels

10-11.9 barrels

12-13.9 barrels

14-15.9 barrels

16-19.9 barrels

more than 20 barrels

States would automatically fall into one of the groups above, and you would look for similarities according to the amount used. Notice that in both cases above, the categories are predetermined using hard-and-fast classes with definite, preset boundaries.

This method of grouping data in preset categories is used quite often in the field of education. The well-known grading scale shown to the left is one of the foremost examples of our inclination to establish the class boundaries without looking at the data, and then let the scores fall as they will into the groups. And we shouldn't overlook the "two-group" model of this method evidenced by "pass/fail" systems. In many places, for a child to get into a program for the gifted, the child would have to score at a certain level on a test; one point less, and the child doesn't qualify! Or a student might have to perform up to a certain standard on a "minimum competency" exam for promotion to the next grade.

90%-100% A
80%- 89% B
70%- 79% C
60%- 69% D
0%- 59% F

The obvious advantage to using such a scheme for grouping numerical information is that no one can claim that you manipulated the groupings because of the data present. In this sense, using preset boundaries seems "fair" and "unbiased," and as such enjoys quite a bit of popularity. Yet at times this method seems inadequate for the purpose at hand. At such times, we frequently resort to our second method of classifying, that of arranging the data so that natural groupings suggest themselves.

THE CLUSTERING TECHNIQUE

As part of its summer offerings, a city recreation department scheduled a 10-week program of exercise classes for people who wanted to maintain or lose weight. According to the plan, registrants would indicate at the first meeting how much weight they wanted to lose during the summer, and a proper program would be established based on that (and other) information. The table on the next page lists the data for one such class.

Weight loss desired, over ten weeks

Allen	25 lb.	Hanson	11 lb.	Martin	22 lb.
Beck	10	Icks	12	Moore	17
Brown	21	James	3	Newton	3
Dukes	9	Jones	19	Owens	12
Evans	2	Karp	16	Stark	4
Gha	20	Lumpkin	8	Tillis	5

The instructor knew that a natural competitiveness can emerge if people with similar interests are grouped together, and that in this situation, such competition would be a healthy factor. To decide on in-class groupings, she used the "clustering" technique. First, she wrote down the entire spectrum of numbers that the registrants might list as the number of pounds to lose. She then recorded the results obtained during the first class meeting as a frequency count, as shown to the right. From this listing, clusters of people interested in losing the same amount of weight could emerge. How many groups are suggested to you by the data as shown? Does this seem like a reasonable way to proceed with such a class?

Weight loss	#	Weight loss	#
2	/	14	
3	//	15	
4	/	16	/
5	/	17	/
6		18	
7		19	/
8	/	20	/
9	/	21	/
10	/	22	/
11	/	23	
12	//	24	
13		25	/

Again, educators make use of this sort of classification scheme quite often, particularly in planning instruction in the classroom. Consider the list of IQ scores below, typical of what you might find in an elementary class:

Sally	85	Robin	99	Sean	118
Bob	115	Brent	103	Melissa	107
Bill	84	Tony	92	Shanna	96
Mary	104	Beth	108	Mark	131
Frank	96	Jody	93	John	94
Angela	81	Eric	114	Vincent	105
Mason	90	Roy	109	Jan	105
Jerry	110	Jill	87	Dee	89
Sabrina	126	Toby	99	Hazel	101

If you were about to teach something that you felt, from past experience, would be best understood if students were "ability grouped" and provided different presentations, you would likely try the "clustering" method of grouping. Starting with a frequency count (as shown on the next page), you could begin a systematic method of determining which students might fare best with a certain level of presentation.

81 /					
82	92 /	102	112	122	
83	93 /	103 /	113	123	
84 /	94 /	104 /	114 /	124	
85 /	95	105 //	115 /	125	
86	96 //	106	116	126 /	
87 /	97	107 /	117	127	
88	98	108 /	118 /	128	
89 /	99 //	109 /	119	129	
90 /	100	110 /	120	130	
91	101 /	111	121	131 /	

You would still have a number of decisions to make, of course—how many groups, what to do with borderline cases, etc.—but at least you would have a sound basis for making an initial attempt at matching ability level to type of material. When combined with a teacher's own knowledge of the "classroom chemistry," this method of grouping students for instruction becomes much more viable than the "preset category" scheme mentioned previously.

It may be worth noting that "clustering" of data is a technique used by many teachers to grade *achievement*. The underlying assumption is that the instrument used to measure achievement will itself make clear distinctions in the achievement-level of the students being measured. This seems reasonable if past experience has shown that, in fact, the test to be used does result in clear lines of demarcation. Could this technique be used with the set of scores on the right? Make a frequency distribution, and then decide.

Scores from a sixth-grade science test			
100	81	69	
98	81	65	
97	80	64	
93	79	62	
93	79	52	
91	76	47	
84	70	39	

Certainly the *A*s above distinguish themselves, but after that the judgments become quite a bit more tenuous. Do you agree?

SUMMARY

The two methods of classifying data mentioned in this section do not exhaust the schemes people use for reducing a set of individual measures to comprehendable groupings. The main issue here is that whatever grouping technique is chosen, the selection should come from an intelligent analysis of what sort of information is desired and what will be produced by the method chosen. In the set of problems to follow, you'll have the chance to make some of those decisions yourself.

SET 1

1. Return to the example of the per capita consumption of motor gasoline for 1978. Use one of the two classification schemes mentioned in the text, or develop a preset method yourself, and apply it to the data given in the table. See if your results lead you to any intuitive conjectures.

2. Return for a moment to the set of scores from a sixth-grade science test (immediately preceding this problem set). Analyze the scores using both the preset technique (90% = A, 80% = B, etc.) and the "clustering technique." Which scores would result in different classifications under your two analyses?

3. The construction industry compiled the following information about materials used in roofing houses. In its "raw" form, little meaning can be derived from the data. Reorganize the data into groups so that a reasonable comparison of the materials can be made. As listed, the first number indicates the life expectancy of the material in years; the number in parenthesis indicates the cost per square foot.

   Slate: Type A (23.4), ($.56)   Type B (27.6), ($.73)   Type C (55.0), ($1.04)

   Asbestos: Grade A (24.5) ($.62)   Grade B (12.3), ($.38)   Grade C (19.2), ($.58)

   Wood: Cedar (12.4), ($.50)   Fir (11.2), ($.62)   Oak (18.7), ($.71)

   Tile: (60.0), ($1.10)

4. On the next page you will find a sample summary for a class that took the Iowa Test of Basic Skills. Within the rectangles in the body of the chart, you will see four numbers. For example, the first rectangle to the right of "Jennifer Anderson" describes her achievement on the vocabulary portion of the test, and looks like:

59	32
8	89

   The number in the upper left, when a decimal point is inserted, gives her grade equivalent; the upper right is her raw score; bottom left is her stanine score; bottom right is her percentile rank. On this part, then, her raw score of 32 gave her a grade equivalent of 5.9, a stanine of 8, and placed her in the 89th percentile.

   Find the column labeled "Problems," which indicates how well students did on mathematical problem solving. Use whichever of the four scores you feel is best for grouping purposes, and make an initial "clustering" of students for instruction in problem solving.

   Do the same thing, but for the column labeled "Computation." Then compare your two groupings. Are there any students high in one classification but low in the other?

Source: © The University of Iowa. All rights reserved. Reproduced with permission of the Riverside Publishing Company.

*Normal ranges of blood pressures
in men and women*

	Systolic			Diastolic	
Age	Male	Female	Age	Male	Female
16	105–135	100–130	16	60–86	60–85
20–24	105–140	100–130	20–24	62–88	60–85
25–29	108–140	102–130	25–29	65–90	60–86
30–34	110–145	102–135	30–34	68–92	60–88
35–39	110–145	105–140	35–39	68–92	65–90
40–44	110–150	105–150	40–44	70–94	65–92
45–49	110–155	105–155	45–49	70–96	65–96
50–54	115–160	110–165	50–54	70–98	70–100
55–59	115–165	110–170	55–59	70–98	70–100
60–64	115–170	115–175	60–64	70–100	70–100

5. The chart on the left shows the normal ranges of blood pressure in men and women, and has already been grouped by age. Study it carefully for a few moments.

What general statements can you make about normal blood pressure from comparing the normal ranges for men and women, or for men and women as they get older? Can you make any other characterizations justified by this grouping of data?

6. In the table below, pick out the forms of exercise that appeal to you. Use the column Cal/min/lb. as the data, and group your activities into meaningful (to you) classes.

How Many Calories Do You Burn in Different Activities?

Figures in column one are calories per minute per pound of body weight. To compute how many calories you burn in any activity, multiply the number in column one by your weight and by the number of minutes. Figures in column two provide an example: one minute of activity for a body weight of 150 pounds.

Activity	Cal/min/lb.	Cal/min/150 lb.	Activity	Cal/min/lb.	Cal/min/150 lb.
Badminton:	.039	6	10 mph (6 min/mile)	.1	15
Bicycling:			12 mph (5 min/mile)	.13	20
Slow (5 mph)	.025	4	Sailing	.02	3
Moderate (10 mph)	.05	8	Skating:		
Fast (13 mph)	.072	11	Moderate (Rec)	.036	5
Calisthenics:			Vigorous	.064	10
General	.045	7	Skiing (Snow):		
Canoeing:			Downhill	.059	9
2.5 mph	.023	3	Level (5 mph)	.078	12
4.0 mph	.047	7	Soccer	.063	10
Dancing:			Stationary Run:		
Slow	.029	4	70–80 cts/min	.078	12
Moderate	.045	7	Swimming (crawl):		
Fast	.064	10	20 yds/min	.032	5
Football (tag)	.04	6	50 yds/min	.071	11
Golf	.029	4	Tennis:		
Handball	.063	10	Moderate	.046	7
Hiking	.042	6	Vigorous	.06	9
Jogging:			Volleyball:	.036	5
4.5 mph (13:30 mile)	.063	10	Walking:		
Judo, Karate	.087	13	2.0 mph	.022	3
Mountain Climbing	.086	13	4.0 mph	.039	6
Running:			5.0 mph	.064	10
6 mph (10 min/mile)	.079	12	Water Skiing	.053	8

Source: Frank Vitale, *Individualized Fitness Programs*, © 1973, pp. 186–187. Adapted by permission of Prentice-Hall, Inc., Englewood Cliffs, N. J.

7. The table below gives the average number of people per square mile, by state, according to the 1970 census. Group these data as you did in the first problem of this set.

When through, compare the two groupings to see if anything "pops out" at you. Does your data lead you to make any conjectures about a relationship between the number of people per square mile, and the amount of gasoline consumed per person?

			People per Square Mile			
Alabama	67.9	Kentucky	81.2	No. Dak.	8.9
Alaska	0.5	La.	81.0	Ohio	260.0
Arizona	15.6	Maine	32.1	Okla.	37.2
Arkansas	37.0	Maryland	396.6	Oregon	21.7
Calif.	127.6	Mass.	727.0	Penn.	262.3
Colorado	21.3	Michigan	156.2	R. I.	902.5
Conn.	623.6	Minn.	48.0	So. Caro.	85.7
Delaware	276.5	Miss.	46.9	So. Dak.	8.8
D. of C.	12401.8	Missouri	67.8	Tenn.	94.9
Florida	125.5	Montana	4.8	Texas	42.7
Georgia	79.0	Nebraska	19.4	Utah	12.9
Hawaii	119.6	Nevada	4.4	Vermont	47.9
Idaho	8.6	New Hamp.	81.7	Virginia	116.9
Illinois	199.4	New Jer.	953.1	Wash.	51.2
Indiana	143.9	New Mex.	8.4	W. Va.	72.5
Iowa	50.5	New York	381.3	Wisc.	81.1
Kansas	27.5	No. Caro.	104.1	Wyoming	3.4

Source: U.S. Government Census, 1970.

Does the picture above appeal to you more than if the title of this section had been simply the words "Presenting data visually?" If so, you're like most of us—visual stimulation enhances your interest in, and understanding of, a presentation. This factor is particularly acute when the presentation to be made relies heavily on numbers, and the audience is not one with a natural affinity for such data.

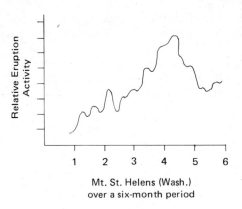

Mt. St. Helens (Wash.)
over a six-month period

You might try to imagine the amount of written description you would need to give the details provided by the graph to the left. Moreover, it isn't likely that the "trend" or "history" of the volcanic activity would be as readily discernible from a written or oral presentation as it is from the graph. Organizing numerical data into a graph of some kind may, indeed, provide a more "holistic" perspective of the situation to the audience at hand. And since clearly communicating our information to others is one of our major concerns in problem solving, it would be worthwhile to spend some time discussing these issues.

In the pages to follow, four types of graphs will be discussed: line graphs, pictograms, bar graphs, and circle graphs. Certainly these aren't new to you, but perhaps you've never been asked to consider the underlying characteristics of each type. Yet this is important, since choosing the right type of graph to display information is not an intuitive decision for many people. These four types of graphs are seen quite frequently in text books for elementary age children, starting as low as the first grade. Perhaps if we teach "graphing" correctly from that point on, future generations will have an intuitive grasp of how to organize and present numerical information in a visually stimulating fashion.

PICTOGRAMS The simplest, most eye-catching form of graph is the pictogram. The display below exemplifies this type of graph:

How the average American uses 60 gallons of water per day	
Drinking water	🝙🝙🝙
Flushing toilets	🝙🝙🝙🝙🝙🝙🝙🝙🝙🝙🝙🝙🝙🝙🝙🝙🝙🝙🝙🝙
Household cleaning	🝙🝙
Kitchen use	🝙🝙🝙·
Washing clothes	🝙🝙·
Washing the car	🝙
Washing and bathing	🝙🝙🝙🝙🝙🝙🝙🝙🝙🝙🝙🝙🝙🝙🝙🝙🝙🝙
Watering the garden	🝙🝙
🝙 represents 1 gallon	

Source: Pollution: Problems, Projects, and Mathematical Exercises. Wisconsin Department of Public Instruction, Madison, Wisconsin.

DESCRIPTIVE STATISTICS   281

Notice that a bucket was chosen above as the symbol for "1 gallon of water," and is an example of choosing the symbol so that it immediately recalls to the audience, at an intuitive level, what the graph is all about.

A pictogram loses some accuracy when fractions are to be represented; notice the "partial buckets" above, and try to guess what they represent. Yet for this same reason, pictograms are actually preferred for many tasks at the primary levels. If a child doesn't succeed in making each symbol exactly like the others, no one gets upset and assumes that a fraction is called for. Other reasons for the popularity of this type of graph at the lower levels are that children enjoy drawing the pictures, they can easily "add on" new data over time, and this type of graph can provide a variety of counting experiences.

BAR GRAPHS

The next step up in graphing is the bar graph, as exemplified below. Notice from this little-known, but unfortunately true data, that an automobile engine gives an output less than 30 percent of the amount of energy put into it. But a large electric motor would give back over 95 percent of the energy put into it! Makes you wonder why electric automobiles haven't taken over the market, doesn't it?

The categories for the data being graphed are generally—but not always—placed on the horizontal axis (perhaps to provide visual "stability") if the bars are darkened in, as shown. The scale most often goes up the left-hand side, but can go up the right-hand side, or even both sides if the graph is so spread out horizontally that the bars in the middle are difficult to read accurately.

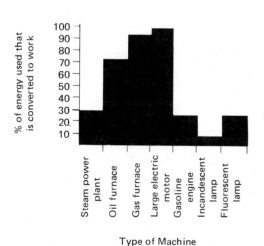

Source: "Mathematics in Energy," National Science Teachers Association, 1977.

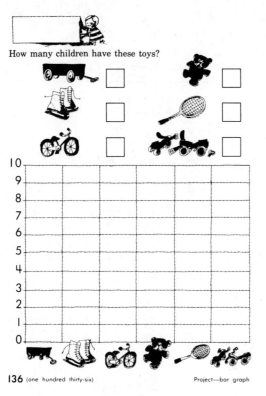

Source: Courtesy of D. C. Heath & Co.

The photograph to the left shows a page from a popular first-grade arithmetic series. As a class project, the students determine the number of children who have the various types of toys shown. Then each student produces his or her own bar graph. Notice the attempt to "personalize" the learning of students by having them obtain and graph data that applies to them instead of data already supplied to them and belonging to someone else.

LINE GRAPHS The most popular graph for showing trends over time is the line graph, as shown by the example below. Again, the categories are usually placed along the horizontal axis, and the numerical scale for the data along the vertical axis. Notice that using this scheme means that the line will go *up* as the data rise in value. While this is not a rigid rule, it does seem to communicate intuitively in most instances. In the example to the right, we subconsciously expect anything representing the price of imported oil during the 1970s to go upward, and quite sharply during the years we remember as those of the "energy crisis." Can you tell from this graph when those years were?

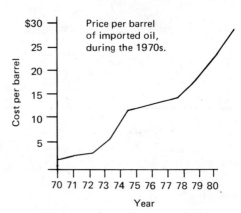

This sort of graph has a distinct advantage over the previous two in that several factors can easily be shown on the same display. In the line graph to the right, for example, Grantz is using a solid line for one of his factors, and a dashed line for the other (with "time" assumed to be the horizontal scale). The one thing that's clear from the graph is that his genetic and environmental factors are both winning and losing at times. Is this generally the case with you?

Elementary students are ready for this sort of graphing in the intermediate grades generally. By then, they can draw the lines fairly well,

Source: Courtesy of Sidney Harris and American Scientist Magazine.

and can understand what the "slope of the line" means. They can also handle the concept of several factors on a graph.

CIRCLE GRAPHS

A circle graph is one of the most useful types in certain instances, but it is also one of the most difficult to construct. Typical of this type of visual display of numerical information are the graphs below:

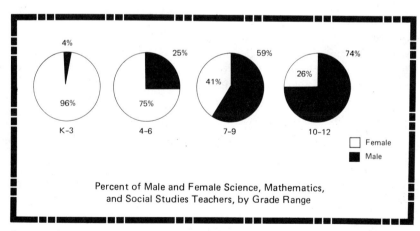

Source: 1977 National Survey of Science, Mathematics, and Social Studies Education (I. Weiss for NSF).

A circle graph is frequently used when all of the fractional parts of a whole are known and need to be displayed. The circle represents the whole, and is divided proportionally to show the various fractions or percentages to be demonstrated.

How could you proceed to divide a circle into various sized pieces—say 30 percent, 47 percent, and whatever is left? The most common method is to first find the central angles that correspond to 30 percent and 47 percent of the central angle of the circle. We make use of the well-known fact that a circle has 360° around its center point, and find 30 percent of 360° (108°) and 47 percent of 360° (169.2°). These are the central angles we need to make inside the circle, and can be done easily using a protractor.

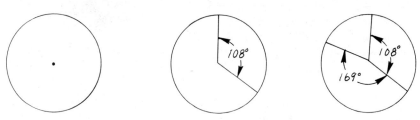

Students can't be expected to construct circle graphs with accuracy until the upper intermediate grades, but they can understand and interpret such graphs earlier.

Circle graphs are not useful if the data are not meant to be interpreted as "parts of a whole picture," if there are more than four or five categories, or if one of the categories turns out to be an extremely small portion of the circle (say 2° or less).

PROBLEMS ENCOUNTERED IN GRAPHING

Selecting an appropriate scale becomes a problem at times, as illustrated by the data on the right. The categories for the data are dates in history; the difficulty for this situation is that the given dates would not be equally spaced on the horizontal axis of a line graph. Intuition would tell most of us that a scale like the one shown below simply would not communicate accurately.

World population figures

Date	Millions
4000 BC	85
1 AD	300
1600	450
1700	700
1800	1,000
1850	1,300
1900	1,700
2000	(est) 6,254

Yet surprisingly, a good many people would use a scale like that instead of one like these:

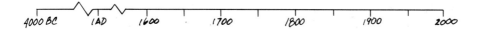

The scale on the previous page has the advantage that the distance between points has been preserved accurately, giving a good perspective of the amount of time elapsing. The one above does not give this same perspective, but the symbol —⌃— alerts the audience to this point, and this scale does have the advantage of the categories not "bunching up" toward the end of the scale.

A second problem illustrated by this example is the one of the data itself (in this case, the population numbers) having such a wide range (from 85 million to over 6000 million). Again we could use one of the two solutions mentioned above: use an unbroken scale that gives the total perspective, or break the scale using —⌃— .

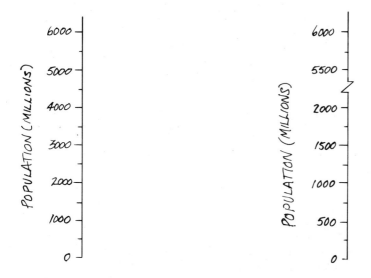

The choice of which solution is best again hinges on the audience. Do they need a total perspective at the cost of some accuracy, or the opposite?

The next problem set will give you the opportunity to solve some of these graphing dilemmas yourself. Keep in mind that the main issue is that of communication of numerical information in the clearest fashion possible.

SET 2

1. Sound is measured in units called "decibels." Typical decibel levels are given below and continued on the next page. Make a graph to illustrate this data.

rock and roll band	120 db.	quiet restaurant	50
jet taking off	150	siren	120
library	35	thunder (close)	100
whispering	30	vacuum cleaner	60

traffic (avg.)	75	leaves rustling	20
motorcycle	110	normal breathing	10
noisy party	80	rocket engine	180
screaming child	92	grass growing	2

Illustrate on your graph the discomfort level of noise (85 db.), which, if inflicted on a subject for about 8 hours, can cause permanent damage. Also indicate the levels at which the eardrum may rupture (125 db.) and death may occur (200 db.).

2. The circle graph below represents categories of accidental deaths among children for 1966. First, estimate the percentage for each category. Then measure each angle with a protractor and calculate the percentages closely. How accurate was your initial visual estimate?

Source: Profiles of Children, 1970 White House Conference on Children (Washington, D.C.: U.S. Government Printing Office, 1970), p. 64.

3. The mileage estimates below come from the 1978 E.P.A. estimates for new cars. Make a pictogram to illustrate how much money the gas alone would cost you to drive each of the cars from this campus to one of your favorite distant spots (home, beach, ski resort, etc.).

Corvette	17 mpg	Triumph	23 mpg
MBG	26	VW Rabbit	31
Fiat X1/9	23	Porsche 911	19
Datsun 280Z	21	Rolls Royce	11

4. A college class at the University of Northern Iowa became interested in the conjecture that a disproportionate number of people die soon after a birthday. The conjecture is supported by the explanation that, since a birthday anniversary is an occasion for special honor and recognition, the anticipation of it constitutes a reason to live. After the celebration, depression is likely to set in, causing a general decline in health. It certainly seems reasonable, then, to consider that more people would die soon after a birthday than at another time of the year.

This class searched the obituary section of the local newspaper for a six-month period, classifying each death according to how many months after the last birthday each occurred. The result is shown on the next page:

Make an appropriate graph to illustrate this data. Also indicate what the graph should look like if the conjecture above were true for these 494 deaths.

How well does your graph of the actual data match the one for the conjecture being true? What is your conclusion?

No. of months survived after last birthday	No. of people
0	47
1	39
2	42
3	35
4	43
5	31
6	42
7	50
8	39
9	31
10	47
11	48

Source: Duncan and Litwiller, "Birth and Death Rates: A Problem for Statistics," *I.C.T.M. Journal* 7 (1978): 14–15.

5. An NSF-sponsored national survey (Weiss 1977) produced the following interesting results. Make appropriate graphs to represent the data visually.

Elementary teachers' perceptions of their qualifications to teach mathematics, science, social studies, and reading (percentages given)

	Math	Science	Social Studies	Reading
Not well qualified to teach	4	16	6	3
Very well qualified to teach	49	22	39	63
Adequately qualified to teach	46	60	54	32
Unknown	1	2	1	2

Date	Millions	Date	Millions
4000 BC	85	1950	2,543
1 AD	300	1970	3,650
1650	525	1975	3,996
1750	800	1976	4,061
1800	1,000	1980	4,817
1850	1,300	2000	6,254
1900	1,700	2007	7,600
1920	1,862	2070	25,000
1930	2,070		
1940	2,295		

6. The table of world population figures given earlier in the text was a subset of the information on the left, which is an official United Nations forecast based on midyear estimates.

You're about to make a presentation to the zero population group, and need a graph. Make one that would communicate appropriately the information to this group.

Population Studies

Numerical information is gathered about all sorts of things, by all sorts of methods, for all sorts of reasons, and in all sorts of ways. Occasionally we even use the "shotgun" method being employed below—that is, asking an open-ended question and looking at the results to see if anything "pops out." Usually whatever pops out (teeth, perhaps) is then investigated in more detail by collecting additional data to answer specific research questions.

Source: B.C. by permission of Johnny Hart and Field Enterprises, Inc.

We will concern ourselves primarily with the latter type of data-collection procedure, that of gathering numerical information to answer very specific questions. In general, the purpose of the data-collection process is to describe a group of individuals. This group of animals, people, apples, or whatever is referred to as the *population* about which we are speaking. Usually the population under study is a relatively large group, but not always.

The cartoon above illustrates a primary concern of dedicated researchers—how to obtain accurate measurements from individuals when the question being asked is one that might evoke passionate, variable responses. When humans make up the population under consideration, personal inhibitions and egos make the data-collection process a real challenge. Even removing these factors doesn't solve all of the problems, since human responses to nonemotional issues change somewhat from one day to the next.

A second concern for researchers is how to draw accurate conclusions about a population when it's impossible or impractical to measure each individual. At such times, a *sample* of the population is taken, measurements are made of this subgroup, and the descriptors of the sample are then generalized to the entire population. As illustrated in the cartoon on the next page, our generalizations are always open to criticism.

How about you? Would you be convinced that "no two are alike" is an accurate description of the population of all snowflakes, if a researcher had examined 800,000 with no duplications? Notice that a sample of some size would have to suffice, since it would be impossible to measure the entire population of snowflakes!

DESCRIPTIVE STATISTICS

Source: B.C. by permission of Johnny Hart and Field Enterprises, Inc.

The first concern mentioned above—that of obtaining accurate, reliable information from individual measurements—is something we can only alert you to. Caution seems to be the watchword here, both when gathering your own data and believing that of others. There are some fairly standard ways of collecting samples of populations, however, and these deserve further consideration.

SAMPLING A POPULATION

In many cases, some important characteristics of the population are well known, and researchers try to account for them by taking a *representative sample.* For example, it seems likely that, in a study of adult self-esteem, the occupation of the adult might be a primary consideration. So a researcher might go to the latest census figures to find the proportion of the adult population classified as professionals, semiprofessionals, businesspeople, farmers, skilled laborers, and so forth. These proportions could then be used to select a representative sample so that this important characteristic was the same in the sample as in the entire population. A *representative sample,* then, is one in which there is a conscious effort to match critical factors in the population within the sample itself.

Another kind of sampling is *random sampling.* Many researchers prefer this method to all others when they can't identify, or match, the factors that might affect the question being studied. One consequence of a random sample is that each individual in the population has an equal chance of being selected for the sample. In such a selection process, it is assumed that if the random sample is large enough, all the critical factors (even though unknown) will appear in the sample just the way they do in the population. The larger the random sample, the more likely this is to hold true by chance alone.

Stratified random sampling combines random sampling with the specific need to have a representative sample. You may know some of the critical factors that would affect the data you obtain, but you may suspect that others exist. So you would randomly select the members of the subgroups that make up your representative sample. For example, to study the opinions of the students at your school regarding abortion, you might want to have a sample that was proportionally the same as the percentage of students in each religious category:

Protestant....... 38%
Catholic........ 27%
Jewish......... 18%
Moslem 11%
Other 6%

Your sample, if you could only measure 200 individuals, would consist of 76 Protestants, 54 Catholics, 36 Jews, 22 Moslems, and 12 "Others." In selecting each of these subgroups, however, you would use a random process. That is, each Catholic student at the school would have an equal chance of being included in the 54 for that group, etc.

The problems in the next set are typical of those that require sampling a population. Be particularly careful and critical of your own work and you'll learn how to judge the statistical claims of others. Failure to be skeptical about claims of random or representative sampling, or generalizing to populations beyond the one from which the sample was drawn, are at the heart of the American public's acceptance of statistical nonsense.

SET 3

1. Devise a plan for determining the approximate reading level of this text, using the "Fry technique" described for a 100-word passage in the chapter on measurement. Describe the size of the sample you would take, what type of sample it would be (random or whatever), how you would select the individuals in the sample, and any limitations of your plan.

2. What criticisms could you make of the data-collection procedures below:

   a. A magazine wants to predict the winner of the next presidential campaign, so it sends out sample ballots to its readers. Analyzing the results, the magazine makes its prediction.

   b. A researcher wants to determine the percentage of education majors that plan to pursue graduate education sometime in the future. The researcher sends a survey to the education students at Ohio State, UCLA, University of Texas, Penn State, and the University of Michigan, and bases his statements on these results.

   c. The student newspaper wants to do an article on alcoholism, but needs to determine to what extent the problem exists on that particular campus. The editor sets up a special booth during registration for the next term, and verbally polls students as they come through the line.

   d. To build up a case for sex discrimination, a woman faculty member sends an opinionnaire to her peers, asking if they feel discriminated against. Based on the fact that 75% of the opinionnaires that were returned said "yes," she reports that "3/4 of our female faculty feel discriminated against."

3. Devise a plan to find out whether male or female students at your school have the best eye-hand reaction times. What size samples would you pick? How would you select the individuals? What factors would be important to control due to human variability (time of day, etc.)?

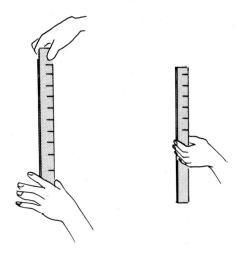

Group Measures

A small school district in Iowa likes to boast about the average salary paid to its employees. Listed alphabetically below are the salaries of the 37 employees in that district (1980 data):

Abernathy	$10,700	Kummer	$13,400	Petry	$13,420
Barnes	10,400	Lathrop	8,800	Reese	7,950
Cyr	8,850	Lukes	9,750	Robbins	10,385
Davis	31,000	Marts	11,230	Sewell	8,780
Dunlap	7,950	Marvin	10,385	Smith	10,500
Gray	8,680	Merkins	29,300	Stevens	10,350
Green	11,240	Miller	7,600	Tanner	12,380
Haskins	15,800	Moncrief	9,675	Tinker	7,975
Hawkins	9,850	Mueller	12,390	Williams, R.	36,000
Hume	11,360	Neimeyer	10,680	Williams, W.	11,235
Iccola	23,470	Nottinger	8,770	Vertuno	10,100
Jaynes	8,060	Olsen	11,240		
Kansky	24,500	Peters	9,890		

The average salary above is $12,812.03, which sounds pretty good as a salary for that particular area. Everyone thinks of themselves as at least an "average" person, and consequently if they work for that school district they should make at least $12,812.03.

Some other questions might come into your mind as you examine the data above:

What is the highest salary? The lowest?

What is the range of salaries?

What is the middle salary?

Do any salaries occur frequently?

Are the salaries spread out evenly between the highest and lowest, or do they tend to cluster at some point?

Questions similar to the ones above, with numerical answers, would give us the essential features of this group of salaries.

Descriptive statistics has developed some fairly traditional ways of describing numerically an entire collection of scores. These "group measures," if you will, allow us to describe with some consistency the essential characteristics of an entire collection of data. This would prove beneficial if you wanted to contrast the group of salaries above to that of another school district, or compare the class of students you have this year to the sort of classes you've had in the past. We need "group measures" just as we need measures of individuals if we expect to make value judgments about phenomena that affect entire populations.

Two aspects of a set of data are of primary concern, and are hinted at by the questions asked previously about the salaries given. First, we would like to arrive at a *representative* number that is, more than any other number, "like" the other numbers that make up the data. Secondly, we need a way to describe how much *variation* there is in the set of scores, i.e., we need to describe numerically whether the individual measures are "close to each other" or "far apart" on the scale. Some possible ways to obtain a representative number are described below as *measures of central tendency;* finding various ways of describing the variability of a set of data will be covered in *measures of variation*.

MEASURES OF CENTRAL TENDENCY

The *mean* (or "average") of a set of numbers is the most familiar measure of central tendency. You're acquainted with the procedure for finding the mean of a set of numbers:

The mean *of a set of numbers is its sum, divided by the number of scores used.*

The average salary ($12,812.03) for the previous example was determined by adding all of the salaries, and dividing by 37.

To get an idea of how good (or bad) the average is as the representative number for the whole set, you might begin by lining up the collection of numbers from smallest to largest, and see where this number fits in. This has been done below for the previous example of salaries:

$7,600	9,890	11,360	
7,950	10,100	12,380	
7,950	10,350	12,390	
7,975	10,385	13,400	← mean
8,060	10,385	13,420	
8,680	10,400	15,800	
8,770	10,500	23,470	
8,780	10,680	24,500	
8,800	10,700	29,300	
8,850	11,230	31,000	
9,675	11,235	36,000	
9,750	11,240		
9,850	11,240		

Does this give you a somewhat queasy feeling about the average always being a good representative for a set of numbers? It should, since only 8 of the 37 employees would make at least an average salary! Can you explain how the mean of this set of numbers is a misleading indicator of what an employee in the district could expect to make?

This particular measure of central tendency is quite useful at times, but its tendency to be affected by extreme scores limits its value as a true representative number. In such instances, we turn to other indicators of central tendency.

The *median* of a set of data is literally the "middle score" of the whole set, and is found without any computation:

The median *of a set is found by lining up the scores from smallest to largest, and then picking the one that divides the list in half.*

In the previous example of 37 salaries, the 19th one splits the list in half. What is the *median* salary of this group of employees then? Is this a better representative number than the mean, for this example?

$7,600	9,890	11,360	
7,950	10,100	12,380	
7,950	10,350	12,390	
7,975	10,385	13,400	
8,060	10,385	13,420	
8,680	10,400	← median	15,800
8,770	10,500	23,470	
8,780	10,680	24,500	
8,800	10,700	29,300	
8,850	11,230	31,000	
9,675	11,235	36,000	
9,750	11,240		
9,850	11,240		

In the event that a set of data has an even number of scores, there are two so-called middle scores, and to determine the median, we merely average these two middle scores. The data to the right is an example—the *median* is the average of 27 and 25, or 26.

$$58, 52, 47, 27, 25, 23, 19, 18$$
$$\uparrow$$
$$\text{median}$$

Both the mean and the median have their day in the sun, and are apt descriptors of what tends to happen in the center of a set of numbers. But at times they fall short as representative numbers for a group of data.

Still a third measure of central tendency is the *mode* of a collection of scores:

The mode *of a set is the most frequently occurring number in the set.*

The set of salaries is "trimodal" in that three salaries—$7,950, $10,385, and $11,240—each appear twice on the list, and no salary appears more than twice.

The mode of a set—or knowing that a set is bimodal or trimodal, or has no mode—certainly might tell us something about how the phenomenon being measured exists in the population. This representative number (or numbers) might give hints about where to begin searching for underlying reasons that the data "bunch up" in certain places.

This set of salaries is one for which the mode would not be an extremely useful measure of central tendency, since between the largest and smallest numbers there is such a wide range of possible salaries. And with only 37 numbers in the set, it would be quite unusual to find many that repeat. Only a few at the lower end of the scale—teachers who just recently were hired at the same base rate—might likely imply that this would not be a good representative number for the group.

In such a situation, sometimes the data are conveniently grouped and dealt with in a similar fashion. A convenient way to group these data would be to use the preset intervals $7,000-$7,999, $8,000-$8,999, and so forth, up to $36,000-$36,999. Then a frequency distribution of the salaries can be illustrated in the form of a bar graph. This technique serves to reduce the list of 37 separate salaries to a much more interpretable, visual form.

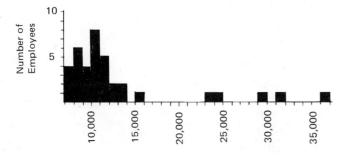

The data has been changed a bit and we might even begin to see a "mode interval." This *mode inverval* would be the category $10,000-$10,999, since that group of salaries occurs most frequently.

Each of these three measures of central tendency—mean, median, and mode—tells us something unique about a group of data. Rather than choose one over the others, it's a good practice to present all three when describing a group of numbers. Then the choice of which one best represents the group can be made in several ways, depending on the situation at hand and the prejudices of the audience.

DESCRIPTIVE STATISTICS

MEASURES OF VARIATION

The simplest measure of the variation of individual scores in a set of data is given by the *range*. Specifically,

> *The* range *of a collection is the difference between the highest and lowest numbers.*

This measure suffers from the same limitation as the mean does in that it is too sensitive to extreme measures. If only one score in a set is far distant from the others, the range picks this up. If a small range is reported, we can feel secure that the data don't vary from each other a great deal; if a large range is reported, we really have no way of knowing if most of the scores are close together and a few far apart, or vice versa. This desire to report an "average variation" number led statisticians to develop another way to measure variation.

The *standard deviation* of a group of scores has become such a commonplace way to report dispersion that it has a universally recognized symbol σ. The process for calculating the standard deviation (also symbolized by SD) is given below:

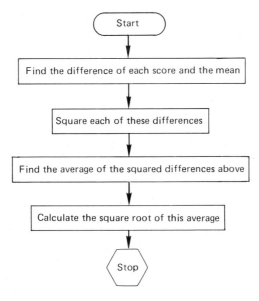

The standard deviation isn't exactly the "average difference" of the scores from the mean of the group, but it was derived from motivation around this idea. And it won't hurt for you to interpret it somewhat loosely in this fashion.

A relatively small SD would indicate that, overall, the individual scores were fairly close together. A larger SD, relatively speaking, would indicate a greater overall spread of the data. In this case, you could conclude that the greater the SD, the more variation in the scores.

Consider for a moment the heights given below for two basketball teams (in inches). Even though the teams have the same mean height (75), one team appears not to vary much from this average, while the other team members seem to vary quite a bit from this score. Using the flowchart on the previous page, run through the computations for both teams, and see how your calculation of the standard deviations compares with what's given below:

Team A			Team B		
Height	Difference from mean	$\left(\dfrac{\text{Difference}}{\text{from mean}}\right)^2$	Height	Difference from mean	$\left(\dfrac{\text{Difference}}{\text{from mean}}\right)^2$
69	6	36	73	2	4
72	3	9	74	1	1
72	3	9	74	1	1
73	2	4	74	1	1
73	2	4	75	0	0
76	1	1	75	0	0
77	2	4	75	0	0
79	4	16	76	1	1
79	4	16	76	1	1
80	5	25	78	3	9

Mean is 750 ÷ 10, or 75.

SD is $\sqrt{\dfrac{124}{10}}$ or 3.52$^+$

Mean is 750 ÷ 10, or 75.

SD is $\sqrt{\dfrac{18}{10}}$ or 1.34$^+$

As you can see, the team with widely varying heights has the largest standard deviation.

This procedure may seem like a lot of work if you have a large number of scores to deal with. Some hand calculators are designed to compute SD automatically merely by entering the data one at a time and following relatively simple steps. It is also helpful to organize the steps of the procedure into a table like that shown above for teams A and B. For statisticians, finding the SD has become one of the routine algorithmic procedures discussed in Chapter 1 as a Level 2 problem solving process!

The standard deviation is also used quite often in examining the degree of relationship (correlation) between two or more sets of scores. For instance, it is possible to use the results of one test (e.g., the Scholastic Achievement Test) to make reasonable predictions of college freshmen grade-point averages for a large number of students. College admissions officers, teachers, and businesses make frequent use of the standard deviation in trying to find the best predictor of a set of scores in terms of future potential. We will not pursue this further, but you might want to keep in mind that the notion of standard deviation serves some important practical and theoretical purposes beyond the scope of this book.

SET 4

1. Return to the standardized test results shown in Problem 4 of Set 1. Find the scores labeled "Reading," and use the *stanines* given to compute the three measures of central tendency for this class. To be sure you're off to the correct start, check to be sure that the first two stanines are 5 and 3.

2. Using this same class profile, but this time using the *grade equivalency* scores instead of stanines, compare the variation in the set of reading scores and the set of total mathematics scores. (Notice that, at the bottom of each column, you'll find the averages already calculated for you—4.62 and 4.49.)

   Does the result surprise you? Would you have predicted that the other subject area would probably have more variation in a fourth-grade class?

3. Engineers who design the braking systems for cars are concerned with two factors—stopping the car in a reasonably short distance and doing so consistently. Two systems are proposed for a new car to be introduced in 1985. Test data on the two systems in their present stage of development are given below:

System	Braking distances in feet
X	98, 99, 102, 100, 97, 103, 101, 101, 100, 99
Y	98, 93, 105, 91, 103, 98, 89, 91, 107, 105

   Use the concepts from this section to support choosing one system over the other, if one of them were to be installed on your personal car next week.

4. The body weights listed in the table below were collected on 110 entering freshmen at an eastern university. Construct a bar graph of the data, and then use the midpoint of each of the categories as the "assumed weight" of each person in that category.

Weight (Kg) interval	Number of freshmen
under 40	4
40 – 50	10
50 – 60	22
60 – 70	35
70 – 80	20
80 – 90	13
over 90	6

   a. Calculate the mean, median, and mode, and mark on your graph where each would fall. Which seems to represent these data best?

   b. Compute the SD of the data, and mark on the scale of your graph on both sides of the mean, how far away one SD from the mean would be.

   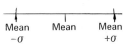

5. Your grade-point average (GPA) should more properly be called a "weighted average" because a grade of B in a five-hour course is worth a different amount in figuring the GPA than a B in a three-hour course.

   Sue attends Auburn University, which employs a four-point system (i.e., A = 4, B = 3, C = 2, D = 1, and F = 0). Her GPA (3.15) for the first term of her first year is computed on the next page.

Course	Grade	No. of hours	Grade points
Mat 104	B	3	9
Eng 243	A	2	8
Edu 300	C	3	6
Art 305	A	1	4
His 216	B	2	6
Mus 305	A	2	8
		13	41

$$\frac{3.153^+}{13 \overline{)41.000}}$$

During the next term, Sue takes three 5-hour courses. Is it possible for her to raise her cumulative GPA to a 3.5 if she makes all A's?

6. Calculate your own GPA (cumulative, if possible, but for your last term anyway), and see if it agrees with what the school registrar says you made.

7. An instructor gave a test to her class in religious practices, collected all the test papers and immediately computed the mean score. She counted all the papers to be sure she had them all, and recomputed the mean a different way, just to be sure. Again she obtained 73 as the class average, and committed it to memory.

When she settled in later that night to finish the grading process by giving letter grades to the students, she found out that a test had been lost, and she had no idea where it might be. She tried but could not find it anywhere. Is there a way that she could assign a letter grade to the student whose paper she lost, legitimately? If so, what would she do?

The Normal Curve

The four graphs below represent just a few of the more "regular" ways in which data from samples may appear visually when graphed:

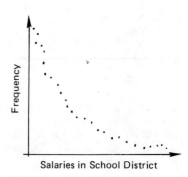

DESCRIPTIVE STATISTICS 299

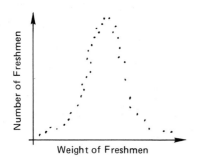

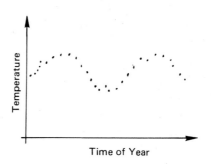

As more and more data from a population are graphed—i.e., as the sample size grows very large—and the categories along the bottom become more refined so that finer distinctions can be made, the points for the frequencies become "tighter packed." Eventually (theoretically) continuous curves replace the individual dots, so curves that represent large population factors are "smoothed out," and might appear as graphs like those below:

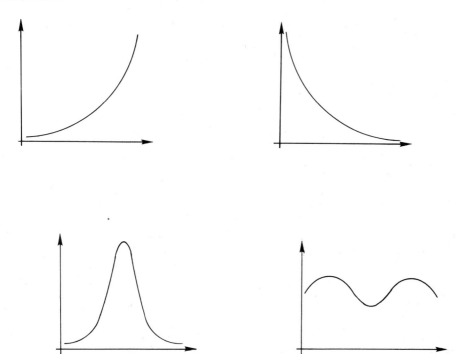

Each of these four illustrate interesting and meaningful distributions, and might be (and in fact *is*) studied in its own right. Note on the next page, for example, where the mean, the median, and the mode might be on each of the four types of graphs:

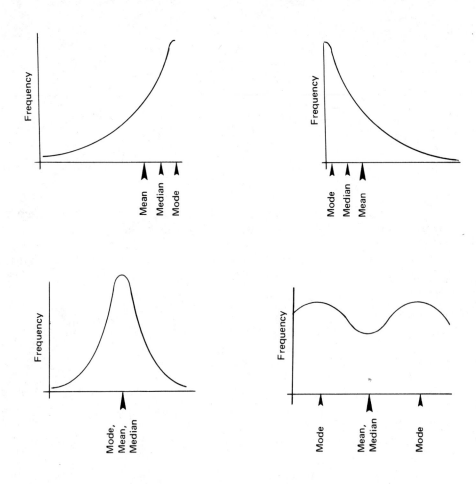

In the third graph of the four—the one shaped like a hat or a bell—the three measures of central tendency are all in the same place. That's nice at times.

This particular shape for a graph is called the *normal* or *bell-shaped* distribution. Due to its utility in the field of education, it deserves special mention when statistics is being related to the classroom. This section, then, will review with you some of the underlying structure of this *normal curve*, and will give you an example of how it affects the standardized-testing market.

STARTING WITH QUETELET Apparently, the Belgian mathematician and astronomer Adolphe Quetelet (1796–1874) first observed and made note of the fact that many anatomical measurements conform to this bell-shaped distribution. His measurements of height, weight, chest size, and even length of limbs, all tended to result in graphs like the one on the next page:

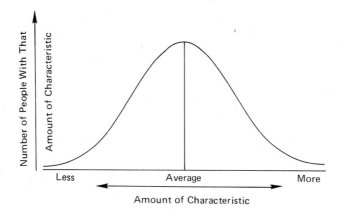

That is, most of the measurements are close to the average; moving away from the average, the measurements occur less frequently, and in about the same numbers on both sides of the mean.

Later on in the history of mathematics, Francis Galton (1822-1911) confirmed that a wide variety of mental attributes of humans also tend to fit this curve. And from there on, the normal curve has assumed the forefront as the most frequently observed, frequently studied type of graph.

STRUCTURE OF THE NORMAL CURVE

One of the ideas from the previous section—that of standard deviation—plays a major role in determining if a normal curve will be:

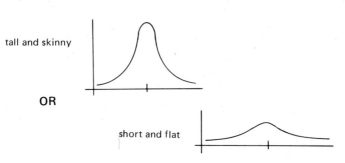

or somewhere between these two extremes. Remembering from the last section that SD represents *the extent of variation from the average,* you can say intuitively that a tall, skinny normal curve has a small SD, since more of the scores are bunched up around the center. Conversely, a relatively large SD means that the scores vary a good bit from the average score, and when graphed they would "spread out more" from the center.

The percentage of the area under the curve out to certain points has been determined by using methods from advanced mathematics. The following graph gives the area under the curve, using the mean and intervals one SD in size.

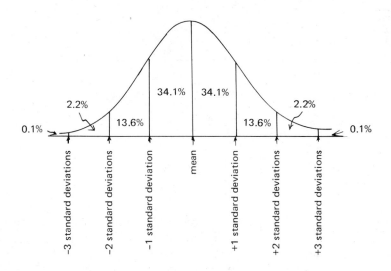

An intuitive idea to keep in mind is that the percentages that give us "the area under the curve" also give us the percentages of the population that would fit into each interval. Look at the curve above as you carefully consider these statements:

> 68 percent of the population would have a measurement that falls within one SD of the average.
>
> About 27 percent of the population would lie between one and two SDs above or below the average.
>
> Almost 5 percent of the population would fall outside of two SDs away from the average.

As you can see, the concept of *standard deviation* is extremely important when discussing the percentage of the population that would have a given amount of whatever is being measured (e.g., IQ or chest size).

You may be wondering about how the *average* and *standard deviation* scores are derived for curves that represent very large populations. Certainly we can't often measure the whole population itself. Typically, a very large, representative sample is chosen and then measured. The average and standard deviation is actually computed for this "norming group," and these numbers are assumed to be good estimates of the population average and standard deviation. In effect, the mean and standard deviation of the sample are accepted as that of the entire population.

An example is in order. Do you recall the traditional way of reporting an IQ score, the method discussed in the last section of the measurement chapter? This is gradually being replaced by a different process for measuring intelligence, one called the "deviation IQ" since it relies heavily on the standard deviation of a "norming sample." A brief return to the former discussion might help you keep what is to follow in perspective and enable you to understand the true differences between the two processes.

An intelligence test is given to a very large sample of people. The only data retained for each person is the chronological age and the raw score (number right) on this test. All the people in the sample are then subgrouped by age, and the mean and standard deviation calculated for each age group. If people that were 10.7 years old had a mean raw score of 53 on the test, and a standard deviation of 12, the basic structure of their normal curve would look like that below:

DEVIATION IQ

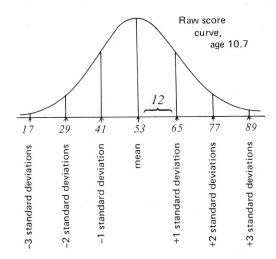

Of course, the rest of the raw scores would be filled in as appropriate. A person who scored 41 on the exam would be exactly one SD below the mean, but would have finished ahead of 16 percent of the others with that same age. To get into the "genius" category of three SD above the mean, a person in this age group would have to get a raw score of 89.

These results are then translated to an IQ normal curve that has arbitrarily been assigned a mean of 100, and an SD of 15 (16 on some tests). So a person 10.7 years old with a *raw score* of 65 would be assigned an *IQ* score of 115, since a raw score of 65 is one SD above the mean, and 115 is one SD above the mean on the constructed IQ curve. The usual cutoff point for mental retardation is an IQ of 70—this would correspond to a raw score of 29 for a child who was 10.7 years of age.

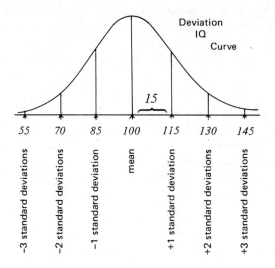

One advantage of the "deviation IQ" over the traditional version is that this measure does not depend on age. People who are 65 years old would be compared only with others of that same age group, and the result would be translated to the curve above. They wouldn't have to continue getting better scores on the intelligence test each year to compensate for an increase in their chronological age, as they would under the traditional formula IQ = MA/CA.

The process discussed above for *deviation IQ* is a typical procedure used for standardized tests of all sorts. We've only meant to give you a "flavor" of what goes on in standardized tests that refer to a norming group. You will certainly meet these same concepts later on as a teacher, and need to understand them before that. This next problem set will give you some practice in using the concept of the bell-shaped curve, both in and out of the classroom.

SET 5

1. The following graph shows the typical monthly cycle of the human female. Compare this normal curve to the others in this text, and approximate where one SD above and below the mean would fall (i.e., on what days of the 28-day cycle). Recall that 68 percent of the area under the curve would fall within this interval. Write a sentence that interprets this information to another student in this class.

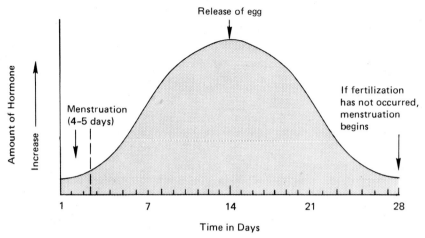

Source: H. K. Wong and M. S. Dolmatz, *Ideas and Investigations in Science: Biology.* Englewood Cliffs, N. J.: Prentice-Hall, p. 114.

2. From your own experience, or from talking to your friends, decide whether each graph below would or would not have a bell-shaped curve. If you have no idea, describe a way you could find out if you had to.

   a. Infant mortality during the first twelve months after birth

   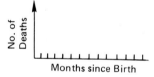

   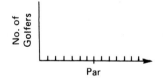

   b. Golf scores from the most recent major tournament

   c. Cost of a college education, during the past 50 years

   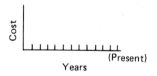

   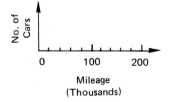

   d. Mileage on cars now being driven in the United States

MATHEMATICS FOR TEACHERS

3. The set of scores to the right represents a college class' results on a biology exam. The maximum possible score was 60. Do several things with this data:

   a. Use the "preset system of 90 percent = A, 80 percent = B, and so forth, to assign grades.

   b. Use the "cluster grouping" method of assigning grades, and see how this compares to (a).

   c. Grade the class "on a curve" by finding the SD, and using it and the scale indicated below to determine A's, B's, etc. The mean of the scores—32—has already been computed for you.

45	47
39	15
32	36
46	24
29	48
30	31
25	28
22	21
57	13
32	20

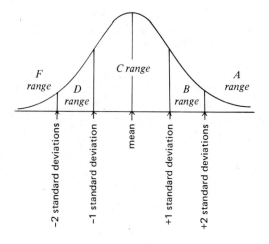

REFERENCES

Tanur, J. M. *Statistics: A guide to the unknown.* San Francisco: Holden-Day, Inc., 1972.

Triola, M. F. *Elementary statistics.* Menlo Park, Calif.: Benjamin/Cummings Publishing Co., 1980.

Weiss, I. *National Survey of Science, Mathematics and Social Studies Education.* U.S. Government Printing Office, Washington, D.C.: National Science Foundation, 1977.

PROBLEM SOLVING, EXTENDED

The curious mind of a four-year-old child is a wonder to behold, particularly if you can do so "at a distance." The question heard so often of children that age is "Why, Mommy?" This is an indication that the child is beginning to grapple with—not merely react to—the underlying structure of his world. Children want to understand what makes the world work the way it does. Hopefully this natural curiosity will be nourished during their formal schooling, producing adults who are naturally inquisitive about their environment—a necessary first step in learning to control it.

The world of mathematics provides fertile soil for developing a person's innate curiosity. At their finest, problem-solving activities can help structure thought processes and encourage originality and creativity in the way we deal with everyday dilemmas. You deserve a chance to interact personally with this expanded view of problem solving so that you can provide similar experiences for your future students.

The activities discussed in this chapter are somewhat different from the rational techniques described earlier (Chapter 1) in that they aren't hard-and-fast strategies appropriate for a particular problem. These are "developmental exercises" that, if indulged in frequently, tend to free your thought patterns by cultivating the habit of continually seeking new perspectives. These activities probably capture the true spirit of mathematics more than anything from the previous chapters. Original thinkers have, throughout history, engaged in intellectual endeavors similar to these.

You'll notice that the dilemmas themselves don't have to be complex or rigorous to prove useful as problem-solving situations. Many of them would translate easily into the elementary classroom. If you get caught up in the exercises, perhaps you'll remember feelings of enjoyment later regarding this important aspect of mathematics.

Other Solutions

When presented with a problem situation, some people are content when they reach a solution. Others feel that finding an answer is only a decent beginning in many cases. These problem solvers appear to have an insatiable thirst for finding more solutions, if they exist, or at the very least other paths to the same answer. It seems that they receive much gratification when they can legitimately feel that they've exhausted all the possibilities connected with that particular situation.

Consider the "magic square" problem given below. Spend a few moments tinkering around to see if you can solve the problem before going ahead.

Place the digits from $\{1, 2, 3, 4, 5, 6, 7, 8, 9\}$ in the squares to the right so that each row, column, and diagonal has the same sum.

Were you successful? If so, give yourself a pat on the back. If you weren't successful, try putting 8, 3, and 4 in the top row, in that order, and proceed from there.

Give the problem to a friend, and see what this person does. Is the solution just like yours? Probably not, unless you give them the hint too. If this problem were given to a room full of people, you'd get a number of different answers. This problem is typical of those for which there are any number of correct solutions (perhaps they all have a good deal in common, but still the answers are different). Many problem solvers would not be content with only one answer to this magic square—they'd want to extend the concept of the problem by trying to find a bunch of solutions, perhaps even all that exist!

History gives us an example of problem solvers coming up with different paths leading to the same conclusion. You probably recall the relationship between the sides of a right triangle, as shown to the left. What you may not realize is that this discovery has evidently been made several times over, by different people in different cultures, each one coming up with a different justification pictorially.

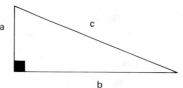

$a^2 + b^2 = c^2$

Along about the time of Christ, the grid in Figure 1 below appeared in the Chinese *Chou Pei*. This established a visual justification for the relationship between the sides of a right triangle, but only for the special case of the 3, 4, 5-sided figure. Can you "see" that $3^2 + 4^2 = 5^2$ from this drawing?

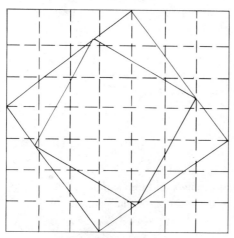

Figure 1

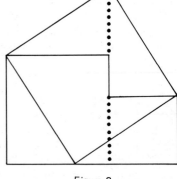

Figure 2

Figure 2 is from Tabit ibn Qorra (Arabian) and came on the scene around 900 A.D. Can you establish the truth of the relationship from this picture of a general right triangle, keeping in mind how to find "areas of squares?"

The Hindu Bhaskara (1150 A.D.) produced the drawing shown in Figure 3 below. As the only clue that he had again demonstrated the truth of the relationship $a^2 + b^2 = c^2$, he left the word "Behold!" Can you observe the same relationship as Bhaskara did?

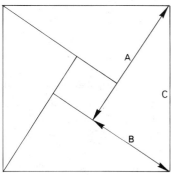

Figure 3

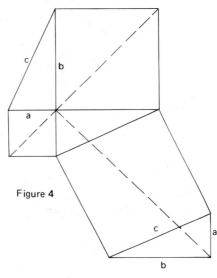

Figure 4

And finally, Leonardo da Vinci got into the act with his organizer to the right, for this same problem, around 1500 A.D.

As you can see from this example, mathematics is an area that clearly should encourage more "personalized problem solving"—different views of the same situation add much to our comprehension of any given situation.

A third example, which you have perhaps seen before, is evident from watching the way different people deal with the problem below. Some would start by writing down all the numbers from 1 to 99, and add them just as if they were doing any other addition problem. A pain-in-the-neck, perhaps, but certainly this is a legitimate solution.

Find the sum of the first 99 counting numbers.
(1 + 2 + 3 + ... + 97 + 98 + 99 = ?)

Others might try to use mathematical induction on this problem—they would solve the first few similar sorts of problems by hand, arrange their work in tabular form, and look for a pattern. If they suspect a pattern, they'd check it out on the next few simple cases by hand, and if it still seems to be a valid rule, they'd apply the rule to the original dilemma to find an answer. Look at the first few steps in this procedure, as shown above— can *you* solve this problem inductively?

#s added	sum	pattern?
1	1	
1, 2	3	
1, 2, 3	6	
1, 2, 3, 4	10	
1, 2, 3, 4, 5	15	$(5 \cdot 6) \div 2$

And yet a clever elementary student (Karl Friedrich Gauss) amazed his instructor a couple of centuries ago by solving this problem using the system below, doing the computations mentally.

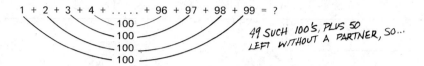

Can you follow his line of reasoning, and finish the problem yourself?

This third example points out quite clearly the advantages of seeing several ways to solve a given problem. If a similar problem were given to someone who had knowledge of the several ways to approach this type of situation, the person would be able to pick the one that was the easiest for the given numbers. A good problem solver might choose a different method for finding the sum of the counting numbers from 1 to 14, than for finding the sum of those from 1 to 317 or from 1 to 9,999.

As you proceed with the problems below, keep in mind the intent of this section —to have you find different solutions, or paths to a solution, for a given problem. Don't be content with just "the answer."

SET 1

1. Return to the magic square problem in the text, and solve it in several different ways. Then answer the questions below:

   a. Do you notice anything that is common to all your solutions?

   b. Do you think you have found all the solutions, or might there be others out there?

2. Return to the problem in the text involving summing the first 99 counting numbers, and solve this problem using both inductive reasoning and Gauss' method. Be sure you get the same answer using both methods. Then answer the questions below:

   a. Can you describe the pattern you noticed for finding the sum of the first n counting numbers?

   b. Do you think that curious elementary students might come up with some other ways of solving this problem? Can you come up with still another method that you could use as a "prompter" in such a situation?

3. One way to write the number *one*, using exactly four 4s and the symbols +, −, ×, and ÷, is like this:

$$1 = (4 + 4) \div (4 + 4)$$

Find some other ways of writing *one* with four 4s, and +, -, x, ÷.

a. 1 = _____

b. 1 = _____

c. 1 = _____

⋮

4. A dozen balls are tightly packed in a box of 3 rows, 4 balls in each row. If each ball has a radius of 2 inches, what are the inside dimensions of the box?

Tip of the Iceberg

You may or may not be aware of the fact that what we see as a gigantic iceberg is actually only about 10 percent of the total thing. The rest is completely hidden by water. All we see is just the tip of what's there!

Some of the most exciting, satisfying, and interesting discoveries in mathematics come not when we find a way to solve the problem in front of us but when we get sufficiently curious about the results to extend the problem in some unanticipated direction. Before we know it, we're deeply engrossed in "problem solving at its best," investigating problems that we structure ourselves. Later, when we come up for air, we realize that the first problem was actually just the tip of the iceberg.

In the last problem set, you considered finding a number of different ways to write the number *one* using exactly four 4s, and +, -, x, and ÷. At the time, you may have been somewhat curious about whether or not there was something "magical" about using *one*—could the same sort of thing be done for *two* and *three*, etc. Would it be possible to write each of the numbers from one to ten, using exactly four 4s and just the whole number operations? You might try doing this now, if your curiosity has been aroused.

2 = _____   5 = _____   8 = _____

3 = _____   6 = _____   9 = _____

4 = _____   7 = _____   10 = _____

Notice some other, similar extensions of the original problem that probably occurred to some in the class:

Could the number *one* be written with two 2s, or three 3s, or five 5s, etc.?

Could the number *one* be written with four 2s, or four 3s, or four 5s, etc.?

You might not want to investigate all of the questions above. Admittedly, most of us would soon get to the "I'm tired of all this" stage. But at least the extensions are possible and there for someone who is challenged by such continuous extensions of a given problem. The rest of the iceberg is certainly there, awaiting the curious mind.

Another example could be obtained from considering logical extensions of the "magic square" concept. Two such come to mind almost immediately:

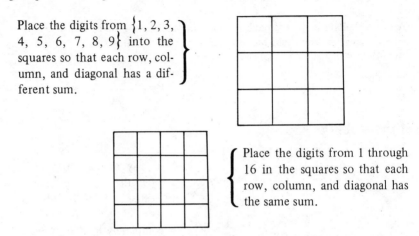

Place the digits from $\{1, 2, 3, 4, 5, 6, 7, 8, 9\}$ into the squares so that each row, column, and diagonal has a different sum.

Place the digits from 1 through 16 in the squares so that each row, column, and diagonal has the same sum.

Notice the first extension above involves changing the rules around, while the second extension keeps the same basic rule, but enlarges the set of numbers to be used. Other convenient attempts to extend the problem might involve going from a 2-dimensional square to a 3-dimensional similar figure, or changing the operation from addition to multiplication. Can you think of others?

A CALIFORNIA EXAMPLE Robert Wertz started toying around with building "brick walls" one day, and soon became interested in building them without "quake lines." A quake line is a line of mortar going all the way up, or across, a wall—this is where the wall would be most likely to crack in the event of an earthquake.

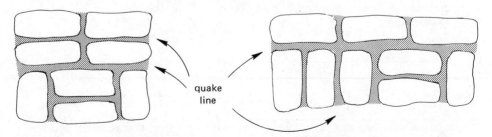

quake line

Notice that the "bricks" he was using (actually, just rectangles cut from cardboard) were all twice as long as they were wide, and he wasn't allowing the luxury of using "half bricks." Here are some more. Can you find the quake lines?

PROBLEM SOLVING, EXTENDED

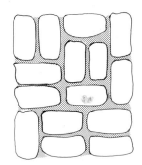
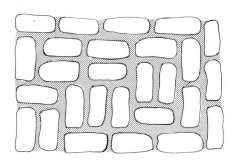

Quite naturally the question arises whether it's possible to build a brick wall without any quake lines at all. This problem can be solved fairly quickly, and leads to an extension of the sort below:

> *What is the* smallest number *of bricks required to build a quake-proof wall (assuming you have to use more than one)?*

Answering the problem above takes some people only a few minutes, and others quite a long time. Those who finish quickly might want to consider further extensions. Can you think of any right off the top of your head? If so, your mind is asking you to explore the rest of the iceberg!

Some of the problems in this next set will provide ample opportunity for you to extend the problem beyond the basic question asked. This is a chance for you to personalize the concept of problem solving by structuring your own activities.

SET 2

1. Return to the "four 4s" problem; extend the problem using one of the suggestions given at the beginning of this section, or perhaps along a path that you saw that wasn't mentioned. Write down any conclusions you reach, and any further extensions you see.

2. Cut out some cardboard rectangles twice as long as they are wide, and see if you can determine the fewest number of bricks necessary to build a quake-proof wall.

3. Place yourself in the role of an elementary teacher for a moment, planning tomorrow's lesson. You'd like to do the "brick wall" dilemma but realize that some students will find the solution to "What's the smallest number of bricks for a quake-proof wall" pretty quickly, while others will take quite some time. What're some extensions you could have in the back of your mind to provide for individual differences?

4. Bob Kansky develops a sequence of "nested numbers" in the following manner:

   a. Put any three numbers (like 4, 5, and 13) in the outer three circles. This is the first nest. Numbers are put into the next inside ring of three circles by addition (4 + 5, 5 + 13, and 4 + 13 go into the next group of circles)—this is the second nest. This procedure is continued ad infinitum.

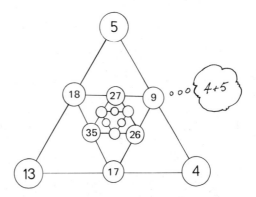

   b. Look for a moment at the *sum* of the numbers in the nests. The sum for the first nest above is 22; the sum for the second nest above is 44; for the third nest, the sum would be 88. Without putting numbers in, you can probably guess the sum of the next nest. What would it be?

   c. Make your own sequence of nested numbers like the one above by starting with a set of numbers other than 4, 5, and 13. Find the sum of the numbers in the 20th nest, using a calculator.

5. Extend the previous problem in some manner, and describe your results. Remember—the results don't have to be fantastic for you to gain from the experience.

Leonardo da Vinci was a prolific inventor, as well as a pretty fair artist. His personal notebooks are full of sketches like the one below. His marginal notes are written in this same fashion, i.e., in a mirror image script. No one knows exactly why!

Practice writing in this manner, and then write a sentence describing the invention shown at the left.

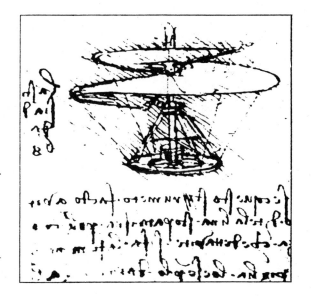

Explorations

Travellers seem to be of two distinctly different breeds—some seem intent only on getting from one city to the next in the shortest possible time, while others feel that time is of absolutely no consequence when you're "on the road." This latter group seems especially adventuresome and inquisitive. They'd much rather take the "scenic route" than the interstate highway. Exploring the backroads has more meaning to these folks than arriving at their final destination.

    The exploratory nature of mathematics has been dealt with throughout this text, but in this section it will receive special attention. The projects mentioned are some of the "sidetrips" that have intrigued mathematical explorers for many years. As you read ahead, don't worry that much about the final destination—relax and enjoy the scenery!

THE MOEBIUS BAND   Named after the mathematician A. F. Moebius, this simple idea has been used to illustrate a large number of geometrical concepts. Surprisingly, it has also found some very practical uses in the "real world." To construct a Moebius band and continue with this investigation, you'll need some newspaper, tape, and a marker of some sort.

The figure to the left represents a strip of the newspaper that has been taped end to end. The figure to the right represents a strip of newspaper that has been twisted $180°$ before attaching the ends together. This is a Moebius band. Make these two shapes, and then we'll proceed with the investigation.

One simple experiment with the Moebius band involves what may seem like a strange question: how many sides does a Moebius band have? To investigate this, use the magic marker and carefully trace a line down the center of the strip of the untwisted band, being careful not to lift the marker until you get back where you started. Since only *one* side of the untwisted band is marked, this shape has *two* sides. Do the same thing with the Moebius strip, and what happens?

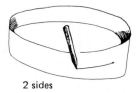 

2 sides

? sides

As you can see, the Moebius band now has a mark down its center all the way around. This means that the Moebius strip has only one side to it! Imagine that: adding a twist has turned a two-sided shape into a one-sided shape!

Did this extension occur to you yet? What would happen if you put in two *twists before taping the ends of the strip together? How many sides would the strip have?*

How many *edges* do you think a Moebius band has? How could you find out?

For another experiment with these twisted bands, you'll need a pair of scissors too. Make several bands: an untwisted one, a Moebius band, a band with two twists, one with three twists, etc.

Now predict what will happen to each strip when you cut it down the center. After making your prediction, go ahead and carefully cut each strip down its center line. You might want to consider such questions as:

> Does the number of twists affect the number of pieces resulting from the cut?
>
> What kinds of bands result from each cut?
>
> How do the lengths of the resulting pieces compare to the length of the original band?

You might want to organize your results using paper and pencil, as suggested in the next problem set.

THE MAP COLORING EXPLORATION

Map coloring has intrigued curious people for years. As a practical problem, the question arose of coloring maps with the smallest possible number of colors (similar to building quake-proof brick walls with the smallest number of bricks). The only restriction was that two regions that meet along a common border can't be the same color. If two regions meet only at a point, they can be the same color, as shown below:

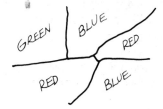

Allowable: Blue meets blue at only one point

Not allowable: Red meets red along a boundary line.

Some maps can be colored with only two colors—they're not too exciting, of course—and some with three colors, as in the preceding examples. Before reading ahead, try coloring the maps on the next page (write lightly in the book) with as few colors as possible:

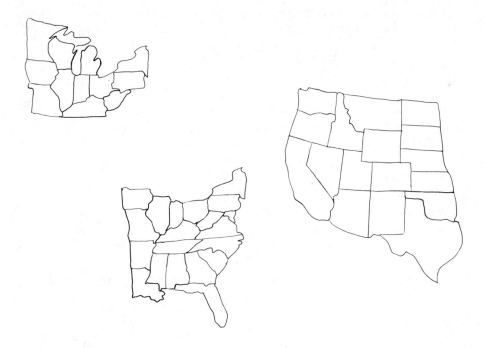

How did you do? Did any of the maps require four colors or five colors?

The map-coloring problem, or "four-color problem" as it is popularly known, is a dilemma that puzzled mathematicians for hundreds of years. The fact that no more than five colors were required was proved years ago, but no one could prove that four colors would suffice for a two-dimensional map. And yet no one could produce a map, no matter how complicated, that needed more than four colors! This phenomenon led most people to believe that only four colors were needed for any such map, but the proof would have to wait awhile!

A computer was used to settle the issue in an exhaustive procedure in the early 1970s (we won't spoil your own fun by telling you the result, of course). Most notable was the fact that this was one of only a few "proofs" that depend on mechanical devices (other than straightedge and compass). You might spend some time on some simple examples to appreciate this coloring exercise, which entertained "grown" mathematicians for ages!

PAPER FOLDING   Much of exploration involves riding one's intuition, playing the hunches, and "going with the flow." Paper-folding experiments are much like this. Through some paper-folding maneuvers it's possible to gain new insights and beliefs about relationships, especially geometrical ones. Here are some notions about points, lines, angles, and perpendicularity that arose earlier in the chapter on geometry.

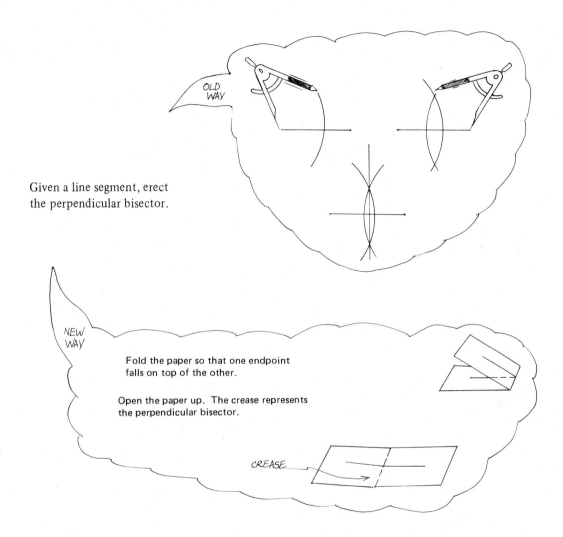

Given a line segment, erect the perpendicular bisector.

Fold the paper so that one endpoint falls on top of the other.

Open the paper up. The crease represents the perpendicular bisector.

Try the exercise above—does the crease really represent the line perpendicular to the given line segment, and does the crease bisect (cut in half) the original? Which of the two methods would be easier for an elementary school child to master?

Can paper folding be used for some of the other constructions we did with compass and straightedge? Consider the next example on how to bisect an angle.

Given angle ABC, find the bisecting line for that angle.

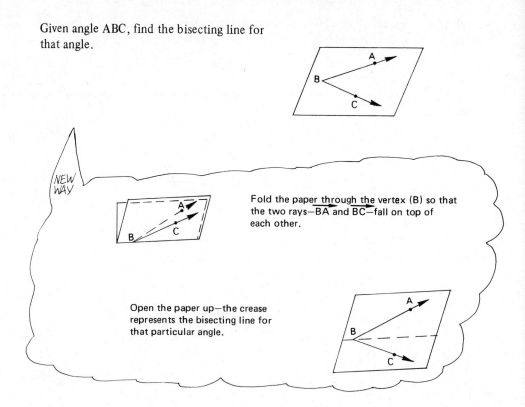

Again, it seems that "paper folding" has produced a much more intuitive, easier-to-remember process for doing this common elementary school construction. Probably you sense the need to explore paper folding further, not merely as a requirement for the next problem set, but in terms of being especially prepared to teach elementary geometry concepts.

In the next problem set, as in most mathematical situations, it seems especially helpful and important to have paper and pencil in hand and ready to go as ideas come to you. Whether you're working at the intuitive level or making a logical organization of the problem, the "nonsensical" marks or doodles on that scratch sheet may evolve into just the notions that lead you to better understandings of the situation. The trick may be just to go ahead, put pencil to paper, and start.

SET 3

1. Organize the experiment with the Moebius band and the magic marker (a line down the center of the band lengthwise) by using the following table. You'll have to find yourself a reasonable way to determine the number of edges the bands will have, and what to do in the case of 20 twists (since it would be difficult to physically make such a band).

Number of Twists	Number of Sides	Number of Edges	
0	2	2	
1	1	1	(verify)
2	—	—	
3	—	—	
4	—	—	
5	—	—	
⋮	⋮	⋮	
20	—	—	

2. In a similar fashion to 1 above, organize the Moebius band experiment and cut down the center of the strip lengthwise. Construct a table like the one above, using: number of twists, number of loops after cutting, number of twists in each loop after cutting, length of the resulting loops, and any other characteristics you think interesting to record.

3. What would happen to a Moebius band if, instead of cutting down the center line of the strip, you made and cut a line 1/3 of the way in from one of the edges? Predict and then see.

4. If you can find an old film-loop that the library has discarded, or one of your roommate's 8-track tapes, carefully take it apart and see if it has a twist in it. What would this be a real-world example of?

5. Suppose you had gone through some intuitive "map-coloring" exercises with your fifth graders, using the steps below:

   a. Students try to color some familiar maps (U.S., etc.) using only four color crayons.

   b. Students make up their own maps, and give them to a friend to try to color in four colors.

   c. The whole class concentrates on coloring any of the difficult ones from (b) above.

   After all this, the class finally concludes that probably any two-dimensional map can be colored in four colors or less. One student pipes up with "Yeah, but how about maps on round things, like a globe. What happens then?"

   Describe what you would do to help a curious student continue this exploration. Be specific—there are some practical problems involved here, since you won't have many globes to practice on.

6. Continue the paper-folding exploration by trying to do some of these common exercises via this "new" method. Use scratch paper. Write down your procedure for each one.

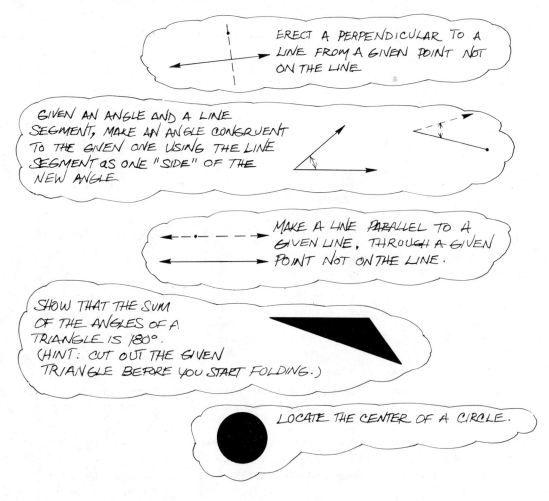

Unique Interpretations and Perspectives

Some problems can be interpreted in more than one way. This possibility can often lead to interesting results, and even broaden our understanding of the problem. The way people choose to represent a given problem indicates their interpretation and, of course, that kind of information is very important to teachers and parents, as well as to the problem solver. In this section we will take a look at problems that can be interpreted in a number of ways, both because the wording of the problem might mean more than one thing, and because "unusual" interpretations meet the conditions of the problem. These are yet other ways to extend problem situations.

THE COIN-CUP PROBLEM

Take out your paper and pencil, or better yet, some coins and cups, and tackle the problem situation below before reading ahead.

> *You have three cups and 10 coins. Arrange them so that you will have an odd number of coins in each cup, using all 10 coins.*

An initial understanding of this problem might include the pictorial representation shown below. Notice that this depiction is not a solution, but could be an example of a stage in a trial-and-error process.

   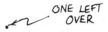

A tabular listing of some possible "ordinary" arrangements is shown below. This table helps us organize an exhaustive list of all possible placements of the coins in the cups (or does it?). As often happens, an attempt to organize and represent a problem may limit our vision about the problem (unless we are very careful). The reason for this is that any organization places certain restrictions on our thinking, because it relies on certain assumptions. In this

Cup 1	Cup 2	Cup 3	Leftover coins
1	3	5	1
1	3	3	3
1	5	1	3
7	1	1	1
⋮	⋮	⋮	⋮

problem, for example, you have probably discovered that no apparent solution is possible under the present interpretation represented by the table.

And yet here is a verbal and visual description of another way to interpret this problem (provided by several elementary students):

> **The problem calls for arranging the coins and cups in such a way as to have an odd number of coins in each cup. One way to do this is to put coins in the cups, and then cups in other cups, so that "odd number of coins in each cup" holds true.**
>
>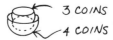
>
> **By putting three coins in cup 1, four in cup 2, three in cup 3, and then placing cup 3 inside cup 2, we have an odd number of coins in each cup!**

The coin-cup problem is a novel one at best in the context of problem solving, but one that could certainly be used in the elementary classroom to point out the tendency for our organizers to have—at times—an unnecessarily restrictive effect on our interpretations.

THE BEAR-HUNTER'S CABIN

Spend a few moments considering this particular problem before reading ahead. You might want to draw a few sketches as you proceed:

> A hunter leaves his cabin and walks a mile due south. At that point he tracks a bear 1 mile due east, whereupon the bear spots him and chases the hunter back to his cabin 1 mile due north. For these conditions to hold, where is the hunter's cabin located?

Below are some geometrical representations that indicate something about the typical first attempts of students to solve this problem. Each of these is a two-dimensional representation of the "real" three-dimensional situation.

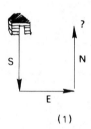 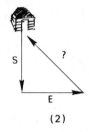 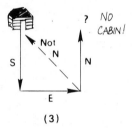

(1)   (2)   (3)

The preceding figures are indeed initial representations and furthermore would lead us to an impasse—it seems that the problem has no solution! Yet let's extend our thinking somewhat.

The representations below provide interesting, nonequivalent perspectives of the problem. Each one represents the problem well enough to be considered a true solution:

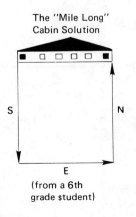 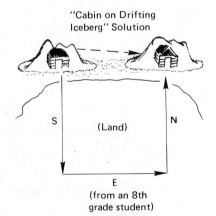 

The "Mile Long" Cabin Solution
(from a 6th grade student)

"Cabin on Drifting Iceberg" Solution
(from an 8th grade student)

"Spherical Triangle" Solution
(from many high sch[ool] and college students)

The first two representations, though not satisfactory from any logical view of reality, demonstrate quite clearly the emergence of original thought about the conditions of the problem.

The third figure, a spherical triangle, ultimately yields the North Pole as an answer to the question. The two-dimensional drawing must be interpreted as a three-dimensional figure, but that's not too difficult! Assuming that the directional arrows represent due south, due east, and due north, respectively, the figure is interpretable as a spherical triangle—thus the North Pole is *a* solution under this interpretation and representation. But is the North Pole the only solution for this realistic situation on the globe?

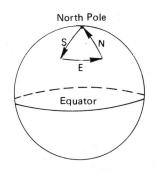

One student experimented with several different shapes and finally used the following one with a great deal of success.

The student later remarked that, after she had derived other solutions, the first solution (the spherical triangle) actually interfered with, or inhibited, her thinking about new possibilities. When you have understood her solution(s), you might agree with her remarks.

The application of her thinking to the globe is depicted in the figure to the right. Point C marks the location of the cabin. The "due east for 1 mile" is a circular path of length exactly 1 mile (somewhere near the South Pole). The proportions are exaggerated to help you visualize it more clearly. From this representation, can you describe an unlimited number of locations for the cabin to conform to the conditions of the problem?

In order to "wrap up" this problem for our purposes of showing different prespectives of a problem, consider this student's verbal description of her mental processes on this problem:

> The directional conditions of the problem are "1 mile due south, then 1 mile due east, and finally 1 mile due north—and the hunter should be back where he started." After struggling with the spherical triangle and trying to make it fit somewhere else on the globe (the Earth), I ended up teaching myself much about directions on the Earth, about what it means to go "due east," for example. I finally was able to break away from the spherical triangle pattern, and was able to make several different kinds of figures on the surface of the globe—this process led me to the key figure, which I think of as a 'headband with a feather in it.' That the hunter might retrace his path finally occurred to me for the due south-

due north part of the problem. The due east circle path is very near the South Pole; then any place exactly 1 mile due north of that circle will be a suitable location for the cabin—of course you probably won't find any bears in such a location on the 'real' Earth!

Kathy Gordon, Junior, F.S.U., 1977

Such admissions make us wonder if there aren't some other solutions out there somewhere, just waiting for a curious mind to find them!

MYSTERY SEQUENCE Determine the next letter of the alphabet in the sequence established by: O, T, T, F, F, S, S, E, ?

Did you answer *E*? If so, you're like most of us adults—we've been conditioned to trust in the fact that most such sequences rest on "numerical sequence" underpinnings. We automatically think that 2 *E*s should follow the 2 *T*s, 2 *F*s, and 2 *S*s, and that there really are 2 *O*s, but one of them has been covered up at the beginning. There's nothing wrong with this sort of conclusion, if we realize that any other conjecture that rests on the same sort of logical reasoning might also be correct.

A primary-school child might say the answer to the question is *N*, resting his or her case on the sequence above being based on abbreviations of number words—*O* for *o*ne, *T* for *t*wo (sorry about that), *T* for *t*hree, etc., on up to *N* for *n*ine. Can you argue that this child is incorrect?

This last example is to point out that "different perspectives" for problems is not the sole province of advanced problem solvers. As a matter of fact, it's sometimes quite refreshing to view the interpretations that uninhibited young problem solvers attach to our somewhat familiar situations.

The process of taking on broad and different perspectives of the same problem situation can be a very useful one for all kinds of problem solving. It requires one to examine all basic assumptions, and to experiment in terms of new or altered assumptions. Explore the exercises in the next problem set by taking on different interpretations or perspectives of the problem.

SET 4

1. You present a slightly different version of the coin–cup problem to a group of fourth graders, as described below:

   > *You have three milkshakes and 30 straws. How can you put an odd number of straws in each milkshake?*

   After thinking for a few minutes, one student says "Put 10 straws in each glass—10 times 3 equals 30, so you've used them all. And 10 straws is certainly an odd number of straws for any milkshake."
   What would you say?

PROBLEM SOLVING, EXTENDED **327**

2. You ask a class of your students "How much is half of 8?" One kid, known for his weird answers, puts on his paper:

   Half of 8 could be either 3 or 0.

   Should you mark his paper wrong, or give him credit for perhaps a unique interpretation?

3. A farmer wants to plant four trees equidistant from each other. Describe how this might be done (use a sketch if appropriate).

4. Can you cause a quarter to fall through a hole the size of a dime? Outline a dime coin and cut out the circle—will a quarter pass through such a hole?

5. So you want to be a pawnshop owner? Suppose you purchased at wholesale a guitar for $80 and sold it for $100. Later you bought this same guitar back from the purchaser for $120, but then sold it again for $140 to another buyer. In reviewing these transactions, to what extent have you gained, lost, or come out even on the dealing? Is there more than one perspective on this situation?

6. There is a game called "petals-around-the-roses" in which five dice are thrown simultaneously and the person who has made up the rule announces an appropriate number (determined by the rule), on each toss of the dice. Here are five situations of tossing, with the number that comes from the rule given on each toss of the dice. Can you determine at least one rule that would give this result?

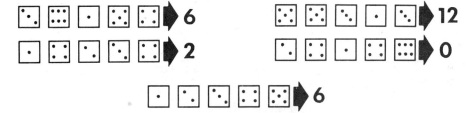

7. Stare at the drawing to the right till you have at least two perspectives about what is represented.

8. Return to the bear-hunter's cabin problem in this section and describe other solutions that match the conditions given. Your solutions can be either of the mile-long cabin/drifting iceberg school, or of the real-world view.

Problems for the Road

We trust that if you have stayed with us this far, the experience has been worthwhile for you. We have emphasized the notion that problem solving involves many processes, including everything from concrete manipulations of objects and figures on a sheet of paper to the internal, mental operations unique to every individual. Your ability to solve problems depends to a large extent on such immeasurable factors as confidence and an inquisitive, adventuresome spirit. These qualities do not guarantee success, of course, but conbined with mastery of certain basic knowledge and logical thought processes, they can go a long way toward making you a better problem solver.

So, we leave you with some problems "for the road." Some fresh ones are scattered among others that are extensions of previous situations. We challenge you to solve them all, using your personal problem-solving style.

SET 5

1. Consider for a moment the nonterminating, nonrepeating decimal
   0.12340123400123400012340000123400000123 ....
   a. What pattern is established by what you can see of this numeral?
   b. Assuming this pattern continues, what would the 200th digit be?
   c. Design one like this yourself, and invent a similar question—be sure to answer your own question!

2. Why are three-legged stools generally more stable than four-legged stools, and therefore easier for the "everyday carpenter" to make successfully?

3. Study this sequence of numbers: 5, 12, 22, 35, ... Can you determine the next two numbers in the sequence above, after establishing a logical relationship that "works" for the first few numbers in the sequence?

4. Pictured to the right is someone's organizer or interpretation for the sequence of numbers in Problem 4. Did your solution fit their interpretation?

5. Study the following sequence of numbers: 1, 3, 6, 10, ... Try to devise a picture, perhaps similar to the one in Problem 5, that would fit the sequence and give you the next numbers.

6. Consider the problem below:

$$\frac{1}{1\cdot 2} + \frac{1}{2\cdot 3} + \frac{1}{3\cdot 4} + \frac{1}{4\cdot 5} + \ldots + \frac{1}{48\cdot 49} + \frac{1}{49\cdot 50} = \square$$

The dots above mean that the pattern established by the first fractions is continued until the last one is reached. Use inductive reasoning (look at the first few simple cases, and search for a relationship) to solve the problem. Hint: reducing sums to "lowest terms" helps.

7. Consider only triangles for a moment—choose any size you want, but you can only use two at a time. Overlap them in various ways and determine how many different numbers of "interior regions" can be produced. The examples below show how two and three interior regions could be formed.

 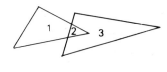

8. What is the usual shape for a manhole cover? Why do you suppose this shape is generally considered the best?

9. A rectangular floor is to be constructed of square tiles, all the same size. The dimension of the floor is 13-1/3 feet by 9½ feet. If the floor is to be covered only with whole squares, find a size of square tile that will work. What number of tiles of this size would it take?

10. Make up a tile problem similar to the one above that could be solved differently and more easily.

11. In Meneghan's "Life, a Fascinating Game" (1976, pp. 56–60), ▪ represents a living cell, and ☐ represents no life (an empty cell). One or more ▪ in a given instalce constitute a colony in a generation of cells. A colony is determined from one generation to the next according to the five "genetic" rules listed on the next page. The arrows in the figure to the right indicate the idea of "neighboring cells" of the one cell in the center.

Rules for going from one generation to the next:

1. Every living cell with at most one living cell as a neighbor dies of "isolation."
2. A living cell having either two or three living neighbors lives into the next generation.
3. Every living cell having more than three living neighbors dies of "overcrowding."
4. A living cell is born an any empty cell with exactly three living neighbors.
5. Births and deaths occur simultaneously in the same generation.

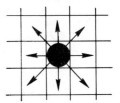

Each cell has 8 neighboring cells, as indicated:

You may start this exploration with any number of living cells, but three is a good one to begin. We will consider some examples and then ask you some questions to get you started on it. After that, you can experiment with "The Game of Life" on your own.

To begin, let's consider what happens in the case of three living cells beginning in cells indicated in the first pattern below. Successive generations are indicated from left to right. When you try this for yourself, you'll need to construct grids similar to the ones shown:

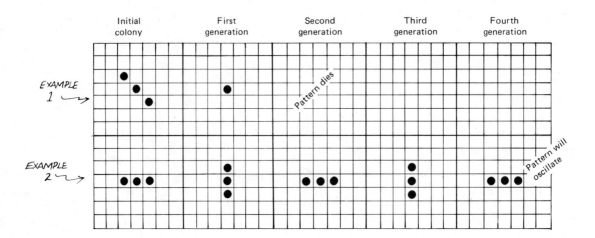

Study the application of the rules to Meneghan's "Game of Life" until you think you know the rules. Then practice on other initial patterns of colonies with three living cells, and investigate what the end result of that kind of colony is. How many different types of end results can you obtain. So far you have "colony dies" and "colony oscillates" as end results.

What about initial colonies with four living cells? Five living cells to start with? And so forth. What other questions can you think of to further the exploration?

12. Recall that a "line of symmetry" for a figure is a line over which you can fold the figure so that the two congruent halves would "match up." Using the figures below, can you find a relationship between the number of sides of a regular polygon, and the number of lines of symmetry it has?

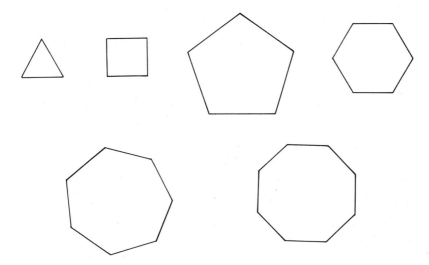

13. A well-known problem to show the dangers of inductive reasoning is given below:

   Is $n^2 + n + 41$ a prime number for all counting numbers n?

   Typically, students begin solving this inductively by taking the simplest possible cases for n first, and after looking at these results (as shown below) conclude that the statement is probably true.

Case	$n^2 + n + 41$	Prime?
n = 1	43	yes
n = 2	47	yes
n = 3	53	yes
n = 4	61	yes
⋮	⋮	⋮

   Try the next few cases, and see if you reach a nonprime number. Can you think of a number that would *not* yield a prime? (Before giving up, try 41 itself!)

14. Can you determine a rule (one of the most common ones ever) that would cause us to arrange the first ten whole numbers in the following fashion: 8, 5, 4, 9, 1, 7, 6, 3, 2, 0.

15. Reason inductively to determine the number of line segments (with end points on the edge of the circle) made by any 300 different points on a circle. See the figures on the next page.

2 pts.   3 pts.   4 pts.   5 pts.

16. Find the area of each figure below by counting, as you learned to do in the chapter on measurement. For each figure, also find the number of "edge points" and "inside points." The first figure has an area of 12½, and 15 edge points and 6 inside points.

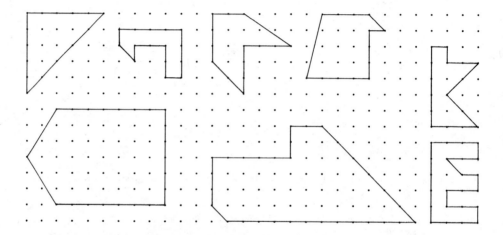

There is a relationship between the number of edge points, inside points, and the area. If you think you've found it, draw a few more such figures and test your conjecture. Organize your work using a chart like the one below. The first one has been filled in for you.

# of Edge Points	# of Inside Points	Area of Figure
15	6	12½ $\frac{15}{2} + 6 - 1$

17. The square to the right has area 64. If it is cut along the dotted lines and rearranged, it will look like the rectangle shown on the next page, which has an area of 13 x 5, or 65. What happens to make the figure lose a square unit of area?

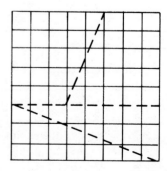

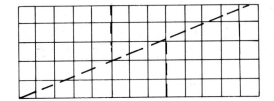

Feel free to cut out such a figure, and play around with it.

18. What letter of the alphabet would be next in this sequence:

    M, V, E, M, J, S, U, N, ?

19. Arrange the digits 1 through 9 into three three-digit numbers so that:
    a. The first number is 1/3 of the last number, and
    b. The middle number is equal to the difference between the first and last numbers.

    ☐☐☐   ☐☐☐   ☐☐☐
    1ST NUMBER   MIDDLE NUMBER   LAST NUMBER

20. When a nurse takes your pulse, she invariably writes it down on her chart as an even number. And yet it certainly seems that your heart would be just as likely to beat an odd number of times per minute! If so, why does the nurse always report a pulse rate as an even number?

21. The Gunning-Mueller Clear Writing Institute developed a "fog index" to determind how much education a person would need in order to understand what has been written. To calculate this number, you find a passage that's between 100 and 200 words long, and find the average number of words per sentence. Then you calculate the percentage of words that have three or more syllables, and add this percent as a whole number to the first number. Then multiply this sum by 0.4, and the result is the number of years of schooling needed to understand that passage.

    Find some such piece of formal writing that you've done lately, and compute your fog index.

22. What's the best possible price you could obtain, to the right, if you needed to purchase 25 pieces of chicken for a party?

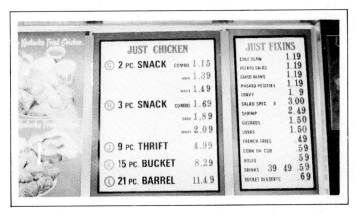

REFERENCES

Ball, F. H. "Posing and solving verbal problems." *The Mathematics Teacher.* 73, (1980): 652–56.

Kilpatrick, J. "Research on problem solving in mathematics." *School Science and Mathematics.* 78 (1978): 189–192.

McGinty, R. L., and Meyerson, L. N. "Problem solving: Look beyond the right answer." *The Mathematics Teacher.* 73 (1980): 501–503.

———. *A Sourcebook of applications of school mathematics.* Reston, Va.: National Council of Teachers of Mathematics, 1980.

Meneghan, Lauren L. "Life, a Fascinating Game." *The Arithmetic Teacher.* Reston, Va.: The National Council of Teachers of Mathematics, 1976, 56–60.

Reeves, C. A., "The Flip Side of Problem Solving." *The Yearbook.* Reston, Va.: 1982, The National Council of Teachers of Mathematics.

Commentary on Selections from the Problem Sets

This section contains comments and suggestions about some of the problems and exercises of the Problem Sets. The authors do not intend to spoil the reader's pleasure and satisfaction in working on the problems. The intent of this section is to provide some leads and suggestions, and, on occasion, some answers to selected problems from each set. Among these comments the reader will find some hints as well as extensions to other ideas related to the problems. Remember that the authors' approach is a "process orientation" and their emphasis is on doing mathematics. In no way do the following comments exhaust all the many interesting ways of thinking about these problems.

Chapter One: Problem Solving Processes

SET 1

1. Experiment with real coins. While experimenting, return your attention often to the target configuration as you move the coins around. To what extent do you focus on the rows? Did you remove the coins at the three corners? Does the resulting shape suggest any additional arrangements other than the target configuration?

2. The first frame of the cartoon means (as we have always known) that the whole banana weighs the same as the peeling + the peeled banana.
   From the second frame it can be determined that "the whole banana weighs the same as the peeled banana + 7/8 oz." So the peel weighs how much?
   Then, since the peel weighs 1/8 of the unpeeled banana, you probably can reason that the unpeeled banana weighs 7 oz.
   The first frame helps directly in answering the question in the third frame.
   Of course, *your* approach to this problem might be more like "trial and error"—making a guess and testing the answer against the description in the cartoon; then, after checking how close your guess is, make another guess for a closer approximation.

4. Should you get the same answer for the following two exercises?

    43   37
    x 37 and x 43   Why do you think so?

5. The method used in this subtraction exercise will always work. The method is different from the one commonly used requiring regrouping (or "borrowing" as it is sometimes called). The method shown here uses two other arithmetic concepts, namely, directional whole numbers and place value.

7. More detail of the pie-slicing problem is provided in the next section of Chapter 1; you need to spend a good deal of time trying the problem on your own to compare your solution to those discussed.

8. There are several different rational explanations for this letter pattern. Visually-oriented problem solvers might see that the letters on top of the line are all made with straight line segments, and those below with curves. A problem solver with strong auditory tendencies might think of the way the letters of the alphabet are spelled. Or perhaps you found another explanation.

10. a. Are you inhibited by sexist words?
    b. Some folks don't always drive their cars in an expected manner; or, "near" the North Pole there is a line of latitude exactly two miles in circumference, so . . . .
    c. Why don't good records skip when played?
    d. Perhaps triplets or quadruplets, or . . . . Also, might the sisters be years apart in age?

11. Think in terms of higher dimensions of space (such as yours!).

SET 2

1. This basketball tournament problem will be discussed in much detail in the next section of Chapter 1, so solve it on your own before going to that section. You'll want a solution of your own to compare to others.

2. The implied "starting point" (the first game) may not be the best way to begin this problem solution. Can you organize the problem by working backwards?

3. There are many ways to organize this problem. Some ways might include fancy algebraic processes, while others require only a little adding, subtracting, and multiplying.

    First, you might think about the differences between rabbits and chickens. You know, the "eyes and feet" thing. Then, you might just make some trial guesses, and test your guesses against the conditions of the problem (chickens and two feet and rabbits and four feet, etc.).
    Algebraically, there are at least two ways to organize the solution:
    First way (the more difficult one!) is with two "unknowns":

    Let R = the number of rabbits and C = the number of chickens. Then, the following two equations derived from the relationship between feet and eyes and rabbits and chickens can be used to obtain the solution:

COMMENTARY

$$4R + 2C = 126$$

$$2R + 2C = 72$$

Do you see what given information is used to derive these equations?

Second way (the easier one!) is with only one unknown:

Given that there are 76 eyes, we know there are 38 animals total! Then, let R = the number of rabbits. We can then represent the number of chickens by (38 − R). Then solve the equation $4R + 2(38 - R) = 126$.

4. Just choose the "right" (left) two numbers and add them.

5. Compare the two situations of fencing in terms of the number of the last post to be installed. Drawing a diagram works nicely here.

6. The next section of Chapter One discusses this circle problem, but you need to solve it on your own before that to internalize how *you* solve such problems yourself.

9. The answer is *not* any of the following: 2-1/8, 4-1/8, or 4-1/4.

10. Estimate, then check by multiplying. For example, a 3 x 3 square has 9 cm^2 of area, so 3 x 3 is too large. On the other hand, a 2 x 2 square has an area of 4 cm^2 which is too small; so, ...

11. If you name the "largest" number, and then add, say 1, to that number, have you created a number greater than the one you first named? Is there a limit to creating larger and larger numbers by such a process?

12. Doesn't this problem sound a little like pie slicing? How many additional sections are obtained as a new straight cut intersects each of the previous cuts?

14. Is the "average" of two numbers always *between* the two given numbers? Would drawing a diagram, and locating the three points, prove it?

SET 3

1. You have studied several ways to approach single elimination tournaments in this section of the text. You may have even discovered a general rule that governs all single elimination tournaments of this kind. If so, apply your rule. If you have not made a general rule discovery, apply any other method you used successfully to solve this problem.

2. All multiples of 3 can be expressed in the form 3 x N, where N is the whole number representing the multiple of 3 you have. For example, 12 is the fourth multiple of 3 because 12 = 3 x 4.

 Now, suppose that M is the "largest" multiple of 3. Then, M can be expressed in the form M = 3 x N, for some whole number N. But, consider the number 3 x (N + 1); is it larger than M? And, is 3 x (N + 1) a multiple of 3?

 This may appear to be a strange way of showing that there is no largest multiple of 3. After all, we began our argument by assuming exactly the opposite of what we thought to be true. (Note: This way of thinking may not be so strange after all, when you consider that in the American court system, the jury assumes that the defendant is NOT guilty and then the jury sits and listens to evidence *against* that assumption! If the evidence is sufficient—that is, if the prosecution's argument is strong enough to remove reasonable doubt—then the jury will decide that the assumption of "not guilty" is wrong, and give the opposite verdict. In a similar process, that is what we did in the argument given above.)

3. Retrace this problem; that is, look at this problem backwards. Or, try to solve it by "guess and check."

4. Another "backdoor" problem. Consider the following line of thinking for the gambler problem: Remember the rule of the game is "loser pays each winner an amount equal to what each winner has on hand."

A	B	C	
$6.00	6.00	6.00	State of affairs at the end of the game
3.00	3.00	12.00	At the end of Game 2 (used the rule above)
?	?	?	At the end of Game 1 (you use the rule)
9.75	?	?	At the start of the game (the solution)

6. Even though most foxes do not like grapes, try working this one backwards, or by "guess and check."

7. First, work from simpler cases: What is the sum of the first two counting numbers? Now determine the sum of the first three numbers. Continue summing in this manner; perhaps a pattern will emerge!

8. Guess and test. For example, 4 is too large, since 4 x 4 = 16 and 3 is too small since 3 x 3 = 9. Next you might guess 3.5. Test with 3.5 x 3.5 = ? Is 3.5 too large or too small, or is 3.5 a suitable answer?

9. One way to organize this problem is in terms of a concrete analysis shown in this section of the text: Use "pairing brackets" and create a "losers' bracket" to accommodate teams that have one loss and play each other until they lose a second game (then they are eliminated).

   You can also use any one of the other three methods of analysis (numerical, inductive, or transformational). Caution: the final solution is conditional; that is, the total number of games required is either 14 games or 15 games. Why?

10. The eraser does *not* cost $.10.

13. To "create" four squares from the five given ones, notice that in all, there are 16 toothpicks, so the resulting four squares must not share any of their sides. Here are several possible configurations which you might consider trying to construct by moving and using exactly two toothpicks in the original setup:

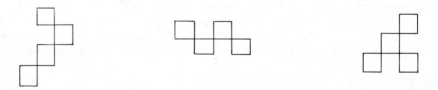

COMMENTARY 339

14. Actually we often think "backwards" when solving problems of time like this one! (How about budgetary problems?)

15. Add 6 + 4 to get 10 on next row.

16. Rule: Strike out the rows in which the halving column has *even numbers* (i.e., 84, 42, 10, 2). Add the numbers left in the double column.

17. What is 5 x 4 x 3 x 2 x 1 = ____? Recognizing that this expression helps you count the number of ways in which five books could be arranged on the shelf should assist you in determining the number of ways six books can be arranged on the shelf. Then what about Peppermint Patty's dilemma?

Chapter Two: The Third R

SET 1

2. a. XLIV b. MMMLXXX c. $\overline{\overline{L}}$ d. CMXCIX e. XXVIII
   f. MCMLXXXIV

3. ```
   1   2    3    4     5      6       7       8        9         10
   /   //   ///  ////  /////  ////\   ////\/  ////\//  ////\///  ////\////
   
   11          12           13            14            ...   35        36         37
   ```

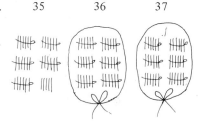

Note: In southern Spain there is a tribe of gypsies known to have among them the common characteristic of six fingers on each hand. These people do indeed have a base six counting system!

4. Base 10: 1 2 3 4 5   6   7   8   9   10 11 12 ... 34 35 36   37
   Base 6:   1 2 3 4 5   10 11 12 13 14 15 20 ... 54 55 100 101

   As word names for the above, you might start with something like:

   one, two, three, four, five, onesi, onesi one, onesi two, etc.

   Notice that the "onesi" would literally stand for "one group of six," just as in our familiar base 10 system "sixty" means "six groups of ten." Continue this process till you get a word name that is 1 more than "fivesi five."

5. a. $334_{six} = 3(6 \cdot 6) + 3(6) + 4 = $?

   c. $1100_{six} = 1(6 \cdot 6 \cdot 6) + 1(6 \cdot 6) + 0(6) + 0 = $?

7. Base 5

   It should be noted that the use of manipulative aids for teaching arithmetic concepts is a very old idea—perhaps thousands of years old. In the first paragraph of the preface of *Elements of Arithmetic* by Charles Davies in 1868 is a poignant suggestion to the reader and teacher of the subject:
   ```

"It has become a settled principle in the science of teaching, that abstract principles and their elementary combinations must be first presented to the mind by the aid of sensible objects."

9. The Mayan calendar year had 360 (or 18 x 20) days.
   a. 10 (18 x 20 x 20) + 11 (18 x 20) + 0 (20) + 3 = ?
   b. 5 (18 x 20 x 20) + 0 (18 x 20) + 4 (20) + 10 = ?
   c. 1 x (18 x 20 x 20) + 1 (18 x 20) + 1 (20) + 1 = ?

SET 2

1. a. Division—partitioning (733 ÷ 70)
   b. Addition (1823 + 145)
   c. Multiplication ($18.25 x 145)
   d. Subtraction—comparison (29 feet, 2½ inches − 17 feet, 7 inches)
   e. Division—partitioning (1900 ÷ 47)
   f. Division—measurement
   g. Subtraction—take away
   h. Addition

2. As an example, choose "partitioning"; then, if we think of dividing a number, say 10, by 0, we would be "trying" to partition 10 things into 0 subsets and then counting the number of things in each of the nonexistent subsets—a situation that has no meaning!
   Now you try it for the "measurement" interpretation for 10 ÷ 0.

3. Some examples of how these exercises might be reorganized for mental computation:
   a. (5 + 68) + 95 as 68 + (5 + 95) = _____
   d. 4 x (25 x 7) as (4 x 25) x 7 = _____
   t. 47 x 110 as 47 x (100 + 10) = 4700 + 470 = 4700 + 400 + 70 = _____
   These exercises make use of the properties of the operations (addition, subtraction, multiplication, and division) for reorganizing.

4. a. Even (how do you know?)
   b. Even (how do you know?)
   c. Odd (how do you know?)
   d. Even (how do you know?)

6. You may have thought of expressions that "go with" the diagrams different from the ones given. If yours are not different you might want to try to find different ones.
   a. 12 ÷ 3   b. 15 ÷ 5   c. 6 + 3 + 5 + 0   g. 275 − 100 + 125

7. The commutative and associative properties of addition.

SET 3

7. a. Mixed multiplication with addition (9 x 7 instead of 9 + 7).
   b. Algorithmic mistake (recorded 4 x 9 = 36 incorrectly).

c. Algorithmic mistake (last '9' should be '10').

d. Algorithmic mistake (37 goes into 340 nine times, not eight).

h. Algorithmic mistake (adding 2 & 3 before multiplying by 7).

8. Recall the commutative property of multiplication for whole numbers and this algorithm won't seem so strange. This problem is similar to one introduced intuitively in Set 1 of Chapter 1.

9. The procedure used in this algorithm for subtraction is consistent with arithmetic principles and it is reliable. This algorithm requires no regrouping ("borrowing"), but it does require knowledge of adding negative and positive numbers (i.e., the integers) and knowledge of subtracting any single digit from any multiple of 10, 100, 1000, etc. This problem is similar to another problem in Set 1 of Chapter 1.

11. One approach to this problem is to compare several different examples. A reasonable comparison is

   987   vs.   876
   x 65   x 95   Which one gives the larger product?
                         Try some others.

SET 4

1. a. Part of a whole   b. Part of a whole   c. Ratio
   d. Division (might also be interpreted as part of a whole)
   e. Ratio

3. Suppose that we tried to make meaningful the expression '10/0' in terms of the "indicated division" interpretation of a fraction. Look back at the explanation of Problem 2, Set 2 of this chapter.
   Also, try one of the other interpretations of a fraction and test out the reasonableness of the expression '10/0.'

4. These exercises use the principle $a/b = (a \times n)/(b \times n)$, for b and n not equal to 0, and a and b whole numbers.
   a. 5/12 and 2/12   d. 1/2 and 3/2   g. 2/3 and 2/3
   b. 25/40 and 6/40   e. 8/30 and 9/30   h. 12/40 and 25/40
   c. 14/12 and 9/12   f. 27/36 and 20/36   i. 220/1715 and 66/1715

5. Having renamed the fraction pairs in each exercise so that they have the same denominator, you need only compare the size of their numerators in order to answer the question.

6. Compare 19/26 to 21/28 in the manner of Problem 5.

7. The metric units, liter and milliliter are slowly replacing the fifth and quart as liquid container units of measurement.

8. "Good ol' boy" type refers to the tools calibrated in the older English unit of inches.

10. How about measuring the thickness of a ream of paper (500 sheets) and then using division! Or find a book around your house.

12. "Cancelling" common factors of the numerator and denominator of a fraction is actually division; that is, cancelling common factors is dividing the numerator and denominator by the same number. In this exercise, *all* of the cancellations are technically *incorrect*; however, by coincidence each of these incorrect cancellations gives a correct answer!

13. a. Equivalent fractional names occur on the same "line."

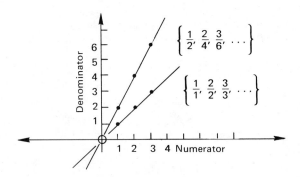

Each line appears to contain the origin of the graph, the point of intersection of the lines. Since two different numbers do not share a common name, it makes sense that the lines (each of which represents a different number) do not actually contain the origin because each point of a line is another name for the same number.

b. If a/b is a fraction larger than c/d, then the line associated with c/d is "steeper" and "above" the line associated with a/b. An inverse relationship.

c. The point (0,0) and other points of the type (a,0) are legitimate points in the Cartesian coordinate system, but their interpretation as fractions would lead to difficulties referred to earlier (see Problem 3 in this section). So the horizontal axis is "outlawed" here as a line for fractions.

14. One way to do this task is, first, rename the two fractions so that they have the same denominator. Then examine the resulting numerators. If the numerators happen to be the same, there is no fraction between them (of course!). If there *is* a whole number between the two numerators, then choose one of them for the numerator of the fraction called for by the problem. Let the denominator of this fraction be the common one you found in the first step.

If there are no whole numbers between the two numerators after you have renamed the two given fractions with common denominators, such as in (c) with 4/5 and 9/10 renamed as 8/10 and 9/10, respectively, then repeat the process by finding a larger common denominator and adjusting the numerators as before. The adjusted numerators will be larger and there will be a whole number between these to choose from for the new fraction called for in the problem.

COMMENTARY 343

SET 5

1. a. 9/8 or 1-1/8 b. 10/15 or 2/3 c. 25/24 or 1-1/24 d. 1/20
   e. 1/7 f. 7/2 or 3-1/2 g. 22/15 or 1-7/15 h. 1/12
   i. 41/88

2. a. 8/15 b. 3/2 c. 2/3 d. 2
   e. 3/4 f. 10 g. 0 h. 6/7
   i. 3/5

3. a. 1/2 b. 21/48 or 7/16 c. 1-3/20 or 23/20 d. 0

4. 30/60 + 15/60 + 6/60 = ___?___, or 30 min. + 15 min. + 6 min. = ___?___ part of an hour.

5. 1/2 + (1/3 of 1/2) + (1/2 of what was left) = fractional part you ate. Drawing a diagram works nicely here to visualize the fractions.

6. 1/10 + 1/4 is taken from you, so how much remains to live on?

10. 1/10 of 30 minutes = 3 minutes, so ___?___ minutes remain for programming. What fraction of an hour is this?

11. The bookstore invested 1/2 of the original price in the book to resell. They resold the book at 3/4 of the original price, so the profit was 3/4 of the original minus 1/2 of the original, or ___?___ of the original price.

13. 1/2 of 4/5 = ___?___. Drawing a diagram helps see this concretely.

14. This problem dealing with teaching time in the classroom is not far from the truth, unfortunately. Two applications of multiplication should give you "from 25 to 16-2/3 minutes" as the solution.

16. Several ways to do this one. Here's one started for you:

    The problem statement says that 1/2 cup sugar gives a three-layer cake. Well, this also means the same as 3/6 cup sugar gives a three-layer cake. Then, shouldn't 2/6 cup sugar yield a two-layer cake? And 2/6 is also 1/3 (the amount we have). What else about this problem? How many one-layer cakes can be made?

    Here is a diagram you might have used in organizing a solution (or just basic understanding of the problem):

1/2 cup three-layer cake   SO   1/3 cup   two-layer cake
or   (3/6 cup sugar)   or   (2/6 cup sugar)
3/6 cup   2/6 cup

17. First, determine the amount the teacher would make in a three-week period of "normal pay." Then could you use 1-1/2 in some way to find the solution? Return to the problem as stated and determine how 1-1/2 fits into the picture.

SET 6

1. a. 22.17 − 9.00 = ___?___ b. 4.008 + 1.319 = ___?___
   c. (188.1 + 27.76) ÷ 50.2 = 215.86 ÷ 50.2 = ___?___

2. a. Think: "15 × $2 − 15 × 2¢ = ___?___"
   b. Think: "639 × 2, or (640 × 2) − 2 = ___?___"

3. a. 5/11 = .454545 ...   b. 63/25 = 2.52
   c. 52/19 = 2.73684210 ... (that should get you started on this one!)
   d. 9/7 = 1.285714285714 ...   e. 10/40 = .25

9. This problem deals with one of the most interesting numbers derived, namely "pi." The Greek letter π is commonly used in representing this elusive number. Even though this number is usually found in the upper elementary grade levels in studies of area, shapes, and geometry in particular, the derivation and explanation of the number is quite complex. For now let it suffice to remind you that there is no exact fractional name for π and even decimal representations are only approximations, though some are better than others. Use $\pi \cong 3.14$ here.

10. Assume that the path traveled by the sensory impulse is from the tip of your toe to the top of your head. A similar path has to be travelled for your brain to tell your leg muscles to start the hopping process!

11. 1/600 millionth of a second, or 1/600,000,000 sec., or .0000000017 sec.

SET 7

1. a. 16.4%   c. 80%   j. 40%   n. 10%
2. a. 9.4   b. $50   d. 16.5 students   l. 12 jobs

5. One way is to think:

   "10% of 48.80 is 4.88, and half of that amount is 2.44. Added together that's a 15% tip of $7.32."

   Or, another way of thinking about this situation is:

   "I'll round off $48.80 to $50, which is half of $100. Now 15% of 100 is just 15, so I'll leave half of that—$7.50."

6. First part: $2500 + $2500 × .09 = ___?___ , or $2500 × 1.09 = ___?___
   Second part: 10/12 × 9% ___?___

7. $2500 − ($2500 × .09) = ___?___
   (or $2500 × .91 = ___?___)

8. The answer here will differ from reader to reader. To think in definite terms you might think relative to 1000 miles for a trip. If you used another person's car, would you think it would be fair for you to pay only 50% of the cost of the fuel (assuming there were exactly two people traveling)? How much above 50% would you think would be fair to all concerned? What other costs relative to his or her car might you need to consider in order to share expenses (e.g., oil, tires, mileage)?

9. Suppose a teacher's 12-month contract had been for $18,000 and the salary was cut by 2/12 or 16-2/3%. Then the new contract for 10 months would be $18,000

x 10/12 or $15,000. But then with the later salary increase described in the problem of 16-2/3% based on the new amount, the increase would be 15,000 x 16-2/3% (or 15,000 x 2/12 = 2500). Therefore the teacher would be making $17,500, or $500 less than the old salary.

11. Let A represent the age of a person, and TH represent the target heartbeat. Then an equation expressing the target heartbeat in terms of the age is:

   TH = .75 x (220 - A).   (What is your equation?)

SET 8

1. The first entry can be represented by $^+316.84 + {}^-250.00 = {}^+66.84$
   The second entry by $^+66.84 + {}^-56.00 = {}^+10.84$
   The third entry by $^+10.84 + {}^-318.00 = {}^-307.16$, and so forth.

3. a. $^-3$ b. $^-10$ f. $^-20$ h. $^+5$

6. For the game of "Hearts" give an analysis of the following results: Jim got the Jack of Diamonds and Andy got the Queen of Spades.

7. $^-10°F = 449.67°$ on the "absolute" scale. You should find it helpful to draw these two scales side-by-side.

   This "absolute scale" is known as the Kelvin scale and is useful in chemistry and physics.

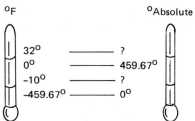

10. If two numbers have the same signs their quotient is a positive number. If two numbers have opposite signs their quotient is a negative number.

SET 9

1. The square indicated to the right has an area of five square units. Study the figure, especially the dotted lines, and try to count five square units—piece-wise—inside the indicated square. The length of the side of this square is the square root of five units.

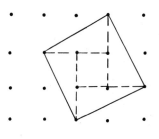

2. One way to approximate $\sqrt{5}$ is by a "guess and test" method. This method is especially appropriate for obtaining some required degree of accuracy. Though impractical, such tables and calculators with square root keys are readily available. The procedure (and tedium) is instructive for learning certain techniques in mathematics.

For your "guess and test" solution the following table may not represent your numbers for approximating $\sqrt{5}$, but the final guess in the table should be reasonably close to yours. A hand calculator will help here!

Approximating $\sqrt{5}$

Guess	Guess x Guess	Difference from 5	Revision
2.5	2.5 x 2.5 = 6.25	+1.25	lower
2.4	2.4 x 2.4 = 5.76	+ .76	lower
2.3	2.3 x 2.3 = 5.29	+ .29	lower
2.2	2.2 x 2.2 = 4.84	− .16	higher
2.25	2.25 x 2.25 = 5.0625	+ .0825	lower
2.24	2.24 x 2.24 = 5.0176	+ .0176	lower
2.235	2.235 x 2.235 = 4.995225	− .004775	OK, at ease!

So, 2.235 is a reasonable approximation for $\sqrt{5}$.

If you did not solve this problem before looking in this section, go back and do the same thing to find an approximation for $\sqrt{13}$.

3. A "proof by contradiction" that the sum of a fraction and an irrational number is an irrational number: We will begin by assuming that the opposite of what we wish to prove is true and look for an eventual contradiction to the assumption.

First, assume that the sum of some fraction, say a/b, and some irrational number, say N, is a fraction (i.e., not irrational!), and can be written as c/d. Then N + a/b = c/d is another way of expressing this fact. But, this equation can be written in the form N = c/d − a/b. In turn, this equation can be written in the form N = (c x b)/(d x b) − (d x a)/(d x b), by using the multiplication principle. Or, N = [(c x b) − (d x a)]/(d x b); that is, this means that N is a fraction—contradicting the original assumption that N is irrational.

Therefore, we have shown that to assume that the sum of an irrational number and a fraction is ever rational (i.e., a fraction) leads us to an impossible situation, for a number cannot be both a fraction and not a fraction.

5. Examples: $\sqrt{2} - \sqrt{2} = 0$ (and 0 is not an irrational number)
$3\sqrt{2} \div \sqrt{2} = 3$ (and 3 is not an irrational number)

7. Isaac Newton's method for approximating the square root of a number with $\sqrt{7}$ as an example: Note that the difference between this one and the one shown in Problem 2 of this section is that Newton's method tells you exactly what to "guess" next. A calculator will help!

"Guess"	Guess x Guess	Difference in 7 and Guess^2	Next Guess
2.5	6.25	.75	2.65
2.65	7.0225	.0225	2.6457547
2.6457547	7.0000179	.0000179	2.6457513
2.6457513	6.9999999	.00000001	FINISHED

Then, 2.6457513 is a very good approximation of $\sqrt{7}$. Try $\sqrt{11}$.

8. Allowing numerals with repeating 9s would mean that some numbers would have two different decimal numerals. For example, the numerals 1 and .999999... represent the same number; .3000... and .29999999... represent the same number, and of course, there are many others similar to these two examples.

   Show that .4400... and .439999... represent the same number!

9. The digit 3 occupies the 1000th place to the right of the decimal point (the authors had quite a deliberation over this miniscule question and finally decided on 3 by a coin toss!).

SET 10

1. There are several ways to set up this problem.

   Let's try organizing it the following way: Consider adding the first term, 1, and 1000th term, 1999; then, the second term, 3, and the 999th term, 1997; the third term, 5, and the 998th term, 1995; and so forth. For each of these pairs we obtain a sum of 2000. How convenient! Now determine the number of such pairs with sums of 2000 and multiply to get the solution.

   Or, we could organize the problem by a different empirical approach: Perhaps a table of the results from simple cases to more complex cases would help us identify a pattern for the sums.

# of odd numbers	Terms	Sum
1	1	1
2	1 + 3	4
3	1 + 3 + 5	9
4	1 + 3 + 5 + 7	16
.	.	.
.	.	.
1000	1 + 3 + ... + 1999	?

   Now go back and compare the first and second processes by writing a rule for finding the sum of the first N odd numbers directly from the application of the processes. Are they the same? Do they provide the same answers? They ought to. Is one of these processes more easily used to derive a general rule for summing odd numbers? Can you find yet another way to approach this problem?

2. If you can think of no other way to do this one, try the processes in Problem 1 above.

4. a. LCM (36, 76) = 2 x 2 x 3 x 3 x 19; GCD (36,76) = 2 x 2

   d. LCM (121, 64) = 11 x 11 x 2 x 2 x 2 x 2 x 2 x 2;
      GCD (121, 64) = 1

   Note: If two numbers have no common factors other than 1—as is the case with 121 and 64—the two numbers are said to be *relatively prime*.

5. a. 5/36 + 3/76 = (5 x 19 + 3 x 9)/(2 x 2 x 3 x 3 x 19) = (95 + 27)/684 = ____?____

   c. 90/115 = (3 x 3 x 2 x 5)/(5 x 23) = (3 x 3 x 2)/23 = ____?____

6. Here are some of the even numbers expressed according to Goldbach's conjecture:

12 = 7 + 5	24 = 11 + 13	38 = 7 + 31
14 = 7 + 7	28 = 23 + 5	40 = 11 + 29 AND
		40 = 3 + 37

   What does the last one suggest to you in terms of the uniqueness of the representation? Can you find several ways for each even number to be represented as a sum of two primes?

Chapter Three: Geometry

SET 1

2. Here are some ideas to consider in describing shape differences between the two sets of figures shown:
   To what extent are the adjectives "convex" and "concave" applicable here? Does the following rule apply to only one of these sets of shapes?

   > *Rule: If we choose any two points "inside" a given figure and notice the line segment determined by the two points, the line segment is completely contained "inside the figure."*

   Can you use this rule in some way to characterize the two sets of shapes? What about the concepts "inside" and "outside"? Can you describe them intuitively, or even more precisely using this rule?

3. Some people perceive that the two line segments appear bent. How do you describe this apparent effect?

4. a. Do the following figure and associated arcs suggest a construction procedure for creating an angle exactly the same shape as a given angle? Study them and think in terms of using your compass and straightedge to develop such a procedure.

   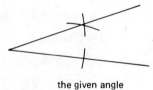

   the given angle

   b. Then, similarly for "twice as sharp":

   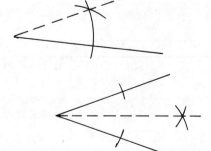

   For "half as sharp" try applying the procedure in (a).

8. The Golden Ratio was referred to in the papyrus of Ahmes, and used in building the great pyramid of Gizeh (ca. 4700 B.C.). Known as the "golden section" among the Greeks, and as the "law of divine proportion" in medieval times, the number (ratio) is obtained by dividing a line segment of any length into two parts so that the lesser is to the greater as the greater is to the whole length of the segment. The most pleasant or desirable rectangle is supposed to be one that has a base-to-height ratio of 1.618 (the golden ratio). Architects and artists have often used this number in designing rooms, buildings, windows, and frames for pictures. In this exercise, the first rectangle from the right is a "golden rectangle."

SET 2

1. ... from three dimensions to two dimensions. (If you consider the light rays leaving the mirror and putting the image onto your retina as another transformation, what would the answer be?)

7. More formal techniques for solving this problem will be dealt with in this chapter; your task now is to explore the properties of the rotation transformation (you can solve this one with your intuition). Here are some things to think about before we attempt to get more formal. The figure should assist you, but keep in mind that this method is only one way. You might devise other ways:

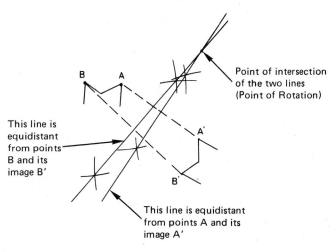

The point of rotation is always the same distance from a point on the figure in the original position as it is from the corresponding point in the image of the figure. Somewhere there is a line that is equidistant from the point A and its image A'. Using a compass and straightedge you can construct this line. The figure to the right provides some clues about using these tools to find the line. Similarly for point B and its image B'. The indicated arcs should suggest the manner in which these lines were constructed. The "special" line for points B and B' is constructed exactly as the one for A and A'. Why should the intersection of those two special lines contain the point of rotation? (Later you will study "midpoint" and "perpendicular" formally; the concepts are meant to be intuitive at this point.)

8. Determine the line of reflection: What properties does this line have relative to points in the original shape and their corresponding image points? The concept of *midpoint* of a line segment might prove helpful!

SET 3

1. Use a mirror.

2. This equilateral (equal-sided) triangle has three lines of symmetry. Do all triangles have three lines of symmetry?
    The square has four lines of symmetry.
    The equilateral (equal-sided, or regular) pentagon has five lines of symmetry. Do all five-sided figures have five lines of symmetry? How do you know?

6. After experimenting with the idea of trying to show the existence of a three-stage composition that cannot be achieved with a simpler two-stage composition, you might try to make a generalization about compositions!

SET 4

1. a. curve   b. region   c. polygon (nonagon)   d. curve
   e. curve   f. quadrilateral
   g. a simple closed curve and its exterior region (shaded)

3. Experiment a little! What happens when you leave point A for point B and cross over the curve? If A is on the outside, do you stay on the outside, or enter the inside, when you cross the curve? Try this method.

4. You could organize this problem of the diagonals into a table.

Number of Diagonals	Number of Sides
0	3
?	4
5	5
?	6

    Is there an apparent general rule for this relationship? What is it?

6. Will this problem have to be thought of as two different cases (even and odd number of sides)? Or can you devise a procedure that includes both situations? Is the concept of midpoint of a line segment useful? Or that of bisecting angles?

9. Of course, you probably "conserve" shape (a Piagetian construct) in this instance, but often the "interaction" of the circles in this configuration would have some people perceive a "bending" effect on the triangle. As you know, the sides are not actually bent, but this exercise might point out the difficulty in always relying on our perceptions and assumptions about what we see in geometrical shapes.

SET 5

1. Suggestion: Trace these angles on another sheet of paper; then, use your compass and straightedge to construct right angles at the vertex of these angles.

2. Suggestion: Trace the two lines below on another sheet of paper; then think about how you might use two perpendicular line segments constructed between the two given lines. (See the figure below, segments $\overline{AB}$ and $\overline{CD}$.) What should be true (at least intuitively) about the two line segments $\overline{AB}$ and $\overline{CD}$? Can you use your compass to compare "approximately" the lengths of the segments?

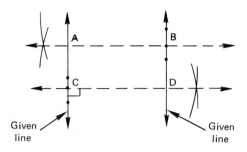

From a different perspective, what should be the relationship between a line segment and both lines, if the segment is perpendicular to one line and intersects the other line?

3. Using the given line segment ($\overline{AB}$ in the figure) as a side of the square you want to construct, you can construct the square from only *one* right angle! First, extend the segment $\overline{AB}$ as shown and construct the right angle (a perpendicular) from point A. Use your compass to mark the same length along the perpendicular as the length of $\overline{AB}$. So, now the length of $\overline{AC}$ is the same as the length of $\overline{AB}$. Then use your compass with this same length and mark intersecting arcs from the point C. Now to complete the square just mark the segments from point C to the point of arc intersection, D, and from point B to point D.

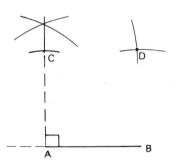

4. Unless you recall the midpoint "construction" from your previous experience, you might approach this problem at an intuitive level. For example, one way to think about this situation is based on the main property of the midpoint—the point that is equidistant from the endpoints of the line segment. What other points are equidistant from the two endpoints? Can you use your compass to locate at least two other such points? How might you use these two points to locate the midpoint of the line segment?

6. One of the two line segments shown is actually longer, though in popular publications recent versions of this configuration focus on the optical effect created by the smaller segments attached at the ends of the two longer segments. The comparison was given here to break the pattern and to tease (possibly) your expectation—real life problems do this to us all the time.

8. What about constructing a regular pentagon by the method of Problem 7, and then using the idea of perpendicular bisector of each of the five sides?

SET 6

2. Start with a cardboard cutout like:

Pick the point C to "go around" by first flipping the triangle over $\overline{BC}$ and then "twist" the figure about the midpoint of $\overline{BC}$. From this standing position flip the triangle down as shown in the figures below.

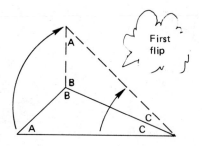
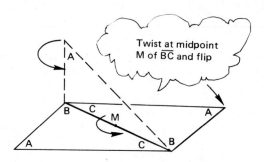

Continue this process, going clockwise around the original point C. After you've completely covered the space around this "turn point," how can you move on another such turn point?

7.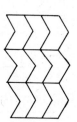

8. Looks like 22 or 23, depending on how picky you are!

SET 7

1. The polygon was an equilateral triangle, which was transformed by the method shown as the second example of a tessellation.

SET 8

1. Is there a polyhedron of which all the faces are triangles? Is the result of constructing a polyhedron of these triangles a polyhedron having the least number of faces possible? How would you construct a polyhedron of five faces? What would the shape of the faces be? Is there more than one kind of polyhedron with five faces?

2.

Figure	Name	# of Faces	# of Vertices	# of Edges
a.	tetrahedron	4	4	6
b.	hexahedron	6	6	10
c.	dodecahedron	12	17	27
d.	pentahedron	5	6	9
e.	heptahedron	7	10	15
f.	octahedron	8	6	12
g.	heptahedron	7	10	15
h.	hexahedron	6	8	12
i.	decahedron	10	16	24

3. From Problem 2 above, let F = number of faces, V = number of vertices and E = number of edges. Various arithmetic operations could be performed on these three columns of numbers as you try to determine whether some pattern or rule exists. You can assume that such a pattern exists in this particular problem, but we wouldn't take away your pleasure of finding it for yourself! You'll know when you do. It's called Euler's formula!

4. A cube (box) can be colored with only three colors. Take a box and try it using letters R, W and B for red, white and blue. But try the same thing on the first figure in Problem 2 above (tetrahedron) and you'll find you must use four colors to prevent adjoining faces being alike in color.
   Look at the other figures in Problem 2 of this set; do any require more than four colors? If so, which one? If not, can you make an inductive conjecture?

5. Has any twisting of planes taken place in any of these shapes? Has any plane been left "stranded," so to speak, such that the planes aren't used as faces? Are any of these two-dimensional representations ambiguous in terms of how they depict a three-dimensional shape? The artist M. C. Escher was a master of this kind of deception.

6. Cubes tessellate three-dimensional space. The measurement we call volume is based on the notion of tessellating our three-dimensional space. What other shapes tessellate the space we live in?

Chapter Four: Answers

SET 1

1. Singular kinds of answers to problems, such as numerical, Yes-No solutions, are certainly important. They often even serve as "rewards" in themselves for having successfully applied some process. Teachers need to be concerned with processes of problem solving because they teach and guide students through processes, not by the answers directly. Knowing merely an answer to a problem in a given instance can rarely provide us the means of solving another problem. Therefore, the emphasis in this book is, and will remain concerned with, all forms of solutions with a primary focus on problem-solving processes.

2. "Conclusion" answer.

SET 2

1. Here are the authors' estimates; yours may differ, but they ought to be in the same "neighborhood." Remember, estimates are often derived from some simple computations but not from exact, detailed computation.
   a. a little over 150 pounds
   b. between 3000 and 4000 dollars
   c. between 37 and 45 miles per gallon
   d. about 250 pounds
   e. about 12 gallons

2. You can get the correct answers for yourself, but here are some suggested operations to use:
   a. Division: $3000 \div 19$
   b. Division: $800,000 \div 229$
   c. Division: $268 \div 6.5$
   d. Add all the numbers (8 of them). Subtract this sum from 9×250
   e. Divide 8.8 by 3 and multiply by 4, *or* multiply 8.8 by 4/3 (Why?)

3. Here are some suggested problems you might have listed from Chapter 1:
   a. The history book problem in Set 2 (Problem 9)
   b. The rational discussion called for in Set 2, Problem 13
   c. The analysis required by Problem 16 of Set 3

   An estimate of the answer to these problems can be described as worthless, because one is a brainteaser and a description or argument is required by the others.

4. a. 2 yds.   b. 3500   c. 3000   d. 300,000   e. 0

5. 0. (Now that's ridiculous, isn't it!) Such an estimate would be based on poor rounding judgment, since multiplication is called for. If you'd been adding the two rounded numbers instead, the answer would be very close to the actual answer. This shows that we must be reasonable when we round numbers.

7. Three ways of rounding are described below; there are many others.
   a. Round each number to the nearest ten before adding to the cumulative sum. For those ending in 5, round alternately up and down.
   b. Round *all* numbers *up* to next ten and add. Then count the addends, take half of that number, multiply by 10 and subtract.
   c. Round *all* numbers *down* to next ten and add. Then count the addends, take half of that number, multiply by 10 and add.

   How do each of these ways compare? Which is the most accurate?

SET 3

1. a. 5 hours
   b. Gold can be hammered into "leaves" so thin that a stack of one million of them would be only four inches thick.

c. About 50 times each minute, or close to one each second.

d. We spent $43 on bread.

e. Blood vessels from an average adult male would encircle the Earth approximately four times.

f. Some computers can perform 10,000 calculations in 1.25 seconds; actually some computers are capable of over 1 million calculations in one second!

g. The mini Mercedes-Benz in this 1979 marathon could have traveled from Maine to the southern tip of Florida on about one gallon of fuel.

2. a. One gram of a "pure" electron mass would have about 10^{27} electrons. Obviously this is a theoretical measurement, since we do not have scales sensitive enough to weigh such a small amount.

b. There is a wide range of differences in heartbeat rate among people, and for the same person from one time to the next.

c. It is doubtful whether anyone can figure their total assets to the nearest ten dollars, much less to the nearest cent! This is certainly true for a span of one year.

d. The world population probably cannot be accurately stated to the nearest 100,000 for a year, much less for a given moment. Some people would die at the given moment, while others would be giving birth.

3. a. 58 x 365 = ? Estimate by rounding. Then, put your answer in a meaningful sentence such as "This year I will use about as much water as (compare to something)."

b. 60 ÷ 1/10 = number of letters per minute, then divide by 5.9.

c. 1/12 x 100 million = ? Estimate by rounding. Communicate answer something like "Last year the people in ___?___ average states gave up their cars."

d. 93,000,000 ÷ 186,282.4 = ? sec. (Don't forget about precision!) Estimate by rounding. Change answer into minutes to make it more meaningful.

e. 2/60 sec. to go 12 feet; so, responses travel at the rate of ___?___ feet per second. This is about ___?___ miles per hour. (Note: compare your results to Problem 10, Set 6, Chapter 2.)

SET 4

2. 56.7 − ⁻88.3 = 145.0 Celsius degrees.

3. 50 + 4 + (4 x 3) − (1 + 4 x 3) + 6 = ___?___

5. Assume that milk, Coca-Cola, and water have equal masses for equal volumes. Also assume from this section that a gallon of milk has a mass of about 4 kg.

8. If you read flowcharts quite well—or want to learn now by looking ahead—you can get a hint by looking at Problem 2 of Set 3 in Chapter 6.

Chapter Five: Chance

SET 1

1. a. Between Ace (low) and 8 are the cards 2, 3, 4, 5, 6, and 7; between 8 and Ace (high) are 9, 10, J, Q, K; so with only this information—that is, no other known cards have been taken from the deck—the probability of the next card being "in between" Ace (low) and 8 is (6 x 4)/50, or 24/50; the probability of it being between Ace (high) and 8 is (5 x 4)/50, or 20/50. Note, we multiplied by 4 because there are four cards of each kind.

    b. From the work in (a) you can determine that a span of six cards is not sufficient to ensure a 50 percent probability. So, how many cards "in between" would be necessary?

      The game of "in between" becomes more challenging when the deck is not reshuffled after each play. It is highly advantageous in this case to remember what has been played previously and recalculate the probabilities from what cards have been removed from play.

4. Only one chance out of 5, or 1/5 probability. Of course, very often when one is taking a multiple choice test, some of the choices can be reasonably eliminated, thus increasing the chance of getting the item correct by guessing. Suppose three of the five choices on an item could be eliminated as being "unreasonable" answers. Then determine the probability of getting the item correct by guessing.

7. Probability of a 5 = 4/52.

    Probability of a red card = 26/52.

    Probability of a red 5 = 2/52. Can you describe at least two ways of thinking about computing this probability?

11. An event with probability one is an event that is "bound to occur."

SET 2

1. Your answer will depend on your state's license plate format. Many states have adopted the one described in the text, so possibly your answer and procedure will not differ from the text's.

3. "Oh, Oh, Oh, Oh, Oh, Oh" is *wrong*. So is "Zero, zero, zero, zero, zero, zero." But some combination of these . . . ?

4. On each of the two dice there are six possibilities, so for each numeral on the red die there are six numerals on the white die to pair with it. Therefore, there are 6 x 6 = 36 combinations, each equally likely.

5. The fundamental counting principle comes in handy here!

SET 3

1. The probability of getting first two heads and then a tail, using the chain rule, is 1/2 x 1/2 x 1/2 = 1/8. Then the probability of getting first a head, then a tail and then a head is 1/2 x 1/2 x 1/2 = 1/8. And the probability of first getting a tail and then two heads on the second and third tosses (or coins) is also, 1/2 x 1/2 x 1/2 = 1/8. Each of these events is mutually exclusive, so the probability of getting two heads and one tail in a toss of three coins, or the successive toss of a single coin is 1/8 + 1/8 + 1/8 = 3/8.

More simply put: There are three ways in which three coins can turn up two heads and one tail, and there are eight ways the three coins can "fall."

2. The three guns are independent of each other and each provides the person aimed at with four chances out of six of surviving. The probability that all the people will survive is 4/6 x 4/6 x 4/6 = 64/216, or 8/27. The chances aren't good at all that all three will make it.

3. For the final question in this item, do you see that there are two possible combinations that would "pay" the top amount? They are three cherries or three lemons. But if you discount these impossible cases, ...

4. First, determine the probability of a person's being born on Wednesday.

8. The probability that the first glass will be different is one. Then, determine the probability that the second glass will be different from the first one. Continue this line of thought, and then apply the modified chain rule. It might help for you to draw a diagram of eight different types of glasses and mark them as you proceed (or use real ones).

SET 4

2. If you don't know exactly how many are in your class, assume a class size of 40. So there would be 40 numbers such as XXX-XX-XX53; ignore all but the last two digits of each one. There would be 40 such numbers to consider and hopefully no pair would be the same. The chance that the first one selected matches none of the "previous" ones is 1; then calculate the chance that the second 2-digit number doesn't match the first one selected, and so forth. Use the modified chain law and subtract from 1 as the final step.

   It might help you to take a look at your own social security card before beginning this problem.

4. The probability that a fertilized human ovum is a male is not exactly 1/2, but closer to 105/205, or about .512. What effect does this information have on your computation?

6. This problem is another example of the old dilemma concerning tossing a coin 10 times, and after nine tosses, all heads, how would you bet on the tenth toss, assuming that the coin is fair? Of course, the "safe" answer is heads in terms of your continued belief that the coin is indeed fair; however, if the coin is not fair, then it likely favors heads in view of the previous nine occurrences of heads. Interestingly, people often forget that the tosses are independent of each other ("Lady Luck has no memory" is the old adage) and under the assumption of fair coin they bet more than they should that the tenth outcome will be a tail, often losing their shirts.

SET 5

4. I. The main fallacy of the coach's strategy for winning the toss lies in failing to take into account the probability of choosing tails and the coin falling tails! Given that the coach was correct in thinking that the probability of the other team captain choosing heads is .75, then the probability of his choosing tails is .25. Then the probability of his winning should be .75 x 1/2 + .25 x 1/2 = .375 + .125 = .50. So, the chance of winning the toss does not depend on

giving the other team the choice, regardless of the probability of their selecting heads! How does the result of your experiment compare to this analysis?

Chapter Six: There Must Be A Better Way

SET 1

2. The number of digits in the dividend for this division problem is larger than most hand-held calculators are capable of handling. All calculators and computers have this limitation to some extent, but some can deal with large numbers more easily than others.

   One of the properties of arithmetic can assist us in overcoming the dilemma posed in this problem. Here is one of the rules that can help us:

   If we want to divide the number N by the number C and we know that N = A + B, for some numbers A and B, then N ÷ C = (A ÷ C) + (B ÷ C).

   Using this rule we can approach this problem by breaking the number 16553847713 into two smaller numbers whose sum is this number. For example 16553847713 = 16,000,000,000 + 553,847,713. Then

   $$16{,}553{,}847{,}713 \div 8 = (16{,}000{,}000{,}000 + 553{,}847{,}713) \div 8$$
   $$= 2{,}000{,}000{,}000 + 69{,}230{,}964.13$$
   $$= 2{,}069{,}230{,}964.13 \quad \leftarrow \text{FROM CALCULATOR}$$

   Can you devise another way to do this?

3. When trying to divide by zero one of several things might happen on a hand-held calculator, depending on the manufacturer or designer, you could get an "ERROR" message, or the numerical display could blink off and on, or the display might just go blank (off).

4. If the 8 key is broken you could key in 34476 – 247 and then add 4000. Can you devise other ways?

6. If the "divide" key is broken, you could use the method of repeated subtraction described in Chapter 2. Or, you might use multiplication in a "guess-check-revise" process, gradually getting closer to the question 239 x ____?____ = 876,984. Can you think of others?

12. Ten steps are required to figure the accumulation in this savings account, beginning with "Amount after one year = 805.437 + (805.437) x .08," or, "Amount after one year = (1.08) x (805.437)."

    To compare the additional amount given by the 7 mill, recalculate these same steps using 805.43 at the initial stage, dropping the "7 mills" portion.

SET 2

3. The flowchart in this problem has many places in it where the right question was raised at the wrong time! For example, at what point in planning a trip should one logically determine whether sufficient money is available?

4. After six years the accumulated amount in the savings account would be $1586.87.

SET 3

2. Converting Fahrenheit to Celsius.

   $32°F = 0°C$; $100°F = 37.78°C$; $212°F = 100°C$

   $131°F = 55°C$; $293°F = 145°C$; $^-15°F = {^-}26.11°C$

SET 5

1. Here is BASIC for summing the first 1000 whole numbers (but your BASIC program might be different from this one):

   ```
   1   S = 0                 OR    1   LET S = 0
   2   N = 1                       2   LET N = 1
   3   S = S + N                   3   LET S = S + N
   4   IF N = 1000 THEN 7          4   IF N = 1000 THEN 7
   5   N = N + 1                   5   LET N = N + 1
   6   GO TO 3                     6   GO TO 3
   7   PRINT "SUM =", S            7   PRINT "SUM =", S
   8   STOP                        8   STOP
   ```

3. For the example on the left, look carefully at line 6. That's one mistake!

   For the example on the right, how would you ever get to line 8?

4. The "dummy number" or "flag" would be a number that would make the program stop by itself. Look at how this would happen in step 7.

5. a. The answer is *not* 100 days, or even 99. This situation is similar to the old "frog jumping out of the well" problem.

   c. Compare your BASIC program for the rabbit reproduction problem to the program shown below. Yours may be different, yet still yield correct results.

   ```
   1    B = 1
   2    Y = 0
   3    A = 0
   4    M = 1
   5    A = A + Y
   6    Y = B
   7    B = A
   8    IF M = 24 THEN 11
   9    M = M + 1
   10   GO TO 5
   11   T = 2 * (A + B + Y)
   12   PRINT T, "RABBITS AFTER 2 YEARS"
   13   STOP
   ```

   Note 1: the LET convention is required by some computer systems for assignment statements in BASIC, as indicated in Problem 1 above.

   Note 2: In this program, 'B' was chosen to keep up with the Babies; 'Y' to keep up with Young rabbits. What would 'A' and 'M' stand for in the program?

SET 6 Most of these problems were used in upper elementary school grades by the authors, and in fact, many of these problems and others like them were created by the students.

3. Your solution to the ultimate thickness of the paper after folding it "theoretically" 50 times will probably surprise you! You might do well to convert the inches to miles in the last stages of your program since the thickness would measure well over half the distance from the earth to the sun.

5. This problem is quite similar to several others in the text and previous problem sets. Find one like it if you need help, and make the necessary adjustments.

Chapter Seven: Measurement

SET 1
2. Wasn't this exercise convenient—you didn't have to use a calculator for converting—just shift the decimal point. (You did, didn't you!) For checking purposes here are some answers to compare with your results.

   a. 5.4 km = 5400 m
   b. 13 dm = 1300 mm
   c. .05 m = 5 cm
   d. 2.47 km = 247 dam

3. Use 1 mile as approximately 1.6 km: e.g., 55 mph = 55 x 1.6 km per hour. You probably have begun noticing the "km per hour" displayed on recent automobile speedometers (alongside miles per hour).

5. Since four laps constitute a mile, one lap is 1/4 or .25 of a mile. Then one lap is 1/4 x 1.6 km, or .4 km. Therefore, two laps = 2 x .4 km = .8 km. Then 2½ laps = .8 km + .2 km = 1 km.

8. Consider this sequence, which doubles after the first term, to get the next term:

   0, 3, 6, 12, 24, 48, 96, 192, 384, . . .

   There's no reason to think this sequence has a relationship to our solar system. But take each number above in turn, add 4 to it, and divide by 10. You get:

   .4, .7, 1.0, 1.6, . . .

   Does this look familiar? It sure did to Bode!

SET 2
3. The figure of the upper left corner has an area of 3½ square units—the figure is shown again to the right. Do these first by counting, and then by applying some formulas used in this section.

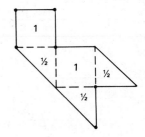

The figure in the lower left corner may be thought of as half a rectangle having dimensions of 4½ x 2½, or as having an area of 11.25. Then the area of the given shape is ½ x 11.25 or 5.625 sq. units.

The figure of the lower right corner has an area of 3½ sq. units. There are several ways to get this one—one way involves the 2 x 4 rectangle containing the given shape. The figure is to the right.

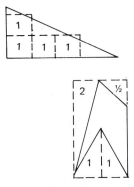

4. a. $1 \text{ dm}^2 = 100 \text{ cm}^2$; b. $1 \text{ m}^2 = 100 \text{ dm}^2$; c. $1 \text{ km}^2 = 1000000 \text{ m}^2$.

If these results are a mystery to you, return to the definitions of dm and cm and study a square with area 1 dm^2. The same reasoning is useful for all of the other square unit conversions.

5. Adults walk about three miles per hour.

SET 3

3. Water (or most other liquids) has three useful properties that enable you to find the volumes of light bulbs and other suitable objects in this problem. Water (1) pours, (2) fits the lateral and bottom shape of its container, and (3) tends to be level on its top surface. That's all you'll need to know to solve this little dilemma.

6. Suggestion (if you have not thought of it already): use centimeters. There are many solutions to this problem, each one requiring two numbers—one for the width and the other for the length.

8. Did you consider weighing the van, both loaded with beans and then empty? You'd then have the weight of all the beans.

SET 4

1. $180° \div 10 = 18°$ for the new 10-based unit angle measure. The "smaller" angle would be $18° \div 10 = 1.8°$.

5. Can a pentagon be divided up into triangles, using diagonals that don't cross? If so, how many? Can a hexagon? Can a heptagon? Would the angles of the polygon then be made up of the angles of triangles?

7. From the formula for the circumference of a circle in terms of its radius you can easily answer this question (the formula is $c = \pi \times 2 \times r$).

8. Use $c = \pi \times 2 \times r$ for the circumference of a circle to answer this one.

SET 5

2. In the quote from Chapter One there are 4¾ sentences, 134 syllables. Then applying Fry's procedure to this information we determined that the quote is on the eighth grade reading level, but just barely.

3. Sam's IQ is 120; Mary's IQ is 93; Kim's MA is 14.

Chapter Eight: Descriptive Statistics

SET 1

3. In selecting roof material many factors would be used to make a final decision. Here are some; you may think of others: availability of materials; amount of money on hand to invest in a roof; length of time expected to stay in the house; safety hazards of the material (e.g., fire hazard).

4. This evidence is typical of most classes and shows that there are three or four students high in one area but low in the other.

5. For young men blood pressure appears higher than for young women, but the blood pressure of older men appears to fall slightly below older women, especially for the systolic pressure. How does the size of the ranges compare between men and women?
    Speculate: What psychological and social factors might be related to these differences among men and women?

SET 3

1. Carry out this exercise for Chapter 7. In your application of Fry's method do you think there are some factors in reading level to which the technique is not sensitive? Does this technique deal with reading matter that contains many uncommon words or phrases that, though they contain few syllables, are not well known by many people? Does the Fry technique depend on the "long run" average over the 100-word passage?

2. a. Who are the readers of the magazine? Do they represent the voters?
   b. To what extent do these schools represent the education majors the research wants to make generalizations about?
   c. What students just happen to be at the booth at that particular time? Do these students reflect opinions representative of the whole campus? Further, would verbal responses be inhibited at all?
   d. How many of the female faculty did not return the opinionnaire, and what were their reasons for not returning it? Could it be that many who didn't return it were satisfied?

SET 4

3. The mean braking distance of System X is 100.00 feet, while the mean for System Y is 98.00 feet. The SD of the braking distance of System X is 1.74, while the SD for System Y is 6.38. So, you can see that each braking system has an advantage (and disadvantage). The question of which braking system is the best is not easily answered: System Y appears to have a shorter braking distance, but System X seems to be less erratic in its braking distance variation.
    This problem typifies a large number of real problems and decisions we have to make. For example, many people have similar comparative results when they try to decide which life insurance policy is best. There are many factors to consider and advantages and disadvantages to all of them.

5. If Sue receives all A's for the 15 hours of course work during the next term, she will have a total of 28 hours credit, and 4 x 15 (= 60) grade points for a total of 60 + 41 (= 101) grade points. Then, her average at the end of her freshman year would be 101 ÷ 28 = 3.61. So, yes, she can even average above 3.5.

   Question for you: What would be her average if she made 2 A's and 1 B during that next term instead of 3 A's?

7. Yes, by a "working backward" technique. What would she do?

SET 5

1. Approximately 68 percent of the area under the curve occurs between days 9 and 19 (estimated: your estimate should be within 1 or 2 of these numbers, but whatever it is, day 14 should be centrally located in the range of days). This assumption suggests that the SD is 5 and it means that approximately 68 percent of the hormone secreted is between the 9th and 19th day of the cycle.

2. (a) Infant mortality and (b) golf scores would best be analyzed with actual data. Use a mortality chart and the sports page of a local newspaper to deal with them.

   For (c) cost of college education, you should consider as unlikely to be distributed "normally," because the cost of education has been on the rise with respect to years for quite some time, and it's more costly now than ever before.

   (d) Mileage of cars in the U.S. is a possibility for a normal distribution. Use a current EPA data report for this one.

Chapter Nine: Problem Solving Extended

SET 1

1. It is helpful to recognize first that the consecutive numbers 1 through 9 are to be inserted in the squares, and thus, the total sum of all rows is 45. Then, you should see that the sum of each row (and consequently, each column) ought to have a sum of 15. A solution is shown to the right. Other so-called "solutions" might be achieved by merely shifting the rows to columns—but you might think this is not really a different solution— and many people would agree with you. Here are some questions to ponder:

   Does 5 always have to be positioned in the center square?

   Are there any other really different solutions to the 3 x 3 magic square? Can you find a 4 x 4 magic square?

3. The number 1 can (as any number) be represented by various combinations of arithmetic operations and other numbers.

   $1 = (4 \times 4) \div (4 \times 4)$ is just an extension of the example shown earlier in the text.

   $1 = (4 \div 4) \times (4 \div 4)$ is another way.

   Can you think of another?

   As an extension, can you represent the number one with exactly three 3's?

4. Tightly packed in a box a dozen balls of the given size could be arranged in several ways:

1. one level with 3 balls x 4 balls—then the box would have dimensions 4" (depth) x 12" (width) x 16" (length).

2. three levels with four balls on each level—then the box would have dimensions 12" (depth) x 8" (width) x 8" (length).

SET 2

4. Did you think of this: You could first construct a flowchart of this nested number sequence, and, as indicated to the right, write a BASIC program to find the 20th set of nested numbers and their sum. The question marks (?) indicate that you can put in any three numbers for A, B and C you wish.

   Note: "A," "B," "C" are storage units to house the three numbers you put at the initial corners. "N" holds the number of the nest you are working on. A fifth storage box, Y, has been used to hold the value for A for a "moment" until it can be used.

```
1    A = ?
2    B = ?
3    C = ?
4    N = 1
5    Y = A
6    A = A + B
7    B = B + C
8    C = Y + C
9    IF N = 20 THEN 12
10   N = N + 1
11   GO TO 5
12   T = A + B + C
13   PRINT A, B, C, T
14   STOP
```

6. This "scribbling" is called mirror writing. Get a mirror if you need to and figure this one out. The authors have found that some people seem to be able to read and write this way—and in other ways (upside down, etc.)—without much trouble. Others seem to find it almost impossible! We wonder why.

SET 3

1. The Moebius band experiment is best done by using plastic strips, e.g., from trash can liners (the thicker the better).
   You can determine the number of sides by pulling the strip under a felt-tip marker to make a continuous mark. Move the strip until the mark returns to the starting point. The number of sides is one in the case in which the strip has no "side" unmarked. The strip has two sides otherwise.

2. Cutting a Moebius band down the center lengthwise with various initial twists in it should result in outcomes related to Problem 1 above. Your prediction of the number of smaller loops and their twists and interlocking properties, based on whether the number of twists were odd or even, should become easy and reliable after this experimentation.

3. You might be somewhat fascinated by the outcome of cutting a Moebius band lengthwise at 1/3 widths—cutting continuously at 1/3 widths (as shown in the Problem Set) until you complete the cut. Odd versus even twists is again pertinent to the question, but a very different result occurs for odd twists situations (large and small loops occur).

SET 4

3. Remember that farm land isn't just flat; there are hills, too.

4. A quarter will surely pass through a hole the size of a dime, if you manipulate the hole. (Can we extend this to the camel and the eye of a needle?)

6. We hate giving this one away in print, but here are some very direct clues anyway:

    "Roses are odd numbered dice,
    Violets are nothing but nice;
    Petals are on the outside looking in,
    So now you know, I'm bettin'."

SET 5

1. As you can see from this example, a pattern exists, but the length of the sequence of digits is not fixed regarding the 1, 2, 3, 4 and 0s.

    Can you use Gauss' idea for summing consecutive numbers (counting the 0s?

    Did you think of constructing a flowchart and perhaps a BASIC program?

    Use trial and error and consider this sum $4 + 5 + 6 + 7 + 8 + \ldots + 20$. Each term in this series can represent the number of digits in a sequence in which each ends in a 4. Then since the sum appears to be part of one you have seen earlier $(1 + 2 + 3 + 4 + \ldots + N)$, you can perhaps use it to answer the question in the problem.

2. From geometry it would be helpful to recall that any three points are always in some plane (that is, three points make their own plane!). The tips of the legs of the stools may be thought of as points. The tips of the three-legged stool are always balanced in "their" plane; however, it is somewhat difficult to construct a four-legged stool so that it has all four tips (points) of its legs in the same plane.

5. The continuation of this sequence would be 1, 3, 6, 10, 15, 21, 28, 36, 45, 55, ... Begin a figure for this sequence of numbers with one point; then use triangles to generate the other numbers in a fashion similar to Problem 4.

9. This problem has many answers, and one method of solving it illustrates that it might sometimes be advantageous to get a common denominator when multiplying fractions (though as you know, this is not usually necessary). Further, reducing fractions can often conceal an answer, rather than making the solution simple! Sound strange?

    Here's one solution:

    The floor is not square, so some calculation appears necessary. Vaguely this problem seems to be related to area, though the main question does not call for it. I think I'll experiment with area and the dimension of the floor as a starting point.

    I'll multiply the length by the width, but for reasons that are only intuitive at this point, I'll get a common denominator for the dimensions and express the numbers as mixed numbers.

> $AREA = 13\text{-}1/3 \times 9\frac{1}{2} = 40/3 \times 19/2$
> $\phantom{AREA = 13\text{-}1/3 \times 9\frac{1}{2} } = 80/6 \times 57/6$
> Then, $AREA = (80 \times 57) \times (1/6) \times (1/6)$
> $ = 4560 \times (1/6)^2.$
>
> *But, wait a minute! There's an answer in the way I've written this. The factor $(1/6)^2$ can be thought of as the area of a $1/6 \times 1/6$ square, and 4560 of these little squares cover the floor because they were derived from the area of the floor. Well, I sure did stumble onto that solution!*

From this answer, you can find other solutions. Can you also find other methods for getting solutions?

14. Alphabetically, using the word names for these numbers.

17. This example illustrates no illusions or tricks; rather it demonstrates one's susceptibility to making assumptions about relationships when they only appear to exist. In reality the figure in this problem, when pieced together, does not fit together exactly; there is a small gap. This result occurs because the angles required by the second figure in the problem are slightly different from the angles in the shapes of the cutout figure. Your eye probably cannot see this small difference. The moral is: Don't be fooled by appearances, for magic is not really a part of mathematics.

18. P.

With that, you should be properly weaned from whatever guidance this book has provided you toward problem solving in mathematics. If you have gotten this far and have worked toward solutions to a substantial number of the problems, you are probably feeling sort of independent and somewhat proud. So be it, but as you know, the challenge goes on, and you have only to take another breath to find another problem awaiting you.

GLOSSARY

Algorithm A step-by-step procedure that can be followed to perform a computation or construction in mathematics. The most common use of *algorithm* in this text is with the step-by-step procedures for adding, subtracting, multiplying, or dividing with whole numbers, fractions, decimals, and negative numbers.

Analysis Taking apart a complex idea, working from "the whole" to its simpler, component parts. The reverse of *synthesis*.

Angle A planar geometric figure composed of two rays with a common endpoint. The common endpoint is the *vertex* of the angle.

Asymmetric Adjective used to describe a geometric figure that has no *line of symmetry*.

Autistic thought Mental processes that appear not to be founded in, or based on, rational thought.

BASIC One of the most common computer programming languages. Acronym for Beginner's All-purpose Symbolic Instructional Code.

Celsius A common scale used to measure temperature. The freezing and boiling points of water (at sea level) on this scale are $0°$ and $100°$, respectively.

Centi A common prefix in the metric system meaning 1/100th.

Circle A geometric figure in a plane, characterized as the set of points a certain distance from a given point. The given point is the center of the circle. The distance of the circle from its center is the *radius*.

Complementary events Two events that are the "opposites" of each other. If one event occurs, then the other can't, and vice versa.

Composite number A counting number with factors other than itself and 1 (i.e., a counting number that can be written as the product of two smaller counting numbers).

Composition A combination of *slides, flips,* and *turns* that produces an image of a two-dimensional figure with the same size and shape as the original.

Congruent Geometric figures having the same size and shape are said to be congruent figures. Being congruent implies that the figures are indistinguishable from each other, except for orientation or position, in a plane or in three-dimensional space.

Counting numbers The infinite set $\{1, 2, 3, 4, 5, \ldots\}$. The counting numbers are the same as the whole numbers, except for 0. Children first learn to count using this group of numbers; this set is particularly important for those who work with preschool children.

Cubic centimeter A cube that is one centimeter on each edge.

Decision Indicates a place in a computer program with two possible paths as successive steps. The computer chooses the path to take after checking its memory bank against the question asked of it—answering either "yes" or "no"—and proceeds accordingly.

Denominator The numeral "below" the bar when a fraction is shown as a/b. In the most common interpretation of a fraction, this refers to the number of equal parts a "whole thing" has been partitioned into.

Dependent events Events that are so related that the occurrence of one alters the likelihood that the others will occur.

Descriptive statistics The methods and techniques of summarizing, graphing, and describing numerical data collected in the study of something.

Deviation IQ An aged-based index of general mental ability. It is based on the difference between a person's raw score on a mental ability test and the typical or average score for that chronological age.

Edges Line segments that form a polygon or border a polyhedron.

Divisibility test A quick method for determining if a fairly small number is a factor of a larger number. Divisibility tests for 2, 3, 5, 7, and 11—the first few primes—are covered in Problem 7 of Set 10, Chapter 2.

Even number A member of the infinite set $\{0, 2, 4, 6, 8, \ldots\}$ or any whole number that is a multiple of 2, i.e., any whole number that can be written as $2 \cdot n$ for some whole number n.

Factor Any counting number that can be used in a multiplication statement to yield a whole number "W" is a *factor* of that whole number W. (E.g., 19 is a *factor* of 57 since 19 x 3 = 57.)

Fahrenheit A common scale used to measure temperature. The freezing and boiling points of water (at sea level) are $32°$ and $212°$, respectively.

Flip Reflecting a plane figure about a given line, producing a "mirror image" of the first figure.

Flowchart A chart that uses arrows to show the flow of steps to follow in the sequence of solving a problem.

Fraction A member of the number system most often represented by the symbol X/Y where X can be any whole number and Y can be any counting number. Any fraction can also be represented by a terminating (e.g., 0.358) or repeating (e.g., 0.358358 ... or $0.\overline{358}$) decimal numeral.

Fundamental counting principle The number of different ways that a sequence of unrelated (independent) events can happen is the *product* of the number of ways each individual event can occur.

Fundamental facts The 390 addition, subtraction, multiplication, and division statements for whole numbers that are the basis for learning an algorithm for that operation. E.g., $54 \div 9 = 6$ is a fundamental division fact, and should be committed to memory at some point in time so that more difficult long division problems can be computed easily.

Grade equivalent score The grade level for which a score on an achievement test is the real or estimated average. (*Anticipated* grade equivalent scores, when reported, are calculated using some measure of the child's mental ability and chronological age. This gives the grade-level score the child should have made, were he or she achieving as expected according to mental ability and age.)

Gram The basic unit of mass. A cubic centimeter of water, at $4^{\circ}C$, is defined to have a mass of 1 gram.

Greatest common factor The largest number that is a factor of several given numbers. This is useful in reducing fractions to lowest terms. Dividing both numerator and denominator by the greatest common factor yields the lowest terms equivalent fraction.

IQ A traditional measure of how "bright" or intelligent someone is—an abbreviation of "intelligence quotient," which gives a hint about how it's derived—by dividing the *mental age* (MA) of a person by the *chronological age* (CA).

Independent events Events that are unrelated in the sense that one occurring will not alter the chance that the others will occur.

Inductive reasoning Generalizing about a phenomenon from considering examples or instances that you have at hand. Inductive reasoning is quite useful in generating new ideas or theories, and goes hand in hand with deductive reasoning in mathematics.

Intuitive thought Mental processes characterized by insightful thinking that initially lacks adequate rational explanation, but for which rational support is sometimes effectively developed.

Irrational numbers Numbers that represent definite points on a number line but can't be expressed in fractional form (a/b). Pi and the square root of 2 are common irrational numbers. In decimal notation, these numbers can only be approximated ($\pi \cong 3.14159$, e.g.), since they are neither terminating nor repeating decimals.

Kilo A prefix commonly used in the metric system, referring to 1000.

Kilogram A common unit used for measuring mass. A kilogram is 1000 grams, or, the mass of a liter of water at $4^{\circ}C$.

Least common multiple The smallest number that is a multiple of several given numbers. This is frequently used when adding or subtracting fractions, and the least common denominator is desirable.

Line of symmetry The line through a symmetric figure about which, if the figure were folded, the two halves would coincide one-to-one.

Line segment A one-dimensional geometric figure consisting of two points in space and all the points "between" them. A *line segment* can be characterized as having length, but no breadth.

Liter A basic unit in the metric system for measuring volume—defined as the volume of a cube that's 10 centimeters on each edge.

Loop A sequence of steps in a flowchart that are repeated over and over again, either for a preset number of times or until a decision step gets an affirmative answer.

Mass A measure of the quantity of matter contained in a physical body. In the metric system, *gram* is the base unit for measuring mass.

Mean A measure of central tendency of a set of scores, computed by adding all the scores in the set and dividing by the number of scores.

Measures of central tendency Numbers so associated with a set of scores that, within the context of their use, they are said to represent the set of scores. Mean, median, and mode are common measures of central tendency.

Median For an odd number of scores, the median is the middle number when the set is ordered from smallest to largest. For an even number of scores, the median is the *mean* of the two scores nearest the middle.

Mental age (MA) The age for which a given score on a mental ability test is average or normal. Most reliable when it applies to young children where mental growth is relatively rapid.

Meter The basic unit in the metric system for measuring length. A meter is about 39.37 inches.

Milli A common prefix in the metric system, referring to 1/1000th.

Moebius band A topological figure made by taking a strip of paper and taping the ends together but inserting a "half twist" before the ends are actually joined.

Mode The most frequently occurring number or numbers within a group of numbers.

Multiple A number that can be written as the product of a second number and some counting number, is said to be a *multiple* of both of the other numbers. For example, 55 is a multiple of both 11 and 5, since 55 = 11 x 5.

Negative numbers The numbers considered the "opposites" of the positive numbers, since a negative number and its corresponding positive number have a sum of 0.

Normal curves A certain family of theoretical curves, associated with specific distributions of scores or measures, that resembles a "bell" or "hat" when graphed.

Norming sample A sample selected for the purpose of providing a standard by which future test results will be compared. The test is given to the norming sample, and the results then become the basic expectations for future test takers with the same characteristics (age, grade level, IQ, or whatever is considered relevant for that particular test).

Number crunching A colloquialism for rapid computations with numbers.

Numeration system A systematic way to symbolize the individual members of a number system. Familiar numeration systems are the Hindu-Arabic for whole numbers, the "fraction bar" system for naming fractions, and Roman numerals for the counting numbers.

Numerator The numeral "above" the bar when a fraction is shown as a/b. In the most common interpretation of a fraction, this refers to the number of parts of a whole under consideration.

Odd number A member of the infinite set $\{1, 3, 5, 7, 9, 11, \ldots\}$ or any whole number that is not a multiple of 2. Still a third definition would be "a member of the set of numbers given by $2 \cdot n + 1$, where n stands for any whole number."

Operation There are many operations associated with numbers; the most common are addition, subtraction, multiplication, and division.

Parallel Lines, segments, or rays in the same plane that would not intersect, even if extended; planes in space that do not intersect.

Percent A term meaning "out of a hundred." Fifteen percent means fifteen out of a hundred and is usually written as 15% (but could also be written as 15/100, or 0.15, or 15:100).

Percentile A reporting technique that shows the percent of equivalent cases at or below the score made by an individual.

Perpendicular A condition of lines, segments, or rays that, if extended far enough, would meet so that the four angles formed are *congruent* (or, are all *right angles*).

Polygon A plane figure that's a *simple closed curve,* but made entirely of *line segments.*

Polyhedron A three-dimensional shape that has an exterior and an interior, and a surface made of polygonal regions.

Population The total group about which one ultimately wants to draw conclusions in a statistical study.

Prime factorization Writing a composite number as the product of its prime factors.

Prime number A number that has exactly two factors (i.e., a number that can't be written as the product of two smaller numbers).

Probability The chance that an event will occur.

Properties of operations Characteristics of an operation (+, -, x, ÷) in a particular number system. For example, addition is *commutative* in the whole number system since $x + y = y + x$ for any two whole numbers x and y.

Radian The size of the central angle of a circle determined by an arc length that's exactly the length of the radius of the circle.

Random sample A *sample* selected in such a manner that each member of the *population* under consideration has an equal chance of being in the sample.

Range The numerical difference between the largest and smallest numbers in a set of data. The range is one measure of the *variability* within a group of scores.

Ratio A way of comparing the relative size of several groups, often shown using the same symbolism as for fractions. The ratio of tires to headlights on a new automobile would be expressed as either 5/2 or 5:2.

Rational thought Mental processes characterized by logical, deductive reasoning.

Ray A one-dimensional geometric figure that results from an unlimited extension of a line segment beyond one of its endpoints, in the direction determined by the segment.

Reciprocal For a given fraction (a/b), the reciprocal is the fraction b/a, assuming neither a nor b is 0. A fraction and its reciprocal, when multiplied together, yield 1.

Recursive relationship A relationship in which each new step is defined in terms of those that came before it.

Regular Refers to having congruent parts. A *regular polygon* has all of its angles congruent, and all of its edges. A *regular polyhedron* has all of its faces congruent.

Repeating decimal A decimal numeral that eventually begins repeating some or all of its digits (e.g., 0.68353535 . . . or 0.68$\overline{35}$).

Representative sample A sample that matches the population with respect to characteristics important to the investigation at hand. There is an attempt to have the sample have the same proportions, relative to the critical factors (e.g., sex, socioeconomic level, or religion) as the population.

Right angle An angle of 90°, or, an angle formed by two perpendicular lines.

Rigid motion Moving a two- or three-dimensional figure without distorting its size or shape. The three basic rigid motions discussed in this text are *slides, flips,* and *turns;* putting several of these three together in a sequence produces another rigid motion called a *composition.*

Rounding off Applying any one of several rules to numbers for retaining only a specified degree of accuracy in calculating or reporting.

Sample A subset of the *population* of a statistical study. Researchers attempt to pick a sample in such a way that their conclusions about the sample will have a high probability of being true statements about the population itself.

Simple closed curve A curve that, if traced with a pencil, would end up back where it started without crossing its path in the process.

Simulation Developing a mathematical model for an event, and making judgments based on the model rather than on the actual occurrence of the event.

Slide "Moving" a geometric figure, without distorting its size or shape, in a specified direction and distance.

Slide vector The specified direction and distance through which all points of a figure are to be moved in the transformation referred to as a *slide.*

Sphere The three-dimensional equivalent of a two-dimensional circle. Can be described as the entire set of points in three-dimensional space that are a certain distance from a given point. The given point is the center of the sphere; the distance of the sphere from its center is called the *radius* of the sphere.

Standard deviation A measure of the variability or dispersion of a set of scores about the mean. The smaller the *standard deviation,* the more the scores cluster around the *mean,* and vice versa.

Stanine A nine-point scale for reporting standard scores. Stanine 1 is the lowest category of scores, stanine 5 is in the "average" range, while stanine 9 is the highest classification in this scale.

Stratified random sample A combination of random sampling and representative sampling. The characteristics and relative proportions of the critical factors are known, but in choosing the sample itself, a random selection technique is employed to fill each of the categories.

Symmetric A geometric figure in which one side is a "mirror image" of the other side.

Synthesis Working with individual component parts, making them mesh together, to form "the whole." The reverse of *analysis*.

Terminating decimal A decimal numeral that has no digits other than 0 to the right of a particular position (e.g., 0.69).

Tessellation Repeating the same shape over and over, with no overlapping and no "holes," so that any region of a plane would eventually be covered completely. The term might also refer to using a three-dimensional shape to "fill up" space in a similar fashion.

The chain law The probability that a sequence of unrelated (independent) events will occur is given by the product of the chances of each one happening individually.

Trace Checking the logic of a computer program by running it through its paces a few times and checking its answer against one calculated by hand.

Turn Rotating a plane figure about some point in that same plane. The resulting figure is exactly like the original, except for its orientation (i.e., the "north pole" might now be pointing toward the "southeast," etc.).

Vertex A point where line segments meet. An angle has only one vertex, while a triangle has three vertices. A cube would have eight vertices (corners).

Whole numbers The infinite set $\{0, 1, 2, 3, 4, \ldots\}$. The importance of this set lies both in its own usefulness and in its being the foundation for other number systems (fractions, irrationals, complex numbers, etc.).

Index

Abacus, 30, 206
Absolute zero, 89
Adding to tens, 44
Addition
    decimals, 68-69
    fractions, 60-61
    positive and negative numbers, 84-87
    whole numbers, 34, 43-44
Algorithms
    decimals, 68-69
    definition of, 10
    fractions, 60-64
    positive and negative numbers, 84-87
    whole numbers, 43-47
Analysis of problems, 11-12, 14-21
Ancient numeration systems, 28
Angle, 106, 258-261
Area, 247-251
    circles, 249
    geoboard figures, 248, 332
    parallelograms, 249
    rectangles, 248
    squares, 91
    triangles, 249
Aristotle, 92
Associative property, 34, 36, 64, 87
Asymmetric, 118
Astronomer's Unit, 246-247
Autistic thought, 3, 6-7, 21

Base ten, 29-31
BASIC, 223-234
Betweenness
    Euclidean, 105
    fractions, 59
Bode's law, 246-247, 360

Cancelling, 58
Celsius, 173-176, 221-222
Chain law, 190-193
Chronological age, 263
Circle, 74, 106, 249
Commutative property, 34, 36, 64, 87
Composite number, 97
Computer
    main components, 223-224
    programming, 223-238
    vs. calculator, 207, 237-238
Concrete analysis, 15, 338
Congruent, 121
Constructions
    bisecting angles, 107
    equilateral triangles, 129
    parallel lines, 135
    perpendicular lines, 132-133
    regular heptagons, 138
    regular pentagons, 137
Corresponding angles, 134-135
Counting numbers, 27
Cubic centimeter, 172
Curves, 124-127

de Vinci, Leonardo, 309, 315
Decimals, 67-69
    related to irrationals, 93-96
    related to fractions, 71-73
    related to percent, 77
Degree
    angular measure, 259-261
    Celsius scale, 173-176, 221-222

Fahrenheit scale, 173–176, 221–222
Denominator, 52, 55–56
Descriptive statistics, 270–306
Dienes Blocks, 30, 32
Dienes, Zoltan P., 29
Distributive property, 36, 64, 87
Divisibility tests, 101
Division
    decimals, 68–69
    fractions, 62–64
    interpretation of a fraction, 52
    positive and negative numbers, 87
    whole numbers, 37–38
Dürer, A., 137–138

Edges
    Moebius band, 320–321
    polyhedron, 152–154
Equilateral triangle, 127
    construction of, 129
    tesselation of, 146–148
Escher, M. C., 138–149
Estimation, 162–164
Euclidean geometry, 103, 105–106
Euclid's fifth postulate, 6
Even numbers, 100

Factor, 58, 97–100
Fahrenheit scale, 173–176, 221–222
Flip, 112–114
Flip line, 112–114
Floccipauchinihilipilification, 166
Flowcharts, 209–222
    as problem-solving processes, 210
    counters within, 217–220, 233
    decisions within, 211–214, 232
    looping within, 213–214, 230–234
    symbolism for, 210
    tracing, 220–221, 232, 234
Fog index, 333
Fold line *See* Flip line
Four color problem, 154, 317–318, 321, 353
Fractions, 51–64
    algorithms for, 60–64
    basic interpretations, 51–53
    comparison, 54–55, 59
    equivalent names, 53–54
    interpreted on Cartesian coordinate system, 58–59, 342
    relationship to decimals, 69–73
    relationship to percent, 77–78
Fry, Edward, 265–267, 290
Fundamental Counting Principle, 186–187
Fundamental facts, 42–43
Fundamental Theorem of Arithmetic, 98

Galton, Francis, 301
Game of Life, 329–330
Gauss, Karl F., 310
Goldbach's conjecture, 101
Golden ratio, 108, 349
Grade-level achievement, 264
Graphs, 279–285
    bar, 281–282
    circle, 283–284
    line, 282–283
    pictogram, 280–281
Greatest common factor, 98–100
Guess, check, revise process, 18–19, 336–338
Gunning-Mueller Clear Writing Institute *See* fog index

Halving and doubling, 23
Harvard Step Test, 74–75
Henderson, K. B., 1
Hindu-Arabic system, 29–31

Identity element, 34, 35, 37, 38, 64, 87
In between, 177, 180–182
Indirect analysis, 20 *See also* Proof by Contradiction
Inductive reasoning, 16, 23–24
Inferential statistics, 271
Intelligence quotient
    traditional, 263
    deviation, 303–304
Intuitive thought, 3–7, 20–21
Irrational numbers, 93–96

Jordan Curve Theorem, 126

Kansky, Bob, 314

Lattice method of multiplication, 45–46
Least common multiple, 98–100
Line, 105
Line of reflection *See* Flip line
Line of symmetry, 117
    for regular polygons, 331
Line segment, 105
Long division, 46–47

Magic square, 308, 312, 363
Mass, 172–173
Mean, 292–293
Measurement interpretation of division, 37
Measuring, 241–269
    achievement, 264–265, 275
    angles, 258–261
    area, 247–251
    central tendency, 292–294
    intelligence, 263, 274–275, 303–304
    length, 243–245
    reading level, 265–267, 290
    variation, 295–296
    volume, 254–257
Median, 293–294
Mental age, 263
Metric system, 29, 51, 172–174, 242–257
    prefixes, 242, 246
    reasons for changing to, 29, 51
    units for area, 249–251
    units for length, 243–245
    units for mass, 172–173
    units for temperature, 173–176
    units for volume, 256–257
Miller, George, 262
Moebius, A. F., 316
Moebius band, 316–317, 320–321, 364
Mode, 294
Multiple, 99
    lease common *See* Lease common multiple
Mayan numeration system, 32–33
Multiplication
    decimals, 68–69
    fractions, 61–62
    positive and negative numbers, 86
    whole numbers, 35–37

Negative numbers, 83–88
    addition of, 84–85
    division of, 87
    multiplication of, 86
    properties of the operations, 87–88
    subtraction of, 85–86
Nested numbers, 314, 364
Newton, Isaac, 95–96, 346
Normal curve, 298–304
Numeration systems
    Babylonian, 29
    decimal, 67–68, 96, 347
    fractions, 51–54
    Hindu-Arabic, 29–31
    positive and negative numbers, 83–84
    Roman numerals, 28
    tally, 28
Numerator, 52

Odd numbers, 100
Opposite angles, 130–131
Optical illusions, 130, 136, 327, 332–333

Paper folding, 318–320, 322
Parallel lines, 143–135
Partitioning interpretation of division, 37
Pascal's triangle, 23
Percent, 76–79
    basic interpretation, 76
    computating percent of a number, 78–79
    decimals to percent, 77
    fractions to percent, 77–78
Percentile, 264–265
Perpendicular, 131–134
Petals-around-the-roses, 327
Pi, 74, 93, 94, 249, 255, 344
Pingry, R. E., 1
Placeholder, 27, 29
Point, 105
Poker hands, 183–184
Polygons, 126–127
    diagonals of, 129

    prefixes for, 127
    regular, 127–130
    tessellating the plane, 139–149
Polyhedra, 150–153
    names for, 152
    regular, 150, 152–153
    relationship of edges, vertices,
        faces, 152–153, 353
    tessellating space, 155, 353
Population, 288–290
Positional principle, 29
Precision, 168–169
Prime factorization, 98–100
Prime numbers, 20, 96–98
Probability of an event, 179–180
    complementary events, 180–181,
        195–197
Problem-solving model, 3–4, 7, 20–21
Programming languages, 207
Proof by Contradiction, 92
    intuitive introduction to, 20, 337–338
    that $\sqrt{2}$ is irrational, 92
    that the sum of a rational and an
        irrational is irrational, 346
Properties
    fractions, 64–65
    positive and negative numbers, 87–88
    whole numbers, 34–39
Protractor, 260–261
Pythagoreans, 92
Pythagorean Theorem, 9, 11, 133,
        308–309

Quake lines, 312–313
Quetelet, Adolphe, 300

Radian measure, 261
Range, 295
Ratio, 51–52
Rational thought, 3, 4–5, 9–12
    Level 1, 9–10, 42, 43, 47
    Level 2, 10–11, 42, 47, 86, 133
    Level 3, 11–12
Ray, 105
Reading level, 265–267
Reciprocal, 64
Recursive relationship, 19
Reducing to lowest terms, 58

Reflection *See* Flip
Regular polygons *See* Polygon, regular
Regular polyhedra *See* Polyhedra,
    regular
Related events, 191–192
Relatively prime, 347
Repeated addition, 35–36
Repeated subtraction
    algorithm, 46–47
    interpretation of division, 37
Repeating decimals, 69–71, 96, 347
    relationship to fractions, 71–73
Right angle, 131–134
Rigid motion, 109–114, 119–121
Roman numerals, 28
Rotation *See* Turn
Rounding off, 46, 48–49, 163–165
Russian peasant method for
    multiplication, 23

Sample, 288–290
    random, 289–290
    representative, 289–290
    stratified random, 289–290
Scale
    for a graph, 284–285
    for temperature *See* Celsius and
        Fahrenheit scale
Setek, William, 89
Simple closed curve, 124–126
Simulation, 201–203
Slide, 110–111
Slide vector, 110
Sphere, 106
Standard deviation, 295–296
Stanine, 264–265
Subtraction
    decimals, 68
    fractions, 60–61
    positive and negative numbers, 85–86
    whole numbers, 35, 38–49
Symmetry, 67–68, 117–119
Synthesis, 12, 14–21

Tally system, 28
Temperature
    absolute zero, 89
    Celsius and Fahrenheit scales, 173–176

conversion, 175–176, 221–222
Terminating decimals, 70–71
   relationship to fractions, 71
Tessellation
   of the plane, 138–149, 351–352
   of space, 155, 353
Translation *See* Slide
Transversal, 134
Turn, 111–112, 349

Unit fraction, 234
Unrelated events, 191–192

Vertex of an angle, 106, 258
Vertices
   of a polygon, 127
   of a polyhedron, 152–154, 353
Volume, 254–257
   of a box, 255
   of a sphere, 255

Weighted average, 297–298
Wertz, Robert, 312
Working backward, 17, 336–339